Bio-Cultural Diversity, Ethnicity, and Local Knowledge

Yos Santasombat
Editor

Bio-Cultural Diversity, Ethnicity, and Local Knowledge

From Lanna to Northeast India

Springer

Editor
Yos Santasombat
Faculty of Social Sciences
Chiang Mai University
Chiang Mai, Thailand

ISBN 978-981-95-4241-3 ISBN 978-981-95-4242-0 (eBook)
https://doi.org/10.1007/978-981-95-4242-0

This work was supported by Ministry of Science and Technology (Fundamental Fund 2020-2023).

This Springer imprint is published by the registered company Springer Nature Singapore Pte Ltd.
The registered company address is: 152 Beach Road, #21-01/04 Gateway East, Singapore 189721, Singapore

Preface

Emerging from a research project hosted by the Center for Biodiversity and Indigenous Knowledge for Development Studies under the Faculty of Social Sciences at Chiang Mai University, *the* monograph aims to expand the boundaries of knowledge by opening up new topics and research areas through interdisciplinary collaboration. Faculty researchers from various departments at Chiang Mai University—such as Social Sciences, Architecture, Sciences, Education, Humanities, and Fine Arts—were working together for 3 years with a shared goal, to generate new knowledge and establish academic networks that connect Thai scholars with those from Northeast India, a region about which we currently have limited understanding.

Although each research project may have its own emphasis and focus, they are all grounded in a shared assumption that knowledge and understanding of bio-cultural diversity and ecosystems stem from diverse ways of thinking, local knowledge, and perspectives that differ across ethnic groups and reflect how humans adapt to various environments and cultural landscapes. Understanding these varied bodies of knowledge and viewpoints allows us to address today's challenges—such as global warming, haze, pollution, ecosystem degradation, biodiversity loss, and sustainable resource management—through more complex, expansive, and holistic approaches.

Traditionally, biodiversity has been seen as the domain of natural scientists like biologists and botanists. But in reality, even natural scientists have limited understanding of tropical biodiversity. We've only just begun truly studying biodiversity as a vital resource, while local communities—who have lived in harmony with forests for generations—have been managing and utilizing biodiversity and genetic resources for hundreds of years. They have developed and selected diverse plant species, accumulating knowledge essential to food and traditional medicine, including methods for use, breeding, and understanding properties of various medicinal and edible plants. This bio-cultural diversity is a pillar of life security—ensuring stable food production not only at the community level, but also nationally and globally.

Biodiversity is not confined to forests; it also encompasses fields, farms, and all local communities that have preserved food and medicinal systems across generations. Local knowledge of biodiversity also holds value for crafts, architecture, tourism, and product development—such as tea, coffee, and herbal medicine —that can generate substantial income now and in the future.

Studying and developing knowledge to harness biodiversity is not only the responsibility of scientists; it is a collaborative mission for scholars in social sciences, humanities, and—vitally—the wisdom and partnership of local communities whose daily lives are deeply intertwined with biodiversity. Biodiversity is closely linked to cultural diversity and indigenous knowledge of ethnic communities who have long lived in harmony with nature. Any form of cultural erosion—intentional or not—undermines opportunities to learn and benefit from that wisdom in ways that could serve the nation and humanity.

This research initiative marks an important beginning in building interdisciplinary teams of scholars. It aims to integrate diverse academic disciplines to jointly explore new perspectives and methodologies for understanding biodiversity and indigenous knowledge. The project also seeks to forge collaboration with communities and scholars in Northeast India to pioneer new research areas, new knowledge, and innovative policies for sustainable resource management—strengthening the academic capacity of both regions.

I would like to express my sincere thanks to all our research team members who spent long years of research in Lanna and northeast India. Our team would like to express special thanks of gratitude to our colleagues from northeast India. Special thanks go to the Department of Philosophy, North-Eastern Hill University (NEHU), Shillong; the Women's Studies Centre, Nagaland University; Department of Sociology, Nagaland University. We would like to express our sincere gratitude to Professor Dr. Lungsang Zeliang and Associate Professor Dr. Khotole Khieya of Nagaland University for their invaluable support in facilitating data collection in the villages of Peren and Kohima districts. We are also honored to have them as co-authors of one of the chapters in this volume. Our deep appreciation further extends to Shri. Supongnukshi, IFS, Chief Conservator of Forests, Department of Environment, Forest & Climate Change, Nagaland, and Dr. Sentitula, IFS, Divisional Forest Officer (DFO) and Head of the Divisional Management Unit (DMU) in Mokokchung, for their generous assistance, provision of insightful information, and facilitation of field visits in Mokokchung and Mon districts. Their contributions were instrumental to the success of this research collaboration between Nagaland and Chiang Mai University. We would like to express our appreciation to the Department of Women's Studies, Gauhati University, Assam; Department of History, Manipur University. Department of History, Sao Chang College, Tuensang, Nagaland; the Divisional Forest Office, Mokokchung; the North East India History Association (NEIHA) and the Manipur Historical Association (MHS). Palaiwan Srisaringkarn deserves special recognition for her untiring organizational assistance to our project.

Chiang Mai, Thailand
October 2025

Yos Santasombat

Introduction

This book highlights the interconnections between biodiversity, cultural knowledge, and resource management in Lanna (Northern Thailand) and Northeast India. By examining traditional practices, economic activities, ethnic identities, gender roles, art and craft, the volume aims to contribute to public understanding and policies that balance conservation with sustainable development. Understanding these dynamics provides valuable insights for strengthening community resilience and promoting equitable, eco-friendly economic systems.

The upper northern region of Thailand and the northeastern region of India share similar geographical features, characterized by alternating mountain ranges and fertile lowlands. These areas are rich in agricultural land and dense forests, which are home to a diverse array of flora, fauna, insects, microorganisms, medicinal plants, and invaluable genetic resources. The forests serve not only as a source of beauty, food, medicine, and tourism revenue but also play a crucial role in maintaining ecological balance, acting as watersheds, and providing sustainable benefits to humanity in both the present and future.

These forests are also home to many local communities and ethnic groups, each of which has developed a way of life closely intertwined with the forest. They have utilized biodiversity for hundreds, if not thousands, of years. Thus, the forest is not only a reservoir of biological diversity but also a repository of cultural diversity and local wisdom. This bio-cultural diversity has evolved within the context of human–forest relationships and has been passed down through generations. The interconnection between communities and forests represents both historical and cultural development. Knowledge related to ecosystems, food and medicinal plants, agriculture, forest and water management is invaluable for biodiversity conservation, genetic resource development, and the future of the food and pharmaceutical industries, as well as biotechnology.

Today, urban and rural communities worldwide, especially in tropical regions, face crises in economic, agricultural, environmental, and cultural dimensions. The breakdown of human–nature relationships has become an urgent issue that people globally are beginning to recognize. Crucial research questions in biodiversity studies are: how can we extract lessons from various field studies to restore local communities' capacity to manage biodiversity? How can we empower communities to regain control over production, utilization, and genetic resource development, including knowledge about food, medicine, and other natural resources?

At the community level, the key questions are: How can communities develop strong organizations capable of negotiating with external forces and managing conflicts effectively? How can they foster a sense of community and sustain cultural diversity, which forms the foundation of knowledge about biodiversity? How can communities establish effective resource management systems that adapt traditional customs to modern needs and contexts? How can communities develop education systems that align with their needs, preserve local wisdom in seed selection and breeding, and integrate traditional knowledge with modern science? How can communities develop financial management systems and enhance their capacity for local business administration to establish a sustainable food and pharmaceutical industry, linking producers with consumers while ensuring community-based production, processing, business structuring, and benefit-sharing?

At the policy level, we must understand the structural issues that undermine communities' ability to manage biodiversity. We need to critically examine policies and laws that restrict local communities' rights to participate in resource management while enabling a privileged few to monopolize and exploit resources.

The fundamental challenge lies in how we perceive and value biodiversity. How can we manage it in a way that benefits the broader society sustainably? How can we develop policies and legal measures that uphold community rights in resource management and protect local wisdom? How can we conduct research to uncover knowledge about our natural resources to foster an "alternative industry" based on biodiversity and indigenous wisdom? This volume aims to synthesize lessons from field studies in Lanna and Northeast India to address these fundamental questions.

Research Objectives

The book *Bio-cultural diversity, Ethnicity, and Local Knowledge: From Lanna to Northeast India* aims to study, understand, and compare the knowledge and wisdom of local communities in northern Thailand and northeastern India. Specifically, it seeks to analyze natural resource management systems, food and medicinal practices, the properties and applications of biological resources, and their relationship with livelihoods, production systems, religious rituals, folk crafts, tourism, and gender roles. The monograph strives to gain insights into the coexistence of biodiversity, nature, non-human and human communities across different time periods. The objectives are as follows:

1. **Comparative Study of Local Wisdom and Cultural Landscapes:** Investigating the natural resource management systems, ecosystems, soil, water, forests, and biodiversity among different ethnic groups in northern Thailand and northeastern India. This includes documenting and analyzing ecological knowledge, cultural landscapes, and ethnobotany, as well as exploring ecotourism with a focus on social and cultural dimensions.

2. **Economic and Social Research:** Examining tangible expressions of local knowledge that can be studied in concrete terms, such as cultural tourism, gender roles, folk crafts, traditions, and belief systems that underpin local ways of life and economic activities. This research seeks to understand how communities and ethnic groups utilize biodiversity and how economic developments, such as commercial agriculture and agribusiness, impact biodiversity conservation and utilization.
3. **Research for Sustainable Development:** Exploring strategies and learning processes to strengthen local communities in conserving, developing, and managing resources. The book also examines the potential for developing community-based products such as tourism, commercial handicrafts, and medicinal plant processing.

Concepts, Theories, and Hypotheses

This research monograph employs the concepts of "Ethnoecology" and "Cultural Landscape" to understand the interactions between biological diversity and indigenous knowledge among different communities in Lanna and Northeastern India.

Ethnoecology is an approach that focuses on the relationship between humans and nature. It emphasizes understanding the worldview, cosmology, and classification systems used by various ethnic groups to interpret their natural environment. This perspective highlights the cultural dimensions of resource management, which directly impact human production and livelihoods within ecosystems that are constantly changing due to economic, social, and political factors.

Since 1954, when Harold Conklin authored "The Relation of the Hanunoo Culture to the Plant World," studies in ethnoecology have begun to expand under various terms such as Ethnoscience, Ethnobotany, Cultural Ecology, and Traditional Ecological Knowledge. Although these fields have different perspectives and theoretical frameworks, their common point of origin lies in the profound and extensive indigenous knowledge systems—particularly in plant and organism classification—of various communities. These systems form the basis of local resource management. For example, Berlin's (1968) research on Tzeltal botany in South America demonstrated that local classification and botanical systems could be comparable to Western scientific taxonomy, such as the Linnaean system. Moreover, research in historical ecology (Crumley, 1994), folk biology (Medin & Astran, 1999), and indigenous ecological studies (e.g., Gragson & Blount, 1999; Nazarea, 1999) reinforce the importance of local knowledge and dispel misconceptions that indigenous wisdom is merely remnants of the past without relevance to present and future sustainability.

Studies on ethnobotanical knowledge, such as those by Clarke (1976), Alcorn (1989, 1994), Bragg (1986), and others, clearly show that local knowledge—especially in botany—is invaluable for drug discovery, breeding new plant varieties, and developing industrial products. However, some scholars argue that this knowledge is just one component of modern science. Figures like Posey and Dutfield (1997), Posey (1997), and Martin (1978) emphasize the importance of preserving indigenous knowledge, considering it humanity's cultural heritage. Comparative studies by Berkes et al. (1993) demonstrate that all societies produce scientific knowledge, albeit with different underlying principles. Indigenous wisdom thus broadens our perspectives and helps us avoid monocultural thinking and the superiority of one knowledge system over others.

Furthermore, indigenous knowledge is deeply intertwined with moral values and ethical systems, offering invaluable lessons on social relationships between humans and nature, as well as among humans themselves. The loss of biological and cultural diversity threatens human security, constraining opportunities for development and diminishing the creative power of humanity.

Biodiversity allows humans to interpret, utilize, and adapt to nature effectively. Ethnological ecology research (e.g., Bragg, 1986; Posey, 1997; Rose, 1996; Sponsel, 1995; Berkes et al., 1993) clearly shows that resource use within cultural contexts involves choices and decisions made by local communities. These communities develop knowledge and lifestyles within "cultural landscapes" they have shaped within their natural environment. Studies (e.g., Blackburn & Anderson, 1993; Chapeskie, 1995; Cronon, 1983, 1995; Santasombat, 2003, 2008) demonstrate that the concept of cultural landscapes is useful for understanding local wisdom and the role of indigenous groups in preserving and promoting biological diversity through knowledge, beliefs, rituals, and resource management practices.

Thus, areas traditionally thought of as forests or natural ecosystems are fundamentally cultural landscapes—products of accumulated knowledge and wisdom passed down through long-term relationships between communities and the environment. Many ethnic groups have transformed valleys, plains, forests, and rivers into repositories of knowledge. These cultural landscapes reflect the achievements and potential of local knowledge systems involved in sustainable resource use, conservation, and management.

This book adopts the concepts of Ethnoecology and Cultural Landscapes as tools to study the relationships between biological diversity and local wisdom among different ethnic groups in Lanna and Northeastern India, hypothesizing that these communities have applied indigenous knowledge and customs to manage and conserve forests, land, and water sources within various ecosystems. It aims to investigate the conditions under which this knowledge and tradition emerged, whether local wisdom retains innovative and adaptable qualities for contemporary and future contexts.

Simultaneously, the ideas of the Anthropocene and its impacts on the environment draw attention to "feral ecologies" and "feral proliferations"—manifestations of destruction and pollution driven by industrial expansion and commercial agriculture (Bubandt & Tsing, 2018; Tsing et al., 2019). Consequently, research areas in Nagaland and Lanna serve as ethnological and biological fields that may help us better understand the spread of these feral ecologies.

Similarly, the concept of a "rupture" in the relationship between the world/society/community and nature helps us to see the breakdown of these connections that may occur in specific places and times. This rupture can result from the interplay between technological change and societal developments that impact the environment, leading to the accumulation of violence and ultimately to environmental crises. The breakdown of these relationships can create a sense of insecurity and anxiety about the future among the public and give rise to various forms of social movements aimed at addressing problems in human–nature relations (Mahanty et al., 2023: 1–2).

"Rupture" is a concept that indicates transformation and uncertainty (Head, 2016). The accumulation of ecological damage reveals crises forming in different ways and intensities across different regions. This concept compels us to revisit the origins of such ruptures, starting from the colonial era through to consumerism in the era of late capitalism, and the epistemological violence (Santos, 2014, 2018) imposed on long-standing social, cultural, and environmental systems. By doing so, we seek to understand the causes and impacts of rupture, as well as the power inequalities that differ across regions.

Lessons from the Northeast India especially Nagaland and Lanna may lead us to question structural changes and explore solutions that could help restore the ruptured relationship between communities and the natural world.

Geographical and Ethnic Context of Northeast India

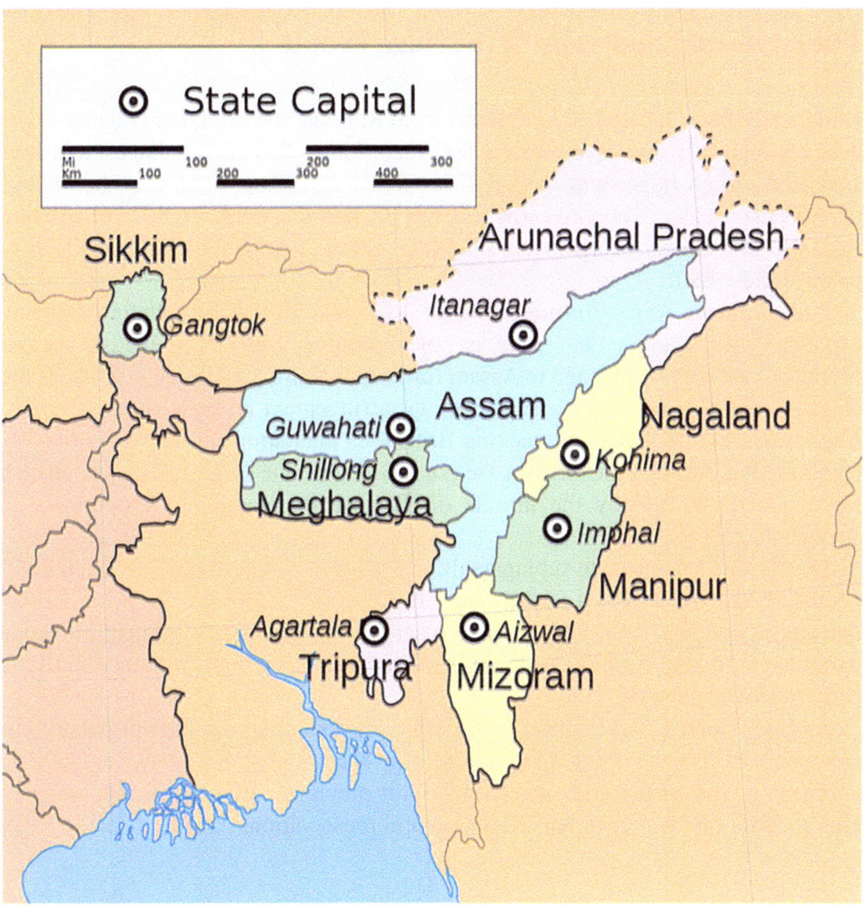

Northeast India comprises eight states: Arunachal Pradesh, Assam, Manipur, Meghalaya, Mizoram, Nagaland, Tripura, and Sikkim. The region is characterized by rugged mountainous terrain interspersed with valleys and is home to numerous ethnic groups, each with unique historical developments and cultural heritage. The Brahmaputra Valley, covering over 56,480 square kilometers, is the largest plain in the region and has been inhabited by diverse ethnic communities for thousands of years.

Most ethnic groups in Northeast India belong to the Mongoloid race, having migrated in waves from different directions, including northern and northeastern Himalayan regions and Southeast Asia. These groups have settled both in the plains and highland areas. Highland communities are generally smaller in population compared to those in the plains. Traditional agricultural practices include terrace farming, shifting cultivation, and livestock rearing. Given that much of the research in this volume focuses on fieldwork in Nagaland, special attention is given to the geographical and cultural landscape of the Naga people, with Khonoma village serving as a representative case study.

The Cultural Landscape of the Naga People

Khonoma village, located 30 kilometers from Kohima, the capital of Nagaland, has been continuously inhabited for over 800 years. Like other Naga villages, it is situated atop a high ridge, with terraced rice fields cascading down into the valleys below. Historically, Khonoma was a center of resistance against British colonial rule, with the Angami Naga people fighting against British forces for nearly fifty years (1832–1880).

Following the Anglo-Burmese War of 1826, the British East India Company expanded into Manipur and what is now Northeast India. The British viewed Nagaland merely as a passage to Assam for tea plantation investments, whereas the Naga people saw colonial expansion as an infringement on their ancestral lands. Naga warriors responded by attacking British tea plantations in Assam, provoking harsh retaliation from the British. Determined to subdue Naga resistance, British forces sought to destroy the morale of the warriors, targeting Khonoma as a stronghold.

Despite British efforts to subjugate them, the Naga people maintained their independence and were viewed as "savages" by colonial authorities. After the British defeated Khonoma, they attempted to resettle outsiders in the village, but none dared to claim land from the original inhabitants. This conflict continued until the early twentieth century.

In 1917, John Henry Hutton, a British anthropologist and administrator, was appointed as Deputy Commissioner of the Naga Hills. His more empathetic approach to the Naga people contrasted with previous colonial officials, shaping later policies on local governance and cultural preservation.

Hutton's policies contributed to the preservation of Naga customs and traditions, fostering a unique governance system that persists today. The integration of traditional governance with modern administrative structures highlights the resilience of the Naga people in maintaining their cultural identity despite historical challenges.

Natural Resources and Economy in Nagaland

Land use and resource management form the foundation of Nagaland's economic, social, political, and cultural systems. Agriculture is the primary livelihood, with most Naga villages practicing two main forms of farming: shifting cultivation (Jhum) and terrace rice farming. Jhum involves clearing and burning forest land for temporary cultivation, rotating fields every 10–13 years to allow regeneration. Despite labor-intensive methods, this system has sustained communities for centuries, maintaining soil fertility and biodiversity.

Terrace rice farming, found primarily near Kohima, requires extensive irrigation infrastructure. Though labor-intensive, it produces higher yields than Jhum farming. The land tenure system in Nagaland varies by tribe, with land ownership distributed among families, clans, villages, and chiefs. Some communities follow hereditary leadership, where land control is passed down through generations.

Resource management in Nagaland is deeply tied to cultural traditions and local governance. Village councils and elder councils oversee land use, water sources, and forest conservation, maintaining community autonomy. Nagaland's community-based forest management differs from northern Thailand, where state-imposed regulations often override indigenous practices. In contrast, over 80% of Nagaland's forests remain under direct community or private ownership, ensuring long-term sustainability.

A comparison with northern Thailand reveals a shift in Lanna agricultural practices toward commercial monoculture, reducing biodiversity and increasing reliance on external resources. The increasing use of chemical fertilizers and pesticides in Lanna agriculture contrasts with traditional Naga farming, which emphasizes self-sufficiency and local ecological knowledge. These differences raise questions about sustainable agricultural policies and the role of local governance in preserving ecological balance.

The Commons, Ethnic Identity, and Resource Management

The term "commons" derives from the traditional English legal term for common land, which are also known as "commons", and was popularized in the modern sense as a shared resource term by the ecologist Garrett Hardin in an influential article called "The Tragedy of the Commons". As Frank van Laerhoven and Elinor Ostrom (2007: 3) have stated; "Prior to the publication of Hardin's article on the tragedy of the commons (1968), titles containing the words 'the commons', 'common pool resources', or 'common property' were very rare in the academic literature."

The commons refer to the cultural and natural resources accessible to all members of a society, including natural materials such as air, water, and land. These resources are held in common even when owned privately or publicly. Commons can also be understood as natural resources that groups of people (households, communities, user groups) manage for individual and collective benefit. In practice, commons involve a variety of informal norms, regulations, and values employed for a governance mechanism not by a state but by a community of users or ethnic groups that self-governs the resource through institutions that it creates. Commons also shed lights into the complexities of human interactions with shared resources, highlighting the balance between cooperation and competition in resource management.

In Chap. 1, Prasenjit Biswas conceptualizes that human motivations, internalized identities, cultures, rituals, and governance systems significantly influence how resources are used and managed, while self-interest often drives individual actions, collective management strategies can lead to sustainablc outcomes. The paper makes use of evolutionary epistemology of experience of lived world in understanding the practices of ascription and deeper realities of valuing nature in both factive and non-factive terms involved in a discourse of affirmation of identity relation between self and other in the context of South and Southeast Asian Communities in their apparent linkages with Northeast Indian communities. Biswas contends that an architectural sharing of an interpretative complex of bio-cultural, transgenic-trans specie and transmigrating symbols, legends and myths with their overflows in worldviews and belief structures mark an unexplored zone on intimacy and extimacy between too many Southeast Asian cultural communities and Northeast Indian ethnic and religious identities. We need to get to the specifics of such societal overflows and subterranean networks of linguistic, ethnic, cultural complexes of common commons arising out as an architecturally shared interpretative complex that exist between communities. For example, Shan-Ahom-Naga-Zo complex, a more dispersed story of successor societies in the landscape of Zomiyas, a continuous movement down Mekong to Brahmaputra and a more-than-human world of specie migration are some of the themes that the paper explores for working out a plausible intellectual project of decodification of this enchanting world of shared cultural worldviews.

Worldviews and corresponding ethnic identities are constructed and reconstructed through a continuous process of life-world experiences, remembrances, and reimagination of relationality between man and nature. A shared notion of natural resources and its patterned meanings mark a bio-cultural logic of conservation and its correlated rituals. Many ritual practices in Northeast India and the Mekong Region are deeply connected to common resources and their representations available in local cultures and their interconnected threads with adjacent Zomia cultural groups.

In Chap. 2, Polly Vauquline examines the relationship of local wisdom, biodiversity, and ethnicity by focusing on food culture of North-East India, with special reference to Assam, from women's perspective. Vauquline contends that local wisdom, especially on food culture, is a foundation of sustainable livelihood, and plays an important role in creating and maintaining a distinctive ethnic identity. Women have been playing distinctive roles in fostering and developing their identity in ethnic communities through food culture. However, the role of women in constituting, storing, and maintaining the local wisdom of food culture, cuisines, and delicacies that emerges from the mundane scape, the kitchen, of the most private, micro space of home, has been overlooked. The paper timely fills this gap by paying the required attention and relevant recognition of women's local knowledge through adequate documentation. It explores the relationship and roles played by women in constituting, storing, and maintaining local wisdom and traditional knowledge of ethnic food in the context of Northeast India in general and Assam in particular. Ecofeminism and Grounded theory were applied to probe deeper and understand the ethnographical knowledge of food through women's perspective. The findings illustrate that women of the region played distinctive roles as the carrier and disseminator of the ethnic knowledge and wisdom of food. They were the practitioner of sustainable use of food in the domestic space of home and community. The wind of changes in the society through the processes and forces of modernization results in a gradual transformation in the role of women as food provider. However, food is still valued as a form of ethnic identity of the communities. Thereby ethnic foods are still valued, practiced, and will continue to exist and cherish by local ethnic communities.

In Chap. 3, Angkhana Inta and her colleagues present a comparative ethnobotanical study of food plants utilized by Tibeto-Burman groups in Lanna (northern Thailand) and Nagaland (northeast India). The paper not only delves into the rich botanical traditions of these two regions, but seeks to identify and discuss cultural similarities and differences in food plant usage, providing insights into their ethnobotanical heritage. A total of 59 key informants from 9 ethnic groups, including Akha, Bisu, Kachin, Karen, Lahu, Lisu, Angami, Ao, and Chakhesang, from 13 studied villages in northern Thailand and Nagaland, were interviewed using a combination of free-listing, semi-structured interviews, group discussions, and the walk-to-the-wood field-survey method. The plant specimens were identified and categorized by food types. Use reports were calculated for Use Value (UV) and percentage of overlap (% overlap). A total of 1,218 use reports were compiled, related to 394 species in 261 genera, 97 families, and 8 unidentified species. Vegetables and fruits were the most common food types in both northern Thailand and Nagaland, with leaves and infructescence being the most frequently used parts. The percentage overlap between the two regions was 24.5% for genera and 12.5% for species. The culturally significant food plants among the Tibeto-Burman groups in both regions include *Allium*, *Alpinia*, *Amaranthus*, *Clerodendrum*, *Colocasia*, *Dioscorea*, *Elsholtzia*, *Ficus*, *Musa*, *Senegalia*, *Solanum*, and *Zanthoxylum*. The cultural significance of food plants among the Tibeto-Burman groups in Lanna and Nagaland is deeply intertwined with their traditional practices, dietary preferences, and environments. The diversity and overlap in plant usage reflect these communities' adaptation to their ecological settings and their ability to integrate a wide range of plant resources into their diets. Understanding these plants' roles offers valuable insights into the complex relationship between culture, ecology, and nutrition, emphasizing the importance of preserving traditional knowledge and biodiversity.

In Lanna and Northeast India, the commons refer to shared forests, grazing lands and, at times, communal swidden agricultural fields. Garrett Hardin's idea of the tragedy of the commons notwithstanding, in Lanna and Northeast India, the community manages and sets the rules of access and use of the resource held in common. The commons do not mean that anyone is free to use the resource as they like. Studies by Arratee Ayuttacorn, Lungsang Zeliang, and Khotole Khieya (Chap. 4) and Sentitula (Chap. 5) have shown that managing forest and land resource as a common property regime often has positive outcomes and avoids the so-called tragedy of the commons.

Arratee Ayuttacorn, Lungsang Zeliang, and Khotole Khieya, in Chap. 4, critically examines traditional and contemporary community forest management practices in northern Thailand and Nagaland. Their paper investigates the processes through which forest-dwelling rural populations establish social networks with non-governmental organizations (NGOs), government agencies, and the private sector. The comparative research paper further explores the resilience of local knowledge systems and their integration with contemporary practices to enhance the effectiveness of forest management. The paper was based on extensive ethnographic fieldwork conducted in five Karen communities in Northern Thailand, recognized as exemplary models of community forest management, and in four villages located in the Kohima and Peren Districts of Nagaland, where ecotourism and development projects are actively being implemented. The findings reveal that both Karen communities in Lanna and Naga communities in Nagaland have developed community-based Payment for Environmental Services (PES) schemes, albeit with distinct approaches. The Karen communities use PES to finance not only conservation activities but also social welfare initiatives, leveraging these schemes to position themselves as forest protectors and strengthen their claims to land rights. In contrast, communities in Nagaland, benefiting from higher income generated through ecotourism, utilize community-based PES primarily to support conservation efforts and fund development projects. Their land ownership reduces dependence on external support; however, the sustainability of these initiatives is constrained by limited resource-sharing across villages. Both Indigenous communities in Lanna and Nagaland effectively leverage their social networks to secure support for livelihood diversification, emphasizing the significance of collaboration with external institutions to sustain their forest management initiatives. Additionally, local peoples in both regions utilize traditional practices and actively construct new indigenous identities to negotiate state policies, foster community cohesion, and promote sustainable forest conservation.

This research paper clearly demonstrates that while Lanna and Nagaland employ different governance systems and management practices, both regions underscore the importance of community-based approaches to forest conservation. Lanna's reliance on community-based PES highlights the potential for locally driven solutions to fund and sustain conservation efforts, even within a state-centric framework. Nagaland's governance empowers communities with autonomy, though it necessitates greater focus on long-term sustainability and resource-sharing mechanisms. Preserving local knowledge, strengthening social networks, and fostering inclusive community-led initiatives are critical to achieving sustainable forest management in both contexts.

In Chap. 5, Sentitula, the head of Mokokchung forestry office, Nagaland, contends that forest management system in Northeast India is inextricably linked to the socio-economic and cultural fabric of ethnic communities. From time immemorial, customary laws and regulations have played a pivotal role in preserving the region's green cover, safeguarding its biodiversity, and supporting livelihoods. Effective forest management strategies must recognize and integrate these traditional systems of governance to ensure long-term success. In Nagaland, Jhum, a traditional swidden agricultural practice, deeply rooted in the region's cultural heritage, has faced criticism for its perceived negative environmental impacts. However, recent studies highlight its potential for sustainable land use management when practiced with appropriate techniques. Sentitula paper explores the historical context of jhum cultivation in Nagaland, its cultural significance, and its role in shaping the region's forests. It also examines the challenges posed by climate change, land degradation, and limited market access for jhum products. Jhum improvement strategies applied in test villages have demonstrated success in fostering community-based forest management and diversifying livelihoods. The ongoing efforts and comparative analyses of the implemented forestry model indicate that slash and burn is beneficial to soil health and agroforestry. Nagaland shows promise in achieving a harmonious balance between jhum cultivation, livelihood and forest conservation, ensuring a sustainable future for the region in the wake of climate change. The success of forest management depends on a coordinated approach that involves government agencies, NGOs, local communities, and individuals working together toward a common goal. In Northeast India, illegal logging, land encroachment, and shifting cultivation practices contribute to the decline of forest cover. Climate change, with its impacts on rainfall patterns and temperatures, also poses a major threat to the region's forests. A range of strategies are being implemented in the region including Community-Based Forest Management (CBFM) for empowering local communities to manage and protect forests sustainably, establishing national parks, wildlife sanctuaries, and community reserves to safeguard biodiversity, promoting sustainable agriculture practices that reduce deforestation and land degradation, raising awareness among local communities about the importance of forest conservation and sustainable livelihoods and implementing effective policies and regulations to address illegal logging, land encroachment, and other threats to forests.

Chapter 6 by Warong Wonglangka and his research team compares local knowledge on traditional agricultural practices between Lanna and Nagaland, focusing on the integration of wet rice cultivation and fish farming in the rice field. Through a case study of Kigwema Village in Kohima city, Nagaland, the paper discloses the evolution and persistence of these practices to better understand Nagaland's cultural landscape and integrated agroecology. The paper captures both environmental and cultural dimensions from an agroecological perspective, using a multidisciplinary approach that combines geographical landscaping, vernacular architecture, and anthropological studies. The comprehensive approach provides deeper insights into the interaction between agricultural practices and cultural landscapes, highlighting how integrated agricultural ecosystems with diverse land use, plant species, and biodiversity have been sustained. For example, the Zabo system, which combines wet rice farming with aquaculture in wet rice fields, has been practiced continuously for over 600 years, demonstrating resilience and adaptability in the face of modern pressures, including urban migration, the commercialization of agriculture, and changing environment. These local practices, rooted in Naga traditions and still in use today, could inform agricultural methods in Lanna, where rice-cum-fish systems are similarly important. The paper addresses the adaptability of integrating local knowledge with the natural environment and offers recommendations that could guide similar practices in Northern Thailand.

Women, History, and Community-Based Ecotourism

History is a systematic study of the past. It employs a narrative to describe, question, and analyze past events, examining their patterns of cause and effect. As an academic discipline, history analyzes and interprets evidence to construct narratives about what happened and explained why and how it happened. Historical narratives often focus on political and economic events, as well as male figures and roles, while there is a significant lack of analysis regarding women and their roles within historical dimensions and socio-cultural contexts.

Chapter 7, written by Pakdeekul Ratana, Bedavati Laishram, and Moirangthem Boby Singh, focuses on the stories of Meitei women, a significant ethnic population who play key roles in shaping society, culture, and the economy in a unique region of Manipur in northeastern India. This paper examines the history and context of Manipur to highlight the role of gender across historical dimensions. It seeks to analyze the relationship between genders and explore gender roles in the unique historical backdrop of the Manipur Kingdom. Additionally, the paper addresses the importance of discussing a space for examining gender roles, which can vary by context, ethnicity, and the specific culture of women in different regions. Furthermore, it proposes opportunities for social and cultural connections and economic development through tourism, which could foster relationships between northern Thailand and northeastern India. This collaboration could open up a new market based on cultural proximity between the two regions.

The rise of community-based ecotourism has transformed gender roles in both northeast India and Lanna, providing women with new economic opportunities and social empowerment. Women in both regions have traditionally been responsible for domestic duties, handicrafts, and food production. However, ecotourism initiatives have enabled them to commercialize these skills, gaining financial independence and greater decision-making power.

In Nagaland, women play an active role in cultural tourism, hosting homestays, crafting traditional textiles, and managing community tourism ventures. Similarly, in Lanna communities, women have leveraged their knowledge of local cuisine, weaving, and hospitality to attract visitors. These initiatives have strengthened community economies while preserving cultural heritage.

Ariya Svetamra, in Chap. 8, employs a feminist political ecology approach to explore the intersections of gender and tourism, examining how gender norms are negotiated within environmental struggles amidst globalization and the cultural commoditization of tourism. The paper provides an intersectional analysis of tourism, gender, power, and the environment. Data were collected through documentary research, field observations, and in-depth interviews with women entrepreneurs who operate homestays in Lanna and Nagaland, Northeast India. This paper argues that cultural commoditization process might reinforce rather than diminish local cultural themes and values.

On the one hand, tourism leads to cultural commodification and a loss of authenticity. This perspective suggests that tourism reduces cultural products and traditions to mere items for sale, often disconnected from their original meanings. For example, festivals and traditions may be transformed into public spectacles for tourists, losing their original significance and meaning to locals. On the other hand, proponents of cultural involution argue that tourism can revive interest in traditional practices, benefiting local communities both culturally and economically. However, this view assumes locals can easily differentiate between sacred and commodifiable cultural elements. It also critiques the idea of authenticity, suggesting that culture is dynamic and fluid, making the notion of an "original" moment in time problematic. By studying these cultural dynamisms, this research seeks to understand how ecotourism can be structured to promote both economic empowerment and cultural integrity.

Rather than focusing solely on the loss of authenticity the paper calls for a deeper exploration of how authenticity is constructed and negotiated within the growing tourism industry, acknowledging the complexities of cultural commodification. This shift moves beyond lamenting cultural loss to examining how people navigate and create meaning within the context of tourism. The process of cultural commoditization can reinforce local cultural themes and values rather than undermine them. Emerging forms of alternative tourism, such as community-based tourism, offer opportunities to challenge gender stereotypes by promoting economic independence and empowerment among women. This type of tourism provides women with more freedom, independence, and income while allowing them to stay in their communities. Through homestay businesses, women have been able to improve their homes, support their children's education, and meet various needs. Interactions with guests have broadened their worldviews, benefiting not only women but also their children, who gain exposure to new cultures and lifestyles. Additionally, women have learned to use social media platforms and engage in the globalized world, extending their experiences beyond their local communities.

Women's engagement in community-based tourism enhances their decision-making power, influencing gender dynamics both within their households and in public life. They leverage networks, including kinship connections, women's networks, and government and NGO support, to empower themselves and manage conflicts, prevent political divisions, and maintain community cohesion. This empowerment through tourism enables women to play significant roles in personal and communal decision-making processes.

From the perspective of feminist political ecology, development and globalization often perpetuate gender inequalities and environmental degradation. Feminist political ecology examines how intersecting social categories—such as gender, class, caste, race, culture, and ethnicity—affect environmental outcomes and influence efforts to maintain environmental sustainability. This framework highlights the importance of recognizing and valuing the environmental insights of women and ethnic groups, who often possess crucial knowledge about sustainable practices.

Community-based ecotourism supports the development of small-scale tourism programs that are culturally and environmentally appropriate. It emphasizes shared leadership, community well-being over individual profit, and the preservation of traditional culture and the environment. This form of tourism provides an alternative to unsustainable development and offers additional income sources for rural and impoverished communities. However, livelihoods encompass more than economic gains; they also include daily interactions, emotions, and attitudes toward the ecosystem. Women involved in community-based tourism emphasize the value of these interactions, they said "the local community benefits and takes pride in their surroundings, environment, and culture. When locals show you their home, they're proud and happy. This interaction not only supports them financially but also strengthens their cultural pride."

A key issue is integrating local knowledge into ecotourism, while addressing gender dynamics and promoting environmental management. Ethnic communities, particularly women, often possess valuable insights into sustainable practices and resource management. However, gender disparities can influence access to and control over resources, affecting decision-making and the negotiation of gender roles. Additionally, ecotourism presents an opportunity to showcase local knowledge, support community-based conservation efforts, and provide economic benefits. By incorporating both local knowledge and gender perspectives, along with promoting sustainable ecotourism, women entrepreneurs can create more inclusive, effective, and equitable strategies for managing natural resources and addressing environmental challenges.

Coloniality, Labor, and Marginality

While women's engagement in community-based tourism enhances their decision-making power within the households and in public life, labor work executed by women in other sectors is generally undervalued, marginalized, and degraded by society's value. In fact, exploitation of women labor in northeast India and Lanna is an integral part of colonialism and modern capitalism. Chapter 9, by Siya Uthai, draws on an ethnographic study conducted in the tea plantations of Assam and Lanna. It analyzes gender relations in the labor market, the role of women laborers in sustaining the structure of the tea plantation economy, and women's status in a capitalist production process deeply embedded in patriarchal and cultural norms that render invisible women's roles in economic processes. It also explores the question of women's location within the existing system of production and traces how their position has changed following transitions in the relations of production and labor processes.

The roots of women's labor in northeast India's tea plantations are deeply intertwined with colonial exploitation. Historical and ethnographic studies reveal how British administrations established tea plantations that relied heavily on cheap, unskilled female labor (Besky, 2014). The colonial plantation system perpetuated gendered labor divisions, with women often relegated to the most labor-intensive and least remunerated tasks, such as plucking tea leaves. This historical legacy continues to influence the socio-economic conditions of women workers in post-colonial societies. Chatterjee (2017) demonstrates how post-colonial governments and plantation management maintained exploitative labor practices established during colonial times. Women continued to face discrimination in wages and job opportunities, reflecting broader socio-cultural norms that devalue women's work. Siya's paper concurs with Harazika (2018) that one of the most pervasive challenges for women in tea plantations is low wages. Ethnographic research in Assam and Lanna highlights that women's wages are often insufficient to meet basic needs, perpetuating a cycle of poverty and economic dependency. This economic marginalization limits women's ability to invest in their children's education and health, further entrenching intergenerational poverty. The paper also focusses on ethnic women and poor working conditions. Ethnic women working in tea plantations often endure harsh working conditions, including long hours, exposure to harmful pesticides, and inadequate access to sanitation facilities.

Findings from tea plantations reveal distinct patterns of occupational segregation. In Northeast India, limited access to education and systemic rights deprivation confine ethnic women primarily to tea-plucking roles. In northern Thailand, while ethnic women are also overrepresented in lower-skilled, informal labor within the tea garden, they exhibit greater adaptability through skill development and gain more occupation options. Meanwhile, female workers in Northeast India faced difficulty in changing their livelihoods. Education and skills training are crucial for empowering women, yet access to these opportunities is limited for women in tea plantations. High rates of illiteracy and low educational attainment among women workers hinder their ability to secure better-paying jobs and participate in decision-making processes. This educational gap is a significant barrier to achieving gender equality and economic empowerment. Consequently, their primary strategy for improving their circumstances focuses on facilitating access to education for the next generation to pursue a better future beyond the tea garden.

In many parts of northeast India and Southeast Asia, profound changes have been brought about by long decades of missionary toil. Christian missionary activity was central to the work of colonialism, providing the colonizers and their supporters with a sense of justice and moral authority. Throughout the history of colonialism, missionary evangelizing offered the British public a model of "civilized" expansionism and colonial community management, transforming conquest and imperial projects into moral allegories. Missionary activity was, however, unavoidably implicated in either covert or explicit cultural change. It sought to transform indigenous communities into imperial archetypes of civility and modernity by wiping out local epistemologies, remodeling the locals through Christian philosophies.

In Chap. 10, Kengkij Kitirianglarp, Pana Kantha, and Sarid Siriteerathomrong trace the multifaceted legacy of the Young family, whose influence spanned from Northeast India to Lanna across seven decades of the 19th and 20th centuries. American Baptist missions operated in Asia since the early nineteenth century, predating direct American political involvement in the region, and often worked with British Empire support in territories including Thailand, Burma, northeastern India, and southern China. By examining their evolution from American Baptist missionaries to military operatives and finally to naturalists who founded Chiang Mai Zoo, this paper reveals how colonial epistemologies persisted and transformed across changing imperial regimes in Asia. The Youngs operated within an intellectual tradition where scientific documentation served both religious conversion and imperial objectives. Their close relationships with ethnic minorities, particularly the Lahu and Wa peoples, proved invaluable as they navigated shifting geopolitical landscapes from British colonialism to American Cold War influence. The establishment of Chiang Mai Zoo represents the crystallization of these intertwined historical forces—where missionary zeal, colonial authority, and naturalist inquiry converged. Institutions like zoos, this paper contends, often presented as neutral educational spaces, emerge from specific historical contexts that reflect and reproduce naturalist knowledge with colonial origins. The significance of this paper lies in its potential to illuminate how colonial processes of knowledge production evolved across different imperial regimes and manifested in institutions like zoos. By focusing on the Young family, whose activities traversed British colonial territories and later areas of American influence, we can trace continuities and transformations in how knowledge about nature was collected, organized, and institutionalized. This paper reiterates that colonial power extended beyond human subjects to encompass control over non-human beings, creating epistemological frameworks that continue to shape our understanding of the natural world.

Art and Cultural Meaning of Lives

Art plays a vital role in society by promoting creativity, fostering emotional expression, and facilitating cultural understanding. Art can be understood from different cultural perspectives and has different meanings depending on the culture of origin. It can help us relate to each other by understanding the differences between our cultures through their traditions, customs, and art forms, which helps us become more tolerant and see things from a different perspective. Through wood carvings, paintings, and attires, we can gain a deeper understanding of emotional expression and attain deeper cultural understanding of others.

In Chap. 11 by Wipawee Panjinda, Rongkakorn Anantasanta, and Thatchai Hongpaeng, wood carving is explored as a form of art that brings together various cultural expressions. Wood carving is the medium to connect common factor; practice of art, cultures, and belief, their similarities and differences of two territories: Lanna and Nagaland. The paper also examines the impact of Covid pandemic on the lives and works of artists and how the pandemic has influenced this art form. Over the past six years, both Lanna and Nagas have faced significant cultural disruptions during the pandemic, including economic challenges for artisans and limited avenues for creative output. The paper highlights the finding that art is a form of resilience, allowing artists to express recovery, hope, and cultural continuity through their carvings. The COVID-19 pandemic illuminates the ways in which art and culture can offer hope, healing, and a sense of connection in times of crisis. A collaboration between Lanna and Naga wood carvers in the contemporary context reflects this broader role of art in addressing global issues.

Iris Odyuo, a historian from Nagaland, argues in Chap. 12 that the Naga world of wood carving has its roots in local epistemology. The paper attempts to describe the involvement of gods and spirits in the creation of woodcarving arts among the Nagas. The Naga tribes hold great reverence for their ancestors and environmental spirits, who embody the unwritten laws, traditions, and customs of their villages and tribes. Prayers were offered to these entities for good harvests, population growth, success in warfare, and protection against sickness, famine, floods, and earthquakes. The art of woodcarving, including the construction of the *morung* (men's house), log-drum, and village-gate, was accompanied at every stage by a series of taboos and ritual performances. Any deviation from the prescribed methods was believed to result in calamity or even death. Despite using simple tools such as chisels, adzes, axes, and machetes (*dao*), and facing the numerous difficulties and insecurities of life, the Nagas' artistic creativity thrived due to their strict adherence to rituals and taboos associated with these arts.

The introduction of Christianity, colonization, and modern education led to the abandonment of many social and ceremonial practices associated with the traditional religion of the Nagas. Modernity significantly reduced their dependence on and loyalty to the *morung* system, as well as the clan and village structures. New sources of prestige, arising from education, professions, and livelihoods, have undermined the traditional prestige system that once supported ceremonial art traditions. Furthermore, with the increase in conversion to Christianity certain traditional songs and dances that invoke the blessings of the village's benevolent spirits, performed by feast givers, are now considered un-Christian and are forbidden. Many Naga villages have destroyed their log-drums in the *morung*, and traditional ornaments, as well as game trophies and Mithun heads stored in individual houses, have been burned or abandoned. Even today, whenever epidemics or unusual incidents occur in the village, people often blame the log-drums or the wood carvings found in the *morung* architecture.

However, some villages in the central and northern parts of Nagaland continue to create new log-drums and village gates and observe ceremonies related to it. Omens and dreams are still taken seriously in these communities. Naga artists who specialize in traditional wood carving possess a deep understanding of the intricate relationship between their craft and the spiritual realm. They recognize that each piece they carve is not merely an object of art, but a powerful expression of their cultural heritage and beliefs. Through meticulous attention to detail and a reverence for the materials used, these artists create works that are intended to bridge the gap between the physical world and spiritual consciousness, reflecting the rich traditions and stories of the Naga people.

In Chap. 13, Temjensosang explores a semiotic approach to comprehend Naga traditional attires. The paper contends that the symbolic meanings of Naga traditional attires and ornaments served as a foundation to mark its ethnic identities and boundaries. The entire concept of attires is centered on the principle of feast of merit and head hunting, the two structural principles of social stratification of Naga society. Attires and ornaments indicate the social distinction of one clan from the other and most importantly attires signify the status and position of the wearer. Therefore, attires are an identity marker. Each pattern and motifs inscribed in the shawls represent a semiotic meaning of the wearer. Tattooing forms an integral aspect of Naga culture and sets a benchmark in the social stratification. The use of hornbill feather signifies leadership and symbolizes high status. The paper illuminates that cultural objects such as animals, birds, human heads signify sociological as well as semiotic meaning. Attires are not just for covering the body but reveal the tribal identity in the context of Naga culture. Naga culture is a celebration of its diverse tribal identities, traditions, and customs and is a testament to the rich and colorful heritage.

Lessons From Northeastern India: Imagined Communities Beneath the Shadow of Doi Suthep

Lessons from comparative studies between Northeast India especially Nagaland and Lanna may lead us to raise pertinent question involving structural changes and explore solutions that could help restore the ruptured relationship between communities and the natural world. Currently, Lanna or northern Thailand relies heavily on tourism as a major source of income. The proportion of household income from agriculture continues to decline, while urban expansion and commercial farming have exacerbated pollution issues, making them more severe and prolonged each year. Forest areas are continuously decreasing, and resource competition is becoming more intense. Local wisdom is being devalued, and uncontrolled environmental degradation is spreading. Social relationships and community cohesion have been replaced by short-term individual interests, and individualism has become a way of life.

However, in Naga society, agriculture remains the primary way of life and livelihood for the villagers. Every Naga farmer is born and raised within the deeply interwoven kinship system known as the clan, which permeates all aspects of life, including society, economy, politics, and culture. Kinship groups form the foundation of identity and belonging. Social obligations among kinship members are reinforced in daily life through communal land use, agricultural labor management, seniority-based hierarchy, traditional learning within the Morung (youth dormitory), and the celebration of major feasts, weddings, and funerals. Kinship structures influence land ownership, preventing land sales or transfers to outsiders and regulating resource use. This kinship system also underpins shifting cultivation or Jhum, which has been the primary agricultural practice of the Nagas for centuries. Around 80% of Naga farmers derive their main income from agriculture, supplementing it with livestock farming, particularly the local Mithun buffalo, as a secondary income or emergency reserve.

Most Naga villagers still reside in their villages, although houses have been modernized using brick and cement due to timber scarcity. Despite these changes, village layouts remain traditionally structured, with homes built on mountain tops and slopes, village gates, security posts, Morung, dance arenas, and coffins lined along the roads. Some villages have developed shops and guesthouses to accommodate increasing tourism.

Terraced rice farming is a common cultural landscape in Nagaland, often accompanied by fish farming within the paddies. However, limited space and maintenance requirements mean that shifting cultivation remains the dominant agricultural practice. Some villages have started converting forest areas into tea, coffee, and fruit plantations in response to rising market demand. Labor shortages in agriculture have led to increasing employment of workers from Nepal, Bangladesh, Assam, and Tamil Nadu in farming and construction work.

Like other Indian states, Nagaland is modernizing its infrastructure, including roads, water supply, electricity, and public utilities. Many young people are migrating to cities to seek non-agricultural occupations. Tourism, garment production, jewelry, and wood carving industries continue to expand, leading to the gradual disappearance of distinct colors and patterns unique to different clans. Education has spread across the state, with all 1112 Naga villages having at least one local school. Kohima, the state capital, hosts several higher education institutions and universities. An increasing number of young Nagas are pursuing university studies, both within India and abroad, making such opportunities more commonplace.

Nagaland's state constitution supports village and tribal councils in managing public affairs and resolving disputes. The customary laws of the 17 tribes across 7 districts are actively upheld in every local community. Village councils play a crucial role in land and resource management, and village chiefs are still respected figures, even though their legal authority has diminished. Chiefs remain the designated heads of village councils and represent their communities in city (district) and tribal councils.

The contrast between urban and rural Nagaland is becoming more pronounced. Cities embody modernity, whereas rural areas preserve tradition. Urban dwellers increasingly adopt Western-style clothing, and traditional tattooing is declining. Even in villages, visible changes are occurring—village gates are no longer maintained as they once were, and many Morung buildings have been abandoned. Feasts, once a means of wealth distribution and social status enhancement, are now rare, except for weddings and Christmas celebrations. Wealth accumulation and luxury consumption have become markers of social status in the capitalist system.

Modernization trends in Nagaland are influenced by religious conversion, market economics, modern education, and communication. Some traditions are adapting while retaining their original significance. Morung youth dormitories have transformed into youth associations that still preserve tribal culture. Loyalty to tribal identity and Naga nationalism is strengthening through organizations like the Naga Federation (Jacobs, 2012: 176). Naga culture and way of life are not relics of history on the verge of extinction due to nation-building and modernization. Instead, Naga ethnic identity and culture continue to evolve and adapt, ensuring their relevance at individual, tribal, and national levels.

The Dynamics of Naga Adaptation

The dynamic adaptation of the Naga people is clearly evident in their daily lives and agricultural production. The Naga shifting cultivation system is not static; farmers continuously improve production efficiency to ensure food security and livelihood, particularly by enhancing soil quality without the use of chemicals (Grandstaff, 1976, 1980). Recent research in agroforestry and ethnoecology (Cairns, 2024) has demonstrated that the Nagas are continually innovating on the foundation of traditional wisdom to improve the efficiency of shifting cultivation and agricultural systems. For example, they employ a composite farming system that integrates various techniques (Rambo 1996, 2006, 2007) by combining shifting cultivation, terraced rice fields, orchards, and perennial crops. In Konoma village, the Angami Nagas plant alder trees in terraced fields and on shifting cultivation plots to serve as firewood, reduce soil erosion, and enrich the soil (Cairns, 2009; Cairns & Brookfield, 2024: 460–461).

In the case of the Konyak Nagas in Mon District in northern Nagaland, farmers use Macaranga trees as pioneer species that help in swiftly rehabilitating fallow fields and maintaining the fertility of the forest (Changkija et al., 2024). This species also possesses medicinal properties in wound healing. Konyak farmers sow Macaranga seeds in their shifting cultivation plots alongside rice and other grains, and they sow additional seeds when the fields are left fallow, thereby transforming them into secondary forests with dense tree cover.

Soil erosion from rainfall is a major issue in the highlands of Northeast India. Erosion affects 60% of the total area, with 81% of agricultural land experiencing moderate to severe soil loss. It is estimated that over 700 million tons of topsoil are washed away each year—especially during the monsoon season from June to July (Nayak & Chauhan, 2001). Traditional techniques based on the indigenous wisdom of the Nagas have increasingly attracted the attention of Indian scholars as effective methods for managing soil erosion (Lotha et al., 2024: 15). These techniques are essential for Naga shifting cultivation; for example, planting alder trees helps stabilize the soil and replenish nitrogen in cultivated areas. The Ao, Konyak, and Lotha Nagas use logs and stones to build embankments along the edges of their shifting cultivation plots to prevent soil erosion. Log bunding is a technique practiced by nearly every Naga group (Humtsoe, 2011); the steeper the slope, the more intensively the log bunding is applied.

Studies on the dynamic shifting cultivation system and indigenous knowledge among the Nagas have led many scholars to propose that this production system not only ensures food security for highland communities but also preserves forests and biodiversity over the long term (Kerkhoff & Sharma, 2006; Nakro, 2009). Shifting cultivation also promotes continuous forest regeneration (Darlong, 2004). The carbon dioxide released from slash-and-burn practices is offset by the rapid carbon uptake of young trees that develop in secondary forests and recovering lands (Nakro, 2009). Continuous and effective innovations in land use, tree planting, soil erosion prevention, and resource management in Naga shifting cultivation have resulted in an agricultural system that underpins sustainable economic, social, cultural, and environmental balance. Decision-making rooted in local relationships facilitates a more equitable distribution of production wealth while maintaining an ecosystem balance based on shifting cultivation principles that prioritize food security. Community-led adaptations in land use and resource management enable Naga agriculture to preserve an ecological mindset that remains balanced—and one that can adapt sustainably in the face of today's dramatic climate change.

Lessons from Cultural Landscapes and Agroecological Systems

Lessons drawn from the cultural landscapes and agroecological systems of the Nagas reveal that many Lanna communities have been molded by state agencies and formal education systems, leading them to abandon their original mindset and overlook the value of traditional indigenous wisdom. Under centralized, authoritarian resource management and with the widespread shift toward monoculture cash crops, local communities have lost control over their natural resources—soil, water, and forests. The rapid spread of degraded ecosystems has been further exacerbated by the increased use of chemicals that are toxic to the environment.

These insights prompt important questions: Should Lanna society revert its focus back to mixed farming systems? How can this key question be addressed when it is intertwined with the structural issues of monoculture agriculture, commercial crop production, and the role of both state and private sectors in managing food production through support for large-scale farming?

Issues such as agricultural trade policies and the corporate-driven development of hybrid crops have stripped farmers of their bargaining power, making it increasingly challenging for them to control production and resource management. The loss of forested areas and biodiversity continues to undermine the food security of Lanna communities. Moreover, neglecting the importance of "knowledge" and local wisdom, as well as overlooking the role of farmers and communities as the primary managers of natural and genetic resources (Santasombat, 2008), has further diminished the potential of Lanna farmers, reducing them to mere producers dependent solely on state aid.

Consequently, Lanna communities have been compelled to depend on external capital, expertise, and technology, which in turn leads to mounting debt and cyclic patterns of disempowerment, marginalization, and environmental degradation. Examples include the contamination of soil and water by chemicals used in the expansion of commercial crops, and the pervasive haze and dust pollution that increasingly affects society. By contrast, the Naga farmers have been able to generate new innovations while effectively blending them with their traditional knowledge in accordance with their cultural landscape. As a result, these local communities continue to maintain their independence and retain effective control over their production and natural resources.

The comparison of cultural landscapes raises further crucial questions: How can we revive a sustainable ecological mindset to counter the spread of degraded ecosystems? How can we restore the once-breakable bonds between communities and nature amid ever-widening economic and political inequality? How might we reinvigorate the role of Lanna farming communities by truly decentralizing power to local organizations—thus enhancing their capacity to produce food securely while conserving critical natural resources and biodiversity? Moreover, how can the balance of power be reconfigured in relationships with state agencies and financial institutions, to reduce dependence on companies that supply technology and production inputs (such as seeds, chemical fertilizers, and pesticides), so that communities can produce their own inputs, generate energy, breed seeds, and raise livestock, ultimately enabling them to invest based on their indigenous knowledge and in collaboration with other agroecological communities?

By establishing community organizations and fostering resource-based relationships at both local and network levels through the direct exchange of experience and shared learning, communities can better manage agricultural lands. This improves the diversity of production—ranging from crop rotations in mixed-use plots to comprehensive animal husbandry and the optimal use of land—and it also supports educational activities that incorporate local wisdom via responsible tourism. Such approaches aim to transform the way land and resources are utilized, ultimately paving the way for an agricultural system that truly achieves economic, social, and environmental sustainability.

Yos Santasombat

References

Alcorn, J. B. (1989). Process as resource: The traditional agricultural ideology of Bora and Huastec Resource Management and its implications for research. In D. A. Posey & W. Balée (Eds.), *Resource management in Amazonia: Indigenous and folk strategies* (Advances in economic botany) (Vol. 7, pp. 63–76). The New York Botanical Garden.
Anderson, B. I. (1978). *We lived in Nagaland: The experiences of a missionary family*. Sema Bible Institute.
Ao, T. (1999). *The Ao-Naga oral tradition*. Bhasha Publications.
Balee, W. (Ed.). (1989). *Resource management in Amazonia: Indigenous and folk strategies* (Advances in economic botany) (Vol. 7). New York Botanical Garden.
Balee, W. (1994). Noble savage or noble state? Northern myths and Southern realities in biodiversity conservation. *Ethnoecologica, 2, 7–19.*
Berkes, F., Folke, C., & Gadgil, M. (1993). *Traditional ecological knowledge, biodiversity, resilience and sustainability*. Beijer International Institute of Ecological Economics.
Berlin, B. (1968). *Tzeltal Numeral Classifiers: A study in ethnographic semantics*. Janua Linguarum.
Besky, S. (2014). *The Darjeeling Distinction: Labor and justice on fair-trade tea plantations in India*. University of California Press.
Blackburn, T. C., & Anderson, K. (1993). *Before the wilderness: Environmental management by native Californians*. Ballena Press.
Bragg, K. (1986). Akha Ethnobotany: The forest resources of a mountain people. In A. R. Walker (Ed.), *The highland heritage*. Suvarnabhumi Books.
Bromley, D. W. (1991). *Environment and economy: Property rights and public policy*. Blackwell.
Brown, A. H. D., Frankel, O. H., Marshall, D. R., & Williams, J. T. (Eds.). (1989). *The use of plant genetic resources*. Cambridge University Press.
Bubandt, N., & Tsing, A. (2018). Feral dynamics of post-industrial ruin: An introduction. *Journal of Ethnobiology, 38*(1), 1–7.
Cairns, M. (2009). *The Alder Managers: The cultural ecology of a village in Nagaland, N.E. India*. PhD dissertation to the Australian National University, Canberra.
Cairns, M. (Ed.). (2024). *Farmers innovations and best practices by shifting cultivators in Asia Pacific*. CABI.
Cairns, M., & Brookfield, H. (2024). Composite farming systems in an era of change in Nagaland, northeast India. In M. Cairns (Ed.), *Farmers innovations and best practices by shifting cultivators in Asia Pacific*. CABI.
Changkija, S., Thakuria, D., & Cynthia, A. (2024). Traditional use of Macaranga trees for soil fertility by Naga shifting cultivators in northeast India. In M. Cairns (Ed.), *Farmers innovations and best practices by shifting cultivators in Asia Pacific*. CABI.
Chapeskie, A. J. (1995). *Land, landscape, culturescape: Aboriginal relationships to land and the co-management of natural resources*. Paper Prepared for the Royal Commission on Aboriginal Peoples. Libraxus.
Chatterjee, P. (2017). Labor movements in Assam's tea plantations: A historical perspective. *Journal of South Asian Studies, 40*(2), 259–279.
Chasie, C. (2017). *The Naga Memorandum to the Simon Commission, 1929*. Standard Printers and Publishers.
Clarke, W. C. (1976). Maintenance of agriculture and human habitats within the tropical forest ecosystems. *Human Ecology, 4*(3), 247–259.
Conklin, H. C. (1957). *Hanunoo agriculture: A report on an integral system of shifting cultivation in the Philippines*. FAO.
Cronon, W. (1983). *Changes in the land: Indians, colonists, and the ecology of New England*. Hill and Wang.
Cronon, W. (Ed.). (1995). *Uncommon ground: Toward reinventing nature*. Norton.

Crumley, C. L. (1994). Historical ecology: A multidimensional ecological orientation. In C. L. Crumley (Ed.), *Historical ecology: Cultural knowledge and changing landscapes*. School of American Research Press.
Darlong, V. (2004). *To Jhum or not to Jhum: Policy perspectives on shifting cultivation, The missing link*. Society for Environment and Communication.
de Sousa Santos, B. (2014). *Epistemologies of the South: Justice against epistemicide*. Routledge.
de Sousa Santos, B. (2018). *The end of the cognitive empire: The coming of age of epistemologies of the south*. Duke University Press.
Drouyer, I. A., & Drouyer, R. (2016). *The Nagas: Memories of headhunters, Indo-Burmese Borderland. Volume 1: History, identity, Architecture and sculpture*. White Lotus.
Gragson, T. L., & Blount, B. G. (Eds.). (1999). *Ethnoecology: Knowledge, resources, and rights*. University of Georgia Press.
Grandstaff, T. B. (1976). *Swidden society in North Thailand: A diachronic perspective emphasizing resource relationships*. Ph.D. Dissertation, University of Hawaii.
Grandstaff, T. B. (1980). *Shifting cultivation in Northern Thailand: Possibilities for development*. The United Nations University.
Hardin, G. (1968). The tragedy of the commons. *Science, 162*(3859), 1243–1248.
Hazarika, D. (2018). Tea plantations and the displacement of indigenous communities in Assam. *Human Ecology, 46*(1), 77–88.
Head, L. (2016). *Hope and grief in the Anthropocene. Reconceptualizing human nature relations*. Routledge.
Humtsoe, L. S. (2011). *Traditional farming and indigenous technical knowledge of the Lotha-Nagas of Wokha district, Nagaland*. Unpublished Master's thesis submitted to the Department of Agricultural Extension, School of Agricultural Sciences and Rural Development, Nagaland.
Hutton, J. H. (1921). *The Angami Nagas*. Macmillan.
Kerkhoff, E. E., & Sharma, E. (2006). *Debating shifting cultivation in the Eastern Himalayas: Farmers' innovations as lessons for policy*. International Centre for Integrated Mountain Development (ICIMOD).
Jacobs, J. (2012). *The Nagas hill peoples of Northeast India*. River Books.
Lotha, L., Khonglah, L., Tripathi, S. K., Singh, N. S., & Upadhyay, K. K. (2024). Log bunding: Laying wood along slope contours in Northeast India and other farmer strategies for minimizing soil erosion on swidden slopes. In M. Cairns (Ed.), *Farmers innovations and best practices by shifting cultivators in Asia Pacific*. CABI.
Mahanty, S., Milne, S., Barney, K., Dressler, W., Hirsch, P., & To, P. X. (2023). Rupture: Towards a critical, emplaced, and experiential view of nature-society crisis. *Dialogues in Human Geography*, 1–20.
Martin, C. (1978). *Keepers of the game*. University of California Press.
Medin, D. L., & Astran, S. (Eds.). (1998). *Folkbiology*. MIT Press.
Nakro, V. (2009). *Traditional agricultural practices and sustainable livelihood: A thematic report*. Nagaland Empowerment of People through Economic Development Project (NEPED) and Project Operations Unit, NEPED Department of Planning and Coordination, Government of Nagaland.
Nayak, R. C. & Chauhan, B. S. (2001). *Soil and conservation measures suitable for northeast India for sustainable production at farm level*. Paper delivered to a National Seminar on Water and Land Management, including CAD, for Socio Economic Upliftment of the NE Region, 22 and 23 November 2001, Tezpur, Assam, India.
Nazarea, V. D. (Ed.). (1999). *Ethnoecology: Situated knowledge/located lives*. University of Arizona Press.
Nienu, V. (2012). *Naga cultural milieu*. Thomson Press.
Nshoga, A. (2008). *Traditional Naga village system and its transformation*. Anshah House.
Posey, D. A. (1997). The Kayapo. In *Indigenous peoples and sustainability: Cases and actions*. IUCN and International Books.

Posey, D. A., & Dutfield, G. (1997). *Indigenous peoples and sustainability: Cases and actions.* IUCN and International Books.

Quinn, N., & Holland, D. (1987). Culture and cognition. In D. Holland & N. Quinn (Eds.), *Cultural models in language and thought.* Cambridge University Press.

Rambo, A. T. (1996). The composite swiddening agroecosystem of the Tay ethnic minority of the northwestern mountains of Vietnam. In B. Rerkasem (Ed.), *Montane Mainland Southeast Asia in transition* (pp. 69–89). Chiang Mai University.

Rambo, A. T. (2006). Human ecology research on tropical agroecosystems in Southeast Asia. *Singapore Journal of Tropical Geography, 3*, 86–99.

Rambo, A. T. (2007). Observations on the role of improved fallow management in swidden agricultural systems. In M. F. Cairns (Ed.), *Voices from the forest: Integrating indigenous knowledge into sustainable upland farming* (pp. 780–801). Resources for the Future.

Rose, D. B. (1996). *Nourishing Terrains: Australian aboriginal views of landscape and wilderness.* Australian Heritage Commission.

Santasombat, Y. 2003 (2013). *Biodiversity, local knowledge and sustainable development.* Regional Center for Social Science and Sustainable Development (RCSD), Faculty of Social Sciences, Chiang Mai University.

Santasombat, Y. (2008). *Flexible peasants: Reconceptualizing the third world's rural types.* Regional Center for Social Science and Sustainable Development (RCSD), Faculty of Social Sciences, Chiang Mai University.

Sponsel, L. E. (1995). Relationships among the world system, indigenous peoples, and ecological anthropology in the endangered Amazon. In L. E. Sponsel (Ed.), *Indigenous peoples and the future of Amazonia.* University of Arizona Press.

Stockhausen, A. V. (2014). *Imag(in)ing the Nagas. The pictorial ethnography of Hans-Eberhard Kauffmann and Christoph von Fürer-Haimendorf.* Arnoldsche Art Publishers.

Tsing, A., Mathews, A., & Bubandt, N. (2019, August). Patchy Anthropocene: Landscape structure, multispecies history, and the retooling of anthropology. *Current Anthropology, 60*(S20), S186–S197.

Van Ham, P., & Saul, J. (2006). *Expedition Naga.* River Books.

Van Laerhoven, F., & Ostrom, E. (2007). Traditions and trends in the study of the commons. *International Journal of the Commons, 1*(1).

Contents

Chapter 1
Understanding an Idea of Common Commons Across Southeast Asia and North-East India

Prasenjit Biswas

Abstract Verrier Elwin had shown in the Chhotanagpur plateau Santal Life-World that even a thorn, a larval worm, an earthquake is sacred, because something is happening—whether the earth is maintaining itself, correcting itself, balancing itself. The case is homologous in India's North East. Rather it is more complicated in the sense that natural phenomena and natural objects are attributed human faculties and vice versa, which is a trans-human perspective. Apatani *KojumKoja*, the Hmong notion of *saibloojmem*, Mizo*thlawmghaina*or Khasi *KaRyngkiew* and many more such notions ascribe personhood by turning things into beings and making communities 'partake' in the rumble of the stone, wind of the mountain and in the crying rain of the sky in more-than-human ways. The paper explores the question of becoming of this ascribed identity and collective personhood across ethno-tribal communities of South and Southeast Asia.

The paper makes use of evolutionary epistemology of experience of the lived world in understanding the practices of ascription and deeper realities of valuing nature in both factive and non-factive terms, involved in a discourse of affirmation of identity relation between self and other in the context of South and Southeast Asian Communities in their apparent linkages with Northeast Indian communities.

An architectural sharing of an interpretative complex of bio-cultural, transgenic-trans specie and transmigrating symbols, legends and myths with their overflows in worldviews and belief structures mark an unexplored zone on intimacy and extimacy between too many Southeast Asian cultural communities and Northeast Indian ethnic and religious identities in late twentieth and twenty-first centuries.

One needs to get to the specifics of such societal overflows and subterranean networks of linguistic, ethnic, cultural complexes of Common Commons arising out as an architecturally shared interpretative complex that exist between communities. To be specific, Shan-Ahom-Naga-Zo complex, a more dispersed story of successor societies in the landscape of Zomiyas, a continuous movement down Mekong to Brahmaputra and a more-than-human world of specie migration are some of the

---

P. Biswas (✉)
North-Eastern Hill University (NEHU), Shillong, India

Y. Santasombat (ed.), *Bio-Cultural Diversity, Ethnicity, and Local Knowledge*,
https://doi.org/10.1007/978-981-95-4242-0_1

themes that the paper shall explore for working out a plausible intellectual project of decodification of this enchanting world of shared cultural worldviews.

Keywords Worldviews · Zomiya · Successor society · Commons · Migration

1.1 Introduction

The pluricultural and non-one-sided existence of religious, ethnic and linguistic diversities in North-East India open up a possibility of experiencing a variety of 'inner worlds' through an ethics of relational, recognitional and shared intersubjective modes of understanding the Other. The Other arises from "origins" of migration of various ethnic and tribal groups who entered in Sub-Himalayan terrains of present Northeast India at various phases of history. A large number of tribal and ethnic groups under the broad rubric of Naga and Kuki-Zo communities carry with them a memory of migration from the Southwestern part of present China, while the Khasi-Jaintia community connects itself to Mon-Khmer groups of Cambodia and upper Mekong. The whole story of migration needs to be historically evidenced and connected through the timeline, yet plural origins, distinctive routes of migration and settlement and evolution into the present sense of becoming and belonging of many of these tribal and ethnic groups present an interpretative complex of bio-cultural, transgenic, transspecies and transmigrating domain of symbols, legends and myths with their universe of discourse constituting the worldviews and belief structures of these very human communities. To be specific, Shan-Ahom-Naga-Zo complex, a more dispersed story of successor societies in the landscape of Zomiyas, a continuous movement down Mekong to Brahmaputra and a more-than-human world of specie migration are some of the themes that needs to be explored for working out a plausible intellectual project of decodification of this enchanting world of shared cultural worldviews.

A pluricultural celebration of worldviews is necessarily an interpretative celebration of otherness of the Other without much ado about a repetitive and a collaborative reconstruction of one's own inner worlds. This is simultaneously qualified and an open-ended manner of self-understanding, self-realization and a negation of dichotomies that arise of cognitive given inside/outside distinction. The situation could be best described in terms of what theorists of decolonization termed as 'epistemic disobedience', which is de-linking constitution of knowledge by nexus of colonial/modern forms of power *from* what is constituted by 'free decisions' of 'free agents'.[1] Rediscovery of such an inner world is what 'celebration' of diversity in the context of North East India would mean. The inner worlds weaved around the percipient agency of inherently relational terms of epistemic stance towards the world

[1] Mignolo (2011). Mognolo argued for a free world conceived by decoloniality of being in which many worlds can fit. My argument is that such plural worlds are part and parcel of decolonized inner worlds of tribal and indigenous communities.

by shared intentionality of the collectively knowing selves gives rise to 'emergent forms of life' produced out of complex encounters of 'a dialectic of reciprocity, relationality, interdependence and de-linking'. Understanding the nature of this relational ontology in the context of India's North East is the task before us.

1.2 The Inner Worlds of Tribes

Tribes are not monolithic constructs, they are an evolving form of heterogeneity and specificity in terms of cultural, ethnographical and possibly in political terms of 'strategic essentialism' (Eide, 2010). All these continually evolved aspects of inner worlds are deeply interrelated that establish the linkage of constitutive relationality and opens it up to a participant/percipient other. This is how auto-ethnographic picturing of this inner world finds its expression in a variety of decolonized and autonomous representation by oneself as well as in the representations by another, by way of a conscious choice of boundary-crossing. A good example of this innerworldly boundary crossing could be given here from Buddhist Monpas cutting the body of their dead members into hundred pieces and throwing it into the river *Tawanchu* in Tawang district. The ritual crosses the boundary between ethnic and religious practices and goes beyond the 'incarcerated consciousness' of the self and the other in a ritual of celebration of dependent body parts and their survival in more complex 'incarnated' material substratum. Similar is the Mao Naga story of origin in which three brothers Tiger, Spirit and Man compete for inheriting mother's land, in which Man wins the contest by hitting the target with a bow and an arrow. This is a kind of an anthropocentric episteme which fulfills the need for human beings' superiority in the race over others such as animals and spirits. But such a contest is far from conclusive in other cases. In the Idu-Mishmi case, members of the tribe believe that they originated from the enchanting place called *Apesha* where death ritual called *Reh* was performed for the first time with the song that stated, as per their belief,

> Oh, the boatman, please take the boat and cross the river
> The soul starts the journey to the other side of the river. (Sarma, 2015)

The death ritual and the presence of goddess Apeh Milli bringing good harvest from Apesha is another way of combining death and harvest as sources of sustainability in travelling to the other side of the river, a cross over that nurtures sustainability in this earth. Similarly in the folk tales centring Loktak, Ibudhou Pakhangba (the Supreme Snake God) incarnating as the wooden boat (locally known as hee) shows a celebration of an ecosystem that nurtures homes on the lake called Phum. Fear and gratitude together marks the making of the *Phum* and *Hee*.[2] Creating a holistic ecosystem of belief and practices by way of combining crucial elements of the lifeworld have been the mainstay of philosophical conceptualization by a variety

[2]Asem Manimala, from her personal communication related to the project entitled, Litoral Community in Loktak Lake: precarity and everyday life.

of communities in Northeast India. Mamang Dai's statement that 'we have long journeys in our blood' or 'we descend from solitude and miracles' conceptualize a certain kind of whispering sounds of words spoken by unknown spirits in one's voice purveys a shamanic symbolism. She recounts it beautifully in her 'songs of rhapsodists',

> In the end we all have *remembrance*. The sword *rattles* and the dancers *sing* in chorus.(…) it is a language that never ceases, and they sing *because hills are old*, older than all sin and desolation and man's *fascination* with blood. (Dai, 2006)

Can remembrance recount the old and the past in a never-ending language? Seemingly music fills the temporal gap between where it all begins and where it all ends. This is a rendering of the ordinary in a different tonality, in a different form of body swaying and silencing that are part of an experiential transition to a desired world that makes entry into one's life-world. It's much more than merely a product of an anthropological machine, but it is a shared, believed and participated form of life that organizes itself around its inner symbols and performs its meanings in multiple ways that need to be celebrated.

There is an inimitable singularity in a collaborative reconstrual of one's own inner world through a variety of performances that simultaneously produces a biography, an individuality and a specificity of a tribe, ethnic group or a community. This reconstrual involves a certain amount of embodiment of other spirits, other elements. Use of buffalo heads, floral designs, dancing bodies and many such aesthetic signifiers create an ambience of an inner world that could be accessed from outside, yet it cannot reach out to elements like 'soul figures', beads of snakes and woodpecker, or the heart of the insect. This inability to reach out to its cultural otherness makes us celebrate them as they give them to us in the mode that they decide autonomously. Celebration of such specific symbolism and its attendant schemes of interpretation are embedded in figurative descriptions and tropes of belief working together in an encoded fashion.

Arnold Berleant (1992) states that the idea of nature is how it affects, which refuses to be constrained by discreet boundaries. Suzi Gablik (2002) discussed the case of tribal imagination in this manner: if something is natural, then it is beautiful. In India, even a thorn, or worm, even an earthquake is sacred, because something is happening whether the earth is maintaining itself, correcting itself, balancing itself. The case is homologous in India's North East. Rather it is more complicated in the sense that natural phenomena and natural objects are attributed human faculties and vice versa, which is a trans-human perspective.[3] Both the ontological and the artefactual are part of these trans-human understanding of human-nature relationship. For example, the cult of recalling the ancestors prevalent among the Nagas, Mizos and Khasis assumed a complex symbolic form. Ancestors are called back through ritualized natural objects, such as bamboo poles, collection of stones and bones and

[3] Stories of 'speaking frog', transformation of human to hornbill, soul leaving or entering the body that presents an impossibility of mind or body or self attaining Nature in terms of presentation of ideas.

by specific water rituals. The complexity of such ritual forms in symbolic practices arises out of construction of 'natural will' (Durkheim, 1982).

Linguistic embeddedness of participation of nature-as-subject in the constitution of the life-world is another feature of tribal subjectivity. In the narrative constitution of the life-world as a Thou (in the second person form), wisdom stands out as an irreducible quality of that life-world. Narrative, in the first-person mode gives rise to the 'hard problem' of making intentional acts of any constitutive subjectivity converge with the horizon of meaning already formed within the life-world.

One example may be drawn from the context of Ethnophilosophy. Ethnophilosophically speaking, these concepts are connected and instantiated only in a worldly way: 'where the world is not construed as a surrounding space, but as the multiple tracing out of the singularity of existence as a tribe'. In the multiplicity of ethnic and tribal plurality in certain parts of the world, a tribe is only a fragment of access, it comes only when the commodities, digits and the formulas of modern world is absent. In other world, tribes present themselves more than a gift of intellect and wisdom, as a surplus of knowing, that lies in its phenomenology of world making. One example of such phenomenology is the Apatani concept of *Kojum Koja*, the lost civilization that is represented in the body of the hills above Subansiri valley, literally the rib of the hills. It is not merely the memory of loss due to deluge, but it is precisely the location of the civilization on the hills. As descendants of Kojum Kuja, the present Apatanis stand against the god of deluge called Biri-Bote. The hill and the river enter into a binary contestation in the narrative refiguration of the self that is never present in the present. This comes quite close to Hmong notion of *saib loojmem*, or 'sighting the veins of the mountains'. There is a kind of geomancy involved in Apatani and Hmong imagination of the earth. Khasis, in a similar manner speak of *Law Kyntang* (sacred forest) as a sacred part of nature, which is a figuration of the self in the depth of nature. In Zeliangrong and Tangkhul tribes, the main pillar (*sana yumbi*) of the house there are human carving, supposedly prepared to signify ancestral descent.

The indignity of non-distinction between the tribal world and the techno-scientific 'modern' world is anticipated in the presence of the tribe that comes in the absence of all the sense-making artefacts of 'our' world. To put it differently, the act of signification is not a catch in the unlimited networks of meaning, as an act of giving meaning begins from the secret of the symbols that act as the 'condition of possibility' of wisdom. This ethnophilosophically empathic statement of wisdom through symbols as condition of its possibility manifests the singularity of the tribe without manifestation. It is an ambivalent expression of the multiple worlds that remain unmanifested in wisdom, while it expresses itself as a secret, a secret of wisdom, a condition of possibility for an interlacing and interfacing between language and praxis, between thought and action. The tribal self arises from this ambivalence of symbols that communicates and shares a secret that remains unmanifest as a singular entity, while its multiplicity of senses do not make any sense unless the tribal self-interlaces itself in the very act of sharing.

1.3 (Bio)Cultural Logic of Diverse Festivities

A shared notion of resources and its patterned meanings mark a biocultural logic of conservation and its correlated rituals. In Khasi-Jaintia tradition, use of rice in important occasions of life is stated in this saying, 'u Khoo na iing; ka doh na trep, u jhur na kper', meaning 'Rice from the granary, meat from the pan, and vegetables from the garden' demonstrate a biocultural legacy of conservation practices in ensuring food to the community (Mohrmen, 2025).

The winter Bihu known as Bhogali Bihu, as opposed to Rongali Bihu of the beginning of the sowing season in summer is marked by several rituals pregnant with very deep faith and meaning. Post-harvest, one needs to reclaim the land and seed and clean up the ecosystem, when water is in short supply. Burning the house of hay in the middle of the field clears up the left over hay after rice crops are collected by the community. The burnt ashes replenish the field with its nutrients. Anthropologically this is an exercise of biocultural rights over land that sustains an agrarian economy in rural Assam.

Young people erecting *Meji* and *bhelaghar* as makeshift huts for the purpose of feasting and then burning it the next morning signify a collective bonhomie. One could also see the symbolic continuance of the practice of erstwhile youth dormitory as it prevailed across the Northeast, where young people were trained in physical and mental exercises and cultural norms and values. Bhogali Bihu celebrated in different names and differently evolved rituals among tribes like Mishing, Deuri, Moran, Bodo, Rabha, in all, some 36 plain and hill tribes assume a colourful form of feasting, dancing, singing and rejoicing coinciding with the progress of the winter season. Interesting recreational and competitive fights involving animals, birds and their eggs, and water sports are organized across Assam to replenish the community after days of hard work in the field. People belonging to various professions too get involved in agrarian rites that reconnect them to their roots.

What is significant at this point is not just a celebration of a traditional ritual, but exercise of biocultural rights in the context of Assam and the entire Northeast. The Kuki-Chin festival of Chavang Kut also brings in a parallel ritual of invocation of ancestors and burning of weeds from the field near the store holes of grains come close to agricultural practices of the hills. Chapchar Kut among the Mizos as a post-harvest festival has a significant bamboo dance where men will only sit and tap the bamboo while women will perform the dance also reminds the matrilineal past of the Mizos, where women were the main cultivators and protectors of the harvest. Among the Southern Angamis of Nagaland, Phousnyi or Sekrenyi festival translated as sanctification or purification festival comes pretty close to Makarsankranti in its style and spirit. Sekrenyi involves ritual observance of *Mi-KI* fire after cleaning the sources of water, absolutely necessary for the next harvesting season. Among the Eastern Nagas like Lothas celebrate post-harvest festival called *Tokhu Emomg* marking a ritual collection of foodgrains and edibles in their traditional baskets, which is redistributed by the community priest.

The winter-festivals celebrated across Northeastern ethnic groups involve a preparation of the next harvesting season. A kind of reobjectification of the rice and maize seeds happen in a variety of ways. In the case of Assam, sowing Boro rice is an effective way of making use of early summer flood carrying sediments and also preserving a good harvest. The cycle of harvesting is regenerated through burning in the field, ritually symbolized in Uruka and Meji burning. In the case of Angami Nagas, for preparation of Zhuto, spouted rice grains along with the hulls is fermented into mash and mixed with bamboo shoot and preserved in a wooden vessel. This preparation during winter allows the rural community to save seeds to be harvested during rainy seasons. Be it winter harvest or preparation for the next season of harvest; the whole process involves rituals meant for preparation of the next farming season.

The entire preparation and processing is guided by a biocultural need and such need is preserved in ritual practices. Such ritual practices have its positive and negative meanings. When rice is turned into feast, it is rice beer that is ritually used to signifiy both success and future sustenance. The variety of rice beers used during winter festivals involve a long chain of grain collection process moving onward into fermentation and observance of strict cultural norms and taboos in ensuring a tasty and healthy drink. Angami and Chakesanga genna (meaning Taboo) guides the community toward certain do's and don'ts that are practically useful in making rice beer and sustaining the food chain.

The large number of rice beer of various kinds across the Northeast also marks the possibility of a material basis for rituals of rejuvenation. Rejuvenation of both ancestral and other spirits living across hills, rivers and fields occupy the ritual imagination of naming them, possessing them and celebrating their release in the form of healing and successful harvest. The interrelationship between enhancing values of harvest in rituals and production of rice beer is an anthropological delight that fascinated many observers. The traditional *Laupani* during the currently held loom festival or hornbill speaks volume about continued significance of an agricultural life-world despite too many contestations in the social world of ethnic and tribal groups.

On the negative side of it, certainly, there is a perpetuation of witchcraft and superstition in certain parts of Northeast. Repeated witch hunting in Assam, Tripura, Meghalaya and other parts shows a kind of failure of state promises and a failure to comprehend rapid technological changes that suddenly dislodges the agrarian land relations and cultural practices. The disruption and displacement affect a variety of significant ritual practices, art, image making and craftsmanship. As some philosophers have argued, this also marks a recession of the sacred and the rise of the so called secular life that abandons quintessential aspects of customary and traditional practices.

In the case of Assam, the internal displacement of minorities and smaller ethnic groups like Tiwa and Lalung along with Bengali Muslims, sandals and Bodos created a situation of lack of inter-cultural empathy and ambience. One of the major reasons for decline in traditional faith is a kind of extreme inward-looking affirmation of culture that is not able to catch up with new challenges coming from new technologies and symbols. The traditional artefacts and their symbolic significance within folk culture and narrative are often given a go by in the upsurge of religious identities that pushes out cultural symbols. One interesting case is a gradual conversion of imageless cults such as Sankari culture, or Doyni-Polo into some iconic Hindu features that internally changes the rules of a culture. Such changes leave their debilitating mark on rituals and festivals like Bihu. Although everyone agrees about its Bihu's great unifying role in Assam's assimilative culture, yet there are signs of depletion that affects its cultural reproduction. The unifying role is both literal and metaphorical as cultural and religious difference are levelled by Bihu by its culinary, musical, ritual and social commonalities that is shared between Hindus, Christians, Muslims and others. This can play a significant role in catalysing public knowledge of each other that feeds back into each other's responses to each other's culture expanding the domains of cultural logic in a democratic manner.

These biocultural rites of Northeast India find too many parallels across whole of Southeast Asia. Such rites share a common structure of belief in sustenance of commons. In upland Laos, what goes by the name of Rmeet culture holds the following as the most important:

> (…)the most important traits of Rmeet tradition (Lao: *hidgong*, R: *riid priim*) were: the sword dance (*wuon*), traditional clothes, singing, drumming on wooden drums, sacrifices at rice granaries, the harvest feast, the sacrifice to the village spirit and buffalo sacrifices in general, as well as visiting ill people. (Sprenger, 2016)

These ritual practices are deeply connected to common resources and their representations available in the very Rmeet culture and its interconnected threads with adjacent cultural groups.

1.4 Sustaining Selves in Migration

The terrains of migrations find its descriptive, symbolic and existential representation within the self-definition or within the core notion of self in many of the tribes and communities. For example among the Kuki-Chin group of tribes, narratives of a common origin somewhere in Southwest China called Chhinlung/Khul and similarly among the Naga groups, a shared common origin called Makhel etc. give rise to an oral narrative of migration through various places in the entire Mekong-Sino-Myanmar terrain. What is more interesting is the dialogical and interactive motifs within these tribes in the form of water-spirit, rock-spirit and hill-spirit that enter into existentially lived self's reflection on its own life. Such spirits assume a variety of forms and functions such as kinship relations, transformations into insects, animals and birds. Spirits act as potential and efficient causes behind natural disasters or gifts and boons that are received from Gods and Goddesses. There is an implicit ordering of all that one experiences in terms of spirit functions and its relationship with higher order beings. One may surmise that concrete objects such as rocks and mountains are situated as spirits that live in the neighbourhood and are contained within a sacred boundary within which the tribal self can define itself. Only on the occasion of death, such a self-travel across the threshold that goes onto land of dead that lies beyond the bounds of the immediate sacred boundaries drawn by concrete hills and rocks. A boundary of the sacred and it's beyond often signifying the inside and the outside that determines the subject position of the communitarian ethno-tribal self.

These boundaries are conceptualized by tribes in terms of belief about past, present and future. Time and boundary lead to abstractions and interpretative surpluses such as memory tales, limits of finite and infinite beings, ritual means of propitiation of these beings, metaphors and images of the supernatural, sickness and healing processes of the mortals etc. These interpretative surpluses open up specific conceptual tokens to a hermeneutic interface with other concepts and beliefs drawn up from other tribes and communities. For example, in a fascinating story of lived experience, there is a competition between sister rivers (who are river goddesses in Khasi tale), namely, KaUmtong, KaTorasa, KaPasbira, KaJani and KaDwara to reach the plains of Sylhet. As all of them take short cuts they lose their way in the hard rocks. Contrastingly the river Umiam quietly flows down the serpentine hills to reach the plains. The flow of the river is construed as a river spirit that gets its representation in competition between rivers with a rider that one who does not take a short cut reaches the destination. River stories and river spirits are absent among those tribes who did not migrate following the course of river. For example, in the Kuki-Chin group of tribes, it is the water creatures such as crabs and insects that assume an anthropomorphic function in determining fate of humans, but there are no significant tales of rivers connected with thick and thin of life. It is rather an issue of survival that arises from such river based animals. It needs to be further explored whether two different kinds of narratives based on two different terrains of migration can find some parallels and comparisons.

On the core concept of self, for example, in the Kuki-Chin context, the concept of Tlawmngaihna/Tomngaina and the concept of Rngiew in the Khasi context that defines the presence of the self/person could be compared in their fine-grained shades of meaning. Any such comparison cannot just be a semantic exercise, but it needs to be correlated to the whole course of migration. How a string of concepts that are available within a specific worldview give rise to a coherent meaning token and how each of these concepts remain connected with some pre-history of migration can provide a better understanding and perspective on conception of self. For example, the Khasi concept of Nongshonoh as the keeper of U Thlen, Palaung (Myanmar and Northern Thailand) concept of snake spirit called I-ran-ti and a place where such a snake spirit lives called *Kok-ya* has a close connection.[4] In the course of migration, Palaungs and Khasis are supposed to have close ties in terms of cultural memory and concepts from one worldview got its repercussions in another. It got altered, transformed and fine-tuned by taking off from each other's specific

[4] My Colleague and co-researcher Basil Pohlong have found this out through his field research among Palaungs of Northern Thailand.

moorings. Similarly, in the Kuki-Chin group of peoples, the *Ciimnuai* genealogy of belonging to a common ancestral tree with tribes such as Miao, Lolo, Moso, Yao, Tung and Lai in Northern Thailand and Southwest China produce an intensional notion of self, a self that derives itself from its relation with other brethren tribes with whom they are separated in the course of migration. An example of this arises in Tlwamngaihna/Tomngaina or the Mizo notion of personhood that finds its close reverberation in Zothang/ Zokang of Lai-Zo and *Tlang* of Zo seemingly travelled from Kham region of Southwest China to move down to Yesago of Northwest Myanmar to Zo-Tlang of the Chin state of Myanmar. Seemingly, Tlang as a locale with its institution of Kuki-Chin group of people assumes the form of Kham-tung-mi, literally, "people on the hill slope" as well as a form of "soul" in the animal or in any other natural object that later gets transformed into the human form of kinship relations and associated norms and codes of behaviour to find itself emerging into *Tlawmngaihna*, as in present Lushei hills. Contrasting people with Zo line of descent further hone the concept as La-Pao as in Kuki culture. Similarly the commonality between Zo and Karen worldviews can be noted in Pa-O and Sgaw Karen belief in *yawa*, the creator who is dispersed in forest, river, field and land and the ancestral female spirit known as *Naka cheng*. The interconnections between tlang, Tlwamngaihna and nakacheng and their mutual transformations deserve exploration into many a hidden contours.

In contrast to Mon-Khmer and Tibeto-Burman tales of origin, ambivalence of origin among the Abo and Tani group of tribes opens up a new dimension of self-conception. Tani conception of Doyni-Polo as the creator involves the Sun goddess and the moon god and their union to give birth to all the living creatures of the world. Further, Tani group's ceremonies such as *Myoko* and *Murung* perform elaborate rituals to regenerate the souls. It is said among the Tani groups, "MiyuSaliang La Gyunyang Santa Tiggo Santa Lingi Du" (man is born with the ceremonies for spirit and therefore we cannot have a prosperous life without these spirits). In a sense, the connection between place of origin and spirit worship is transformational, the former transforming into the latter. In case of such a transformation, there is a creative appropriation of the idea of origin in symbolic forms. Tai speaking people such as Hmongs, normally settled in lower valleys intimate and indicate the emergent social formations in upper hills of Nortwest Laos. This marks a transition from Hmong to successor societies such as Tai Nua to Khmu communities and their settlement. Karen ritual of au brae in Mae M'Lau as an animated landscape of spirits comprising of land of the dead (pblukau) as well as domesticated wild that are invisible located at Doi Inthanon in Northern Thailand. In their agreement with Hmongs that gave rise to the ritual called 'eating the head of the rice' and 'conserving old bamboo groves' that essentially meant continuation with rai munwian system of swidden farming. Below the area of swidden farming is the taboo area called *doo ta* area. Similarly, Tani ambivalence towards water-spirits and a greater attachment to hill and celestial spirits mark an aversion towards riverine routes, routes which are occupied with more ancestral or other communities such as Akas, Khampas, Padamyongs and/or Tibetans and Bhutiyas. Indeed water-spirits in the Tani tradition assume the form of *wiyu* or evil spirit. Territorially water-spirits come from trans-border of Southern Tibet, marking a kind of descent into lower depths and finally they are supposed to reside in the nether world. This is a non-hierarchical horizontal construction of self that purges out alien spaces and spirits from the inner core of self-definition.

1.5 Successor Societies

Such lived experiences of shared cultural concepts throw up a history of migration that has been captured fundamentally in two distinct frameworks: one by James C. Scott's model of non-state spaces inhabited by stateless peoples who lived beyond the frontiers of organized state in upland Southeast Asia and lived a migratory nomadic life (Scott, 2009) and the other framework by Richard A. O'Connor speaks of 'ethnic succession' in relation to agro-cultural complexes involving subordination of lowland ecology by a succession to complex agro-cultural systems on the upland leading to a "closure" for later migrants (O'Connor, 1995). These two frameworks throw up logical and paradigmatic models to understand empirical details and systematize these details by going beyond local knowledge of the subjects. One of the major debates in such systematization is the constitution of the State as a relationship between subjects and power, between agents and structure. James C. Scott provides an account of "liminal zones" that are neither inside nor outside the State that reduces the authority and the functional control of the State. Such zones are elevated uplands that pose physical barrier to the entity called State. This physical barrier served those communities that wanted to flee or resist the State and in the process these communities avoided governance under the State and choose to live in a stateless condition. Examples of migrating communities from the Upper Mekong to lower Mekong delta, Northern Thai and Burmese communities including major communities like Mon, Khmer, Han and numerous nomadic tribes such as Chin, Lisu, Karen, Akha, Hmong, Kayah to name just a few, whom Scott labelled as Zomiyas. Zomiyas are, in Scott's (2009: ix–x) inimitable words:

> *Zomia* is a new name for virtually all the lands at altitudes above roughly three hundred meters all the way from the Central Highlands of Vietnam to northeastern India and traversing five Southeast Asian nations (Vietnam, Cambodia, Laos, Thailand, and Burma) and four provinces of China... It is an expanse of 2.5 million square kilometers containing about one hundred million peoples... at the periphery of nine states...virtually everything about these people's livelihoods, social organization, ideologies, and... even their largely oral cultures, can be read as strategic positionings designed to keep the state at arm's length... to avoid incorporation into states and to prevent states from springing up among them.

The Zomiyas remained illegible as they escaped the abstract categorization by the State and remained elusive to the processes of settlement and subjection within the territorial jurisdiction of the State. As legibility implies a viewer whose vision is synoptic and who can see from outside, the Zomiyas not only remained outside the pale of institutionalized land tenure, legally appointed chieftainship and such other forms of systematized constructions of identity. Indeed the illegibility of the

Zomiyas prevented the State from coming into being. Zomiyas could follow a parallel path of livelihood and belief system that does not require an arbitrator or an authority over and above their social organizations.

These illegible Zomiyas of Scott turned out to be singular and specific agro-cultural complexes marked by a set of agrarian and cultural practices such as home-garden or wet-rice cultivation in O'Connor's understanding. Wet rice cultivators left the lowland delta and inhabited the highlands as successors to sedentary agriculturalists. The highlanders carried the practice of wet rice cultivation by shifting from flood management to flow management. This transition is explained by O'Connor (1995: 970) in the following way:

> Now let us map peoples and contrast, say, 700 A.D. and 1700 A.D. In the early era Pyu, Mon, Khmer, and Cham rule the mainland, while to their north are Burmese, Tai, and Vietnamese. Ten centuries later the northerners dominate everywhere. In irrigable niches the southerners are gone (Pyu) or in decline (Cham), and their remaining strength is in the unirrigable lower deltas where Mon and Khmer flood farmers live in states under northern overlords. Juxtaposing these two eras reveals an epochal change. The date, pace, and completeness of the change vary but not the larger fact of a succession.

The larger fact of succession from lowland to highland in the form of a disjunction between cultivating practices and practices of rule or state acts as the source of "social contrast" between lowland settlers and highland successors. If Scott's Zomiyas are the highland successors who always moved away from the state and lived in the periphery of centralised states, then Zomiyas also are the successors of settled agro-cultural complexes from the lowland. This contrast of state and non-state spaces in terms of highland and lowland gives rise to interesting cultural contrasts as well. Lowland communities practice a kind of collective burial of the dead in their fields, who ensure land right and collective belief in such spirits, while highlanders have no such ownership attachment to their land and community spirits become lineal spirits. Transformation of community spirits into lineal spirits also marks a kind of transformation from patrilineage of community spirits to matriliny of lineal spirits, a contrasting transformation that marks the story of migration to highland. Matriliny goes with highland as successor community could see itself as a lineage not from the pastoral lowland but from the mountainous upland that held them on its laps and contours. Mountains connecting the world above are conceived as maternal nature, while the plain land below is conceived as paternal. Lineages and its imaginative domains formed around mountains imagined themselves to be freed from patrilineal regulations and tended to remain more matri-local and matri-lineal in its pristine connection with sky and the world beyond. This matrilineal tendency in-built around settlement in mountainous terrains and spaces separated by deep gorges and steep hills often converged in water-heads and watersheds. The Brahmaputra River system in the Northwest of Southeast Asia, the Irrawaddy on the North and many other rivers act as common meeting grounds between hilly terrains and adjacent plains.

1.6 Enchanting World of Shared Worldviews

There is a reciprocal and alternating convergence and divergence between physical and cultural spaces with the basic elements of self-conception. The initial organization of a dwelling space brings out the underside of tribal imagination: the presence of an alter subjectivity to whom experience could be related in the mode of conscious self-relation. The idea of KojumKoja and saibloojmem is a form of everyday cognition that gives rise to a various level of conscious understanding of how the self cognizes itself. Such a cognition is relational that arises through specific representations.

Another form of representation is construction of conceptual artefacts that it is not only that the reflective consciousness of the world is modified by the pre-reflective in the form of a transformation of pre-reflective into reflective, but it also produces a 'phenomenological datum' within any expressive, metaphorical, rhetorical and literary rendering of the experiential. In the phenomenology of the tribal world, any rendering of the experiential takes the form of geomancy or some kind of spiritual explanation of the phenomenon.

In the Thai-Ahom context of shared worldviews, such conceptual artefacts find a representational token in artefacts found in Ambari excavation in Guwahati in two different periods of pre-Ahom era spanning between seventh and thirteenth century AD and also of Ahom era spanning between thirteenth and eighteenth century AD. One of the artefacts was 'a peculiar type of pottery made of kaolin clay' in the shape of globular jars with flaring rim and short neck, flat based goblets, sharp rimmed cups (Buragohain, 1994). Seemingly Kaolin clay used in Southwest China travelled to the Brahmaputra valley with certain common artistic motifs embossed over these globular jars. One of the most common motif is that of a dragon like small animal available among art objects made and used by communities like Ahom, Khamti, Phake, Aiton, Turung, Khamyang, all of whom had their roots in Southwestern part of China. Conceptually speaking, the practice of building elevated wooden-floored houses among Tai-Ahoms called *Chao-Phas* and *Karengs* and inscribing floral designs on potteries with motifs of dragon show the kind of transmission of artefacts.

Such a construction of self-combining the lived space and the phenomenological datum along with a variety of conceptual artefacts draw up its resources from beyond, which is the larger terrain of Southeast Asia. Apart from the tribes, there is major cultural inter-relatedness between Shan people and Ahoms, Meitei people and other Tibeto-Burman groups of people as indicated in the previous paragraph.

As far as Naga tribes are concerned, they have their close kins with Mao, Kham and Akha people of Southwestern China. There are significant linguistic and cultural similarities that can be understood on a closer examination. Kuki-Chin group of people has significant and shareable cultural similarities with Hmong, Miao, Kham, Lai, Tai Lue communities. Indeed there are cross-cutting beliefs, folk motifs and shared stories of origin in space and time that opens up possibilities of wider comparison. For example, Hmong origination from the hills and their settlement to the plains point to a reverse migration that happened in search of water, which is similar to Naga migration in search of sources of water for swidden cultivation.

This comparison would open up various core notions of self as it is available in the worldviews of various communities and tribes. An important constituent of these worldviews is the religious beliefs and practices that throw up a number of ideological categories. For example, among the Tai Lue community of Southwest China, there are conception of self that sink deeper below the surface level praxis, adjusting to a subterranean world of secret alliances and multiple meanings. Tai Lue community adjusts itself to dry plains, to the spirits of shrubs and groves and flower valleys that are invoked in festivals such as *Wat Phrathat Sob Van* performed to bless the community. Similarly in the context of Northeast India putting down roots in a deeper substratum of personal and social networks will lead us to intricate ways of world-making that various tribes engage in.

The account brings into play religious belief and rites of passage from one territory to the other in a larger function of a 'grain producing community' that keeps copying itself, from plains to hills and from hills to plains in relation to fields, water, mountains and spiritual well-being (Scott, 2009). Taking this as a point that undergirds much of the Southeast and South Asian formation of successor communities and their boundaries within the landscape, one needs to inquire how nation-spaces with its logic of territorialization actually subverts the sense of belonging of those who own a fuzzy history of migration of their life-worlds. Such a subversion is more metaphysical and ideological, as it consists in a project of appropriation of these nascent communitarian forms of life into a legitimized narrative of national identity with its force of unification. Depending on the level of tolerance of plurality and difference, various nation-states attempt to draft a variety of policies of recognition and assimilation of these essentially minority-ized fuzzy ethnic identities. Scott and van Schendel's description of these identities as northwestern "borderland" identities of Southeast Asia (Van Schendel, 2005) Graham Chapman's description of these identities as 'resource owning and resource sharing de-territorialized identities' (Chapman, 2009) are more liberal than that of the 'singularities' of nation-state. Singularities that create a binary opposition between native vs. settlers, indigenous vs. non-indigenous actually subvert both the common space of belonging as well as the idea of 'indigenous citizenship'. The conflict is patently reflected in territorial claims beyond one's lived spaces that result into certain singularities. The question is, why such claims are made? Do such claims create a conflict between ethnically defined citizenship and civic citizenship? How borderland identities deal with these questions of adaptations, struggles and cross-border alliances is a major question.

Such adaptations give rise to altered conceptual schemes and frames of interpretation of one's own culture in its proximity and distance from other cultures. It also creates a nexus of lifeworlds in the context of North East India. Decoding such a nexus in terms of multiple shared roots and routes is still a cultural project, which is like a work in progress. More reflections on such possible sharing of meanings and ideas of life in its variety of manifestations and in its existential quest for authenticity unravels dialogic subjectivities that clamour for recognition and difference.

References

Berleant, A. (1992). Human being and natural world. In R. E. Hart (Ed.), *Ethics and the environment* (pp. 25–28). University Press of America.
Buragohain, R. (1994). *The lost trails: A study of the Tai peoples of North-East India* (Vol. 1, p. 76). Eastern Tai Literary Association.
Chapman, G. (2009). *The geopolitics of South Asia: From early empires to the nuclear age.* Ashgate.
Dai, M. (2006). *The legends of Pensam* (p. 55). Penguin India.
Durkheim, E. (1982). Rules of sociological method (pp. 41–42) (W. D. Hall, Trans.). Free Press.
Eide, E. (2010). Strategic essentialism and ethnification: Hand in glove? *Nordicom Review, 31*(2), 63–78.
Gablik, S. (2002). *Living the magical life: An oracular adventure* (p. 45). Phanes Press.
Mignolo, W. D. (2011). Epistemic disobedience and the decolonial option: A manifesto. TRANSMODERNITY: Journal of Peripheral Cultural Production of the Luso-Hispanic World, 1(2), 44–66. https://doi.org/10.5070/t412011807.
Mohrmen, H. H. (2025, April 7). Rice connection with rituals and traditions. *The Shillong Times.*
O'Connor, R. A. (1995). Agricultural change and ethnic succession in Southeast Asian States: A case for regional anthropology. *The Journal of Asian Studies, 54*(4), 968–996.
Sarma, R. (2015). *Disappearing Dialect: The Idu-Mishmi Language of Arunachal Pradesh (India).* International Journal of Intangible Heritage, 2015 no.10, pp.61–72 https://doi.org/10.35638/ijih.2015..10.008
Scott, J. C. (2009). *The art of not being governed: An Anarchist history of Upland Southeast Asia* (p. 9–10, 23). Yale University Press.
Sprenger, G. (2016). The connectivity of ethnic displays: New codes for identity in northern Laos. *Asian Ethnicity, 18*(1), 95–116. https://www.tandfonline.com/doi/full/10.1080/14631369.2016.1148428#d1e290
Van Schendel, W. (2005). Geographies of knowing, geographies of ignorance: Jumping scale in Southeast Asia. In P. H. Krastoska, R. Remco, & N. S. Henk (Eds.), *Locating Southeast Asia: Geographies of space and politics of knowledge* (p. 275). Singapore University Press.

Chapter 2
Relationship of Local Wisdom, Biodiversity and Ethnicity on Food of North-East India with Special Reference to Assam: Probing and Understanding from Women's Perspective

Polly Vauquline

Abstract Globalization and modernization have brought in many positive changes and has also transformed the nuance of traditional habits and behaviours though diffusion of knowledge across the globe. It has also transformed the food we eat. So has knowledge and culture which has evolved over time, which is inevitable. North-east India is enriched with rich biodiversity and a mosaic of diverse communities and ethnic groups. In the recent years aggressive growth of digital media has brought NEI closer to mainland India and is now emerging as a tourist destination for the rest of India.

Local wisdom is one form of cultural wealth and life value system inherited from previous generations (Hartati and Chandra, Local wisdom in the form of traditional food, Corn in Karo People tradition, in 2nd International Conference on Social Science and Character Educations (ICoSSCE 2019), pp. 212–216. Atlantis Press, 2020) and enhanced through one's experiences and deep knowledge. Local wisdoms of ethnic groups is a pertinent way to build sustainable livelihood. And food is one of the basic elements of subsistent livelihood. It plays an important role in creating and maintaining a distinctive ethnic identity (D'Sylva and Beagan, J Gender Stud 20(3):279–289, 2011). And women have been playing distinctive roles in fostering and developing their identity in ethnic communities through food culture. However, the role played by women in constituting, storing and maintaining the local wisdom of food culture, cuisines and delicacies that emerges from the mundane scape, the kitchen, of the most private, micro space of home required attention and relevant recognition through adequate documentation. Thereby, this study explored the relationship and roles played by women in constituting, storing and maintaining local wisdom and traditional knowledge of ethnic food in the context of NEI in general and Assam in particular. Ecofeminism and Grounded theory

P. Vauquline (✉)
Department of Women's Studies, Gauhati University, Guwahati, Assam, India
e-mail: pollyvauquline@gauhati.ac.in

Y. Santasombat (ed.), *Bio-Cultural Diversity, Ethnicity, and Local Knowledge*,
https://doi.org/10.1007/978-981-95-4242-0_2

were used to probe and understand of the ethnographical knowledge of food through women's perspective.

The findings revealed that women of the region played distinctive roles as the carrier and disseminator of the ethnic knowledge and wisdom of food. They were the practitioner of sustainable use of food in the domestic space of home. Because of the changes in the position of women in the society through the processes and forces of modernisation there is gradual transformation in the role of women as food provider. However, food is still valued as a form of ethnic identity of the communities. Thereby ethnic foods are valued, practised and will continuity to exist.

Keywords Local wisdom · Women · Assam · Ethnic food · Ethnic knowledge

2.1 Introduction

Globalization and modernization have no doubt brought in many positive changes in the society, alongside, it has also transformed the nuance of traditional habits and behaviours though it's extensive diffusion of knowledge across the globe. Be it the commodities we use, the dresses we wear and the food we eat, they are all tinted by globalization to some extent. Evidences of such transformations in the food we eat are distinctively visible with the flourishing MNC food outlets like KFC, McDonalds, Starbucks, Subways, Domino, Pizza Huts, etc. in India as well as in Northeast India (NEI). Along with global intrusion, Indian delicacies and foods have also influenced the food habits of people of NEI. 'Tandoori chicken'[1] a popular delicacy of Punjab, chicken kababs[2] a dish of north India, idli[3] and dosa,[4] delicacies of south India, momos,[5] a Bhutanese delicacy, etc. are common food served in the restaurants in the urban places of NEI. Some of these foods have also diffused to a certain extent to the rural areas and individual households too. With the uprising of the open economy, digitization, easy access to the virtual space and affordable digital services, accessibility to such knowledge has become easier and more convenient. These transformations are inevitable in the era of globalization and digitalization. At the same time knowledge and culture are not constant. It evolves and transforms with the progress of time. As such, at this juncture it becomes utmost important to preserve the local wisdom and ethnic knowledge which have been passed on from one generation to another in a conventional way. Northeast India is a site of the two

[1] It is a cuisine of Punjab where marinated chicken is baked in earthen oven.

[2] It is a cuisine of northern and central part of India where minced meat are baked skewer in earthen oven.

[3] It is a special dish of southern part of India where fermented grounded rice and special lentils are steamed.

[4] It is a special dish of southern part of India where fermented battered rice and special lentils are spread in hot pan and fried as thin as paper.

[5] It is a dumpling.

biodiversity hotspots, the Himalayas (Easter Himalaya) and Indo Burma, of the total four hotspots of India and the 30 hotspots recognized by National Geographics across the globe. It is the genetic treasure of diverse flora and fauna, animals and micro-organisms (Prabha & Jain, 2016). Together with its nature-bestowed endowment, the region is a mosaic of diverse communities and ethnic groups with its rich customs, traditions, heritage and their knowledge associated with it. According to Vikaspedia (a Government of India initiative) there are more than 140 indigenous communities who identifies themselves as tribals in NEI. Within these tribes there are again sub tribes which makes NEI one of the most culturally diverse region of the country. Some scholars even considered NEI as one of the most diverse ethnical regions of Asia (Marchang, 2019). Of the seven NE states, Arunachal Pradesh, Nagaland, Mizoram and Meghalaya are tribal dominated state with more than 80% of tribal population (Marchang, 2019; Census, 2011). Each of these tribes have their own ethnic uniqueness and certain similarities too in terms of their language/dialect, outfits, rituals, dialect, food habits etc. Their rich culture, ethnic practises, traditional knowledge and wisdom remained intact for a long period of time. The bottleneck location of the region and policies on restriction of movement of people to certain parts of the region can be considered responsible for its preservation and also blamed for its economic backwardness. These have kept the region physically distant and isolated from the rest of the country for a long period of time.

But in the recent past, the aggressive growth of digital media has reduced this isolation of NEI in the virtual space. With the coming in of political stability in the region, NEI is gradually emerging as a tourist destination for rest of India. With the thrust on Act East Policy by Government of India various initiatives for the economic upliftment of the region have taken it momentum. In such situation, it is utmost important to move forward valuing sustainable development taking along with it the local wisdom and ethnic knowledge.

2.2 Rationale of this Study

Local wisdom is one form of cultural wealth and life value system inherited from previous generations (Hartati & Chandra, 2020, February) and enhanced through one's experiences and deep knowledge. It may be in the form of religious practices, culture norms, traditional customs and norms, traditional practices and habits etc. which are generally passed on from one generation to another in oral form and are regarded as a part and parcel of a society's social structure. For the continuity of the community, it's sustainable growth and local management local wisdom plays a crucial role (Abas et al., 2022; Kuasa et al., 2015). It supports the community by offering practical and feasible solution for sustainable growth and development. The traditional knowledge gained from local wisdom helps to maintain the balance of one's life with its surroundings, nature (Mungmachon, 2012) and its preservation too. This knowledge comes from one's experience with deep understanding.

Local wisdom is closely associated with indigenous groups (Amri, 2018). They are the original and the noble knowledges evolved from the indigenous groups that they subsequently accepted as traditional values. Most of this local wisdom belonged to the indigenous groups (Ginting et al., 2020). These traditional knowledges are maintained by the ethnic groups as customs, believes and norms all across the world as well as in NEI. Local wisdom of the ethnic groups is a pertinent way to build sustainable livelihood.

One of the basic elements of subsistent livelihood is food. Food plays an important role in creating and maintaining a distinctive ethnic identity (D'Sylva & Beagan, 2011). Food has its own symbolic denotation. It is a medium through which ethnic identity and cultural norms is transmitted (D'Sylva & Beagan, 2011). Diverse ecosystem of the tropical rainforest, within which NEI is located, is a huge repository of food. It generates varied food culture and diaspora of ethnic cuisines (Deka et al., 2023). Unique food diversity of a region supports and sustain the food diversity at the community level (Wahlqvist, 2003). This diversity can be explored through the study of the food ingredients used in the traditional foods and cuisines. The way the traditional foods and ethnic cuisines are prepared can unfold a pandora box of knowledge. The processes involved in the preparation of the varied cuisines will definitely illustrate the local wisdom and traditional knowledge attached with it. Much of this Indian local knowledge was orally 'constituted, stored and maintained' (Kapoor & Singh, 2005: 12) and much of this knowledge of food was transmitted from one generation to another. As such, the prevalence of traditional knowledge and beliefs of which food to be avoided or preferred during different condition, situation (Kapoor & Singh, 2005) and gender (during childbirth, pregnancy, etc.) are still common practices in communities and ethnic groups. Indian Knowledge System (IKS) is exploring the different means to document knowledge about food which exists in proverbs, songs, etc. Thereby the role played by women in constituting, storing and maintaining the knowledge of food culture, cuisines and delicacies that emerges from the mundane scape, the kitchen, of the most private, micro space of home required attention and relevant recognition through adequate documentation in IKS too.

Women have been playing distinctive roles in fostering and developing their identity in ethnic communities through food culture. Even though, women have been the founders of varied technologies related with food like, 'food-gathering, processing, cooking, storing, feeding' (Wahlqvist, 2003: 307) women's association with food is often trivialised because of the very mundane nature of food (D'Sylva & Beagan, 2011), it's relation with women's role and the space within which it is associated, the private and domestic space (Bowlby et al., 2023; Brickell, 2012).

Again, Sustainable Development Goal 2 (SDG 2), aims to achieve zero hunger by 2030. One of the targets of SDG 2 is to ensure sustainable food production systems. It also emphasizes that the significance of sustainable agriculture is to maintain ecological balance. The role of women in this process of sustainable production of crop and food remained invisible and marginalized for varied socio-cultural factors. Even in the recent documents of IKS (Pollock, 2000; Kapoor & Singh, 2005) these roles played by women are not mentioned. In the patriarchal society knowledge is androcentric. This androcentric knowledge does not see the intrinsic value of nature's diversity whereas promotes capitalist patriarchy, where nature is esteemed on the basis of commercial gain and economic benefits (Shiva, 1992). On the contrary, the very core of local wisdom and traditional knowledge of ethnic communities are based on feminine principles whereof the essence of sustainability (production of crops and thereafter food), diversity and ecological balance are kept intact (Shiva, 1992; Samal & Dhyani, 2006). As such, women can be regarded as the true custodians of indigenous knowledge (Samal & Dhyani, 2006) of food. Thereby, this study intends to explore the relationship and roles played by women in constituting, storing and maintaining local wisdom and traditional knowledge of ethnic food in the context of NEI in general and Assam in particular. This will further lead to their understanding of their biodiversity and ethnic foods from women's perspective. Keeping these arguments in mind the following objectives are taken for the study.

2.3 Objectives

The following objectives are taken for the study

1. To study the different types of locally available food ingredients used to prepare ethnic cuisines both at commercial food joints and at individual households
2. To explore the ethnic knowledge and local wisdom of women in the process of preparation of traditional food cuisines.
3. To study the role of women in dissemination of local wisdom and traditional knowledge of ethnic food within its community.

2.4 Research Question

1. Are local wisdoms intact with the ethnic knowledge of food in Northeast India?
2. Are local wisdom and ethnic knowledge impacted by factor and agents of globalization and modernization?
3. Are women the true bearers of local wisdom and ethnic knowledge of ethnic food in Assam?

2.5 Database and Methodology

Database The study is based on both primary and secondary data. These two sets of data are used to explore two perspectives. The secondary data is used to explore the regional perspective, the North East India, NEI and the primary data is used to explore the in-depth understanding of ethnic knowledge and wisdom of food of a specific state, Assam.

Secondary Data To study the locally available ingredients used to prepare ethnic foods of NEI, secondary data from different sources were collected. Data which has information of all these three aspects; the ethnic communities of the region, the ingredients used to prepare the ethnic food and the ethnic food are only taken into consideration. The data are then segregated on the basis of the seven states (excluding Sikkim), of NEI. Information of all the 140 major ethnic communities of the NEI could not be reflected in this study due to lack of data. These data were then minutely analysed to find out the most common ingredients used across the different ethnic communities of the state. Some specific ingredients used by certain ethic communities were also identified from this set of secondary data.

Another set of secondary data were collected to study the different cuisines of the region. This information was collected from 6 travel blogs and tourist guides with information about NER food. The information collected are then described according to the popular ethnic foods of the respective states.

Primary Data To explore the knowledge of ethnic food and the local wisdom associated with it, two Focus Group Discussion with three participant each and five in-depth interviews were conducted. Effort was made to interview participants from different ethnic communities and age groups. The participants were from nine ethnic communities, three religious background and five regions of the state of Assam. They are referred according to the name of their ethnic community, such as Bodo, Rabha etc. Of the eleven participants, ten were women and one was a man. The participants were purposively selected and it is a non-probability and availability sample survey. An unstructured interview framework was carried out focusing only on the objectives and the research questions of the study.

Data Analysis Both qualitative and quantitative methods are used in this study. ***Content analysis*** was used to analyses the data of ingredients used by the ethnic communities of NEI. ***Descriptive analysis*** was used to interpret the data of the food bloggers and tourist guides. ***Thematic analysis*** was used to analyse the interviews of the participants. The findings from the interviews were first analysed as broader themes and subsequently sub-themes were derived to have a deeper understanding of the narratives.

Theories Used Ecofeminism and Grounded theory is used in this study. Grounded theory is used as in this study it is required to supports and shapes the methodology (Holt et al., 2022) of ethnographical knowledge through women's perspective. These theories were pertinent to this study as they generated new perspectives which lead to discovering new theory of women's wisdom and ethnic knowledge from the data in social science research rather than forcing data into a few existing theories (Urquhart, 2022; Gogoi et al., 2023; Gogoi & Vauquline, 2023). Ecofeminism promotes women's indigenous knowledge (Warren, 1988) which have been ignored by the mainstream knowledge and led to mal-development (Shiva & Mies, 2014). In this study ***feminist research methodology*** is used to conceptualize women's indigenous knowledge of food in NEI.

2.6 Findings

2.6.1 Locally Available Food Ingredients Used to Prepare Ethnic Cuisines

Rice Rice is the commonest ingredient used across all the ethnic communities taken for this study. Varieties of rice are available on the basis of the texture, aroma, size, shape, colour, etc. It is the main staple food of almost all the tribes of the region. It is grinded, dry fried, flattened, puffed etc. to make different cuisines and delicacies. Rice is also brewed to prepare rice beers of different forms and their names differ according to the ethnic communities.

Meat Almost all the ethnic communities of the region are non-vegetarian. Pork is the commonest meat used to prepare ethnic food across these ethnic communities of the state. Beef is consumed only by few ethnic communities. Along with these, chicken, flaws and mutton are other forms of meats used to prepare their ethnic food. Meat of deer and wild boar are also consumed by the Anals of Manipur. Mithun and sheep meat are common among the Gangte of Manipur. Dog meat is consumed among few tribes of Nagaland and the Reyangs of Tripura. Consumption of fermented meat is common among Uchai tribe of Tripura. Smoked meat is common among ethnic communities too. Meat is usually grilled more so among the tribes of Assam. Roasted meat is popular among the Mogs of Tripura.

Fish Fresh water fish of different sizes are a common food of all the ethnic communities. They are usually consumed fresh but are also fermented, powered, dried, grilled and also stuffed in bamboo hollows and used as ingredients for varied delicacies. Fish soup with local herbs is common among Khamtis of Arunachal Predesh. Fish stew among the Mogs and Tripuris of Tripura. '*Sidol*', a form of fermented fish and shrimp cuisine, among the Chakmar of Mizoram, Noatia and few other ethnic communities of Tripura are common foods. Crabs, turtles, prawns and frogs also forms a part of their food among the Reyangs of Tripura.

Bamboo Shoots They are the tender shoots of bamboo which is used as ingredient across almost all the ethnic communities of the region. The softer part (upper portion) of the bamboo shoot is usually used fresh as ingredient to prepare meat, fish or just boiled with other vegetables. The lower part of the bamboo shoot is fermented and is consumed among all the communities. At times it is dried after a short spell of fermentation. These bamboo shoot with pork and fish is a common food culture of the region but it is favourite among the tribes of Nagaland.

Leafy Vegetable and Herbs Wide verities of edible leaves and herbs form a major part of ethnic cuisines across the communities of the region. Some of these leafy vegetables are seasonal and some perennial. In the rural areas they are usually not cultivated but are collected from the wild. Banana flowers, bulbs, rhizomes, roots, buds, stems, tubers, flowers and shoots of some native plants are common supplementary food of the communities of the region. Singphos of Assam are observed to use wider varieties of plants products. Elephant yam leaves, brinjals, berries (bitter), some specific types of leaves etc. are fermented for prolong use or to be consume during their lien seasons.

Maize and Millets The Akas of Arunachal Pradesh, takes maize and millets as their staple food. Millet is observed to be used more among the tribes of Arunachal Pradesh, Koms of Manipur, Chakmas of Mizoram, Jamatia of Tripura and Angamis and Pochuries of Nagaland. Beer is also brewed from millets and maize in Arunachal Pradesh and in Nagaland.

Exotic Foods and Specialty Since a major portion of this region lies on the tropical rain forest area, small animals and insects are found in plenty mostly during the rainy season. These have become a part of the diet and food among some ethnic communities. Red ant larvae and silk pupa as food are relished by the Ahoms and the Kacharies of Assam. Bettle nuts along with piper leaves is consumed in plenty in Assam, and Meghalaya. Some says that it helps in digestion of food. Chille of varied size and colours are favourite across the communities of the region. Intake of salt is lesser in the foods of the region. During the older days people of this region never used salt and the intake of iodine was supplemented through different herbs and leaves. Garlic and ginger are the only ingredients used as spices in most of the cuisines of the region.

Even though few of the ingredients were specifically mentioned as used by certain ethnic community, however, they are also used by the other communities of the region.

2.6.2 Ethnic Cuisines of Northeast India

The cuisines of NEI have its own distinctiveness because of its geographical isolation, the way it is prepared, the ingredients used and the process involved. As such the region has unique food culture which has evolved over a long period of time. In general, the foods of NEI are described as bland, pungent and healthy. It is also described as hot because lots of use of chillies, bland because of no use of spices, it is considered aromatic because of the different herbs used and it is considered healthy because of very less use of oil. Majority of the foods are oil free. They are either steamed or boiled. The foods of the region are protein rich because of high intake of meats, poultry products and fresh water fishes. They are also rich in minerals because of regular intake of leavy vegetables and locally available herbs.

Following are the types of ethnic dishes of the seven states of NEI recognised by the food vloggers and tourist guides.

2.6.2.1 Assam

Khaar *Khaar*, which means 'ash' is considered as a common food of Assam where the filtered water of banana ashes is used. It is used to prepare dishes with raw papaya and pulses.

Kumurat Diya Hanhor Mangxo Duck meat cooked with ash gourd is another popular dish of Assam. It is prepared with mustard oil and spices like black pepper and crushed cardamom.

Pitha Pithas are Assamese smack which are available in different forms and shapes. The main ingredient of *pitha* is grounded rice, both sticky as well as plain rice. They are made both salty and sweet. Jaggery is used to make it sweet. Some of the *pithas* are deep fried and some are pan roasted without any oil. They are rolled with crushed coconut and black sesame seeds inside it and are elongated in shape. *'Suya pithas'* are made in bamboo hollows where full grained rice is steamed inside it.

Masor tenga It means fish cooked in sour gravy. Fresh water fishes of different sizes are the most common food across the states. They are prepared in different ways, but cooking fish with sour gravy is the favourite among the people of this state. The most common way of preparing the tangy is by making gravy out of elephant apple (Ou tenga), dried tomatoes, fresh tomatoes and lemon juice. Local herbs are used to spice it up.

Delicacies made from rice such as *'kumool chawul'* (soft rice), *'Chira'* (flattened rice), *'akhoi'* (puffed rice), *muri* (puffed rice), and *xandoh* are common in Assam. These are taken with curd or milk and jaggery. During Bohag Bihu, a festival of Assam, a distinctive dish is made with 101 herbs and leafy vegetable.

2.6.2.2 Arunachal Pradesh

The travel blogs and tourist guides do not have much information of the ethnic dishes of Arunachal Pradesh. Since millet, maize and wheat is produced in this part of the region, most of the dishes are prepared by these ingredients.

Zan It is a porridge made from millet flour and boiling water. This is common among the Monpa tribes of the state. It is taken with boiled meat and fresh vegetables. Local herbs are added to enhance the flavour of the dish.

Wungwut ngam It is a chicken dish which is prepare with rice powder. Locally grown pepper is used in this dish to give it a distinctive taste. Native peppers are used to give chicken a distinctive taste.

Apong It is a locally brewed beer prepared both from rice or millet.

Pasa It is a spicy soup made with fish.

There is no mention of names of the dishes made with pork, poultry or Mithun. However, consumption of meats in sizable quantity and that it is cooked with local herbs are acknowledged. Again, rice of varied indigenous variety which are steamed, boiled and shallowed fried are taken as staple food of this state.

2.6.2.3 Mizoram

Sanpiau It is a dish made with powdered rice, coriander paste, onion, black pepper, and tangy fish sauce. It is also prepared with egg/s with other ingredients remaining the same. Scrambled eggs is garnished over it. It is a common household as well as street food of Mizoram.

Vawksa Rep This is a special dish prepared during the time of festivals. It is a smoked pork dices softly cooked with local herbs and fresh green vegetables. Most of these herbs are grown in their kitchen garden.

Zu It is tea, a popular drink of Mizoram.

Sa-um This is fermented pork fats added with vegetables to make a unique dish of its own taste.

Dawlrep Bai It is a spicy curry made with beef or pork.

Because of the existence of rich flora and fauna, edible leaves, roots, nuts and mushrooms they are commonly used as ingredients to prepare unique cuisines in Mizoram. In addition to that people also grow edible plants in their kitchen garden, which is a common practice. To ensure easy accessibility of food products throughout the year ethnic communities of Mizoram sun dry or smoke dry (more common) their meats like pork, chicken and *mithun* and ferment vegetable like bamboo shoot and yam leaves. They are again dried after fermenting them.

2.6.2.4 Nagaland

Axoni It is a side dish (chutney) served along with the main dish. It's main ingredients are fermented soyabeans and ginger along with roasted fresh green chillies and tomatoes.

Galho It is like porridge made with rice and meat or vegetable. Herbs and roots are added to create its own taste and aroma.

Thevo Chu It is a dish cooked with pork and fermented bamboo shoots where fresh red chilies, garlic and ginger are added.

Most of the Naga foods are prepared with no oil. It is boiled, steamed, smoke dried (over kitchen fire) and fermented. Steaming is done in locally made bamboo and wooden utensils. The cuisines are served with boiled rice. 'Bhoot Jolokia' (ghost chilli), one of the hottest chillies of the world is commonly used as an ingredient along with locally available herbs and seeds. Dog meat and hornet larvae are also part of Naga delicacies. There are 15 tribes in Nagaland. Each tribes have their own cooking style and specific essential ingredient; Akhuni (fermented soy bean) of the Seema tribe, bamboo shoots for the Lothas, Anishi, dried yam leaves for the Aos etc. are some of them.

2.6.2.5 Manipur

Nga Atoiba Thongba It is a traditional dish of the Meitei community of Manipur which is a fish dish specially prepared with fresh fishes of Lokthak lake. It is cut into small pieces and mashed a bit and prepared with boiled potatoes, fermented lentils, fresh bay leaves, onion, cumin, chillies, and chives.

Iromba This is also an ethnic dish of the Meitei community prepared with fermented fish which has a strong pungent smell. The common ingredients used are mashed potatoes, locally available vegetable and red chilies with no oil with it. It is taken as a side dish with rice of local variety.

Kabok It is made with rice and vegetables.

Hawai-uri thongba It is a bean stew.

Bawngsha It is a beef curry.

Ooti It is a dish made with yellow pea, garlic leaf, ginger and garnished with dill.

Hanggam chamthong It is kind of a stew made with boiled mustard greens and fermented fish.

Khamen Thongba It is made with brinjal, green chillies garnished with coriander.

Kheer It is a sweet dish made with rice, milk and sugar. It is served as dessert. It is prepared both with white and black rice. It turns purple and has a sticky texture when cooked.

Chahao kheer It is a black rice pudding served as dessert in Manipur.

Another interesting aspect of Manipuri delicacies is that they serve them in culinary of different shapes and sizes made from banana leaves. There can be as many as 30 items in a Manipuri delicacy served in plates and bowls made of banana leaves. Manipuri cuisines are spicier, more evolved, elaborate cooking methods and processes are involved with use of oil. The vegetable options are also wider. Chillies are commonly and elaborately used in almost all the Manipuri dishes.

2.6.2.6 Meghalaya

Jadoh It is rice and meat (pork) based cuisine prepared primarily by the Khasis, and Jayantias ethnic communities of Meghalaya during festivals and important ceremonies. The common ingredients to cook this dish are onion, garlic, turmeric, pepper, ginger and oil. Some also used pork liver to prepare this dish. It looks like Biriyani[6] but has its own taste and aromatic flavour. *Jadoh Snam* is it's another version where raw pork blood is mixed with the rice and cooked in pork fat. Again, *Ja Mardoh* is mixed together with soft tomatoes and crunchy onions.

Tungtap It is a dry fish side chutney prepared with dry small fish which is mixed with onion and fresh chillies. It is a popular both among the Khasi and the Jayantia communities.

Pukhlein It is a dish made with rice flour, sugar and oil. It is prepared in the other states of the region but the Khasi's claims it to be their traditional sweet delicacy. It is first made into batter and then deep fried in edible oil. It is prepared on a daily basis as snack and also during festivals.

Dosyear It is a chicken dish made with lots of pepper to make it spicy and consumed when one suffers from flu and cold. The pork is boiled and then shredded into tiny pieces. It is then mixed with pork fat and onion together.

---

[6] It is an Indian dish which is very spicy, lavishly made with highly seasoned rice and meat, fish, or vegetables.

Nakham bitchi It is a popular Garo soup made with dry fish.

Kyat It is the rice bear prepared during festive occasions of Meghalaya.

Seasoning and use of animal fat is used occasionally in the cuisines in Meghalaya. Most the food consumed by the ethnic communities of this state are found in the hilly areas and are grown locally. Mushrooms which sprout abundantly during the rainy season finds their way into many local cuisines of these communities.

2.6.2.7 Tripura

Chikhvi It is a lavish delicacy of Tripura where the main ingredients are pork and bamboo shoots. Only home-grown herbs and locally prepared spices are used in this dish. It is served with boiled rice.

Mui Borok It is another traditional dish of Tripura. Its main ingredient is 'berma'.[7] It is ubiquitous in almost all the dishes of Tripura. This dish is also prepared with local herbs and spices but no oil.

Chuwarak It is the traditional drink of Tripura. It is prepared with special a variety of rice, pineapple and jackfruit which creates an exotic taste of its own. Mui Borok is generally eaten with Chuwarak.

Shigal It is a paste made with fermented fish. The fresh water fish of small size is processed in a unique way in clay pots using salt and mustard oil till it releases strong pungent flavour. It is also used with other vegetable to prepare other dishes.

Traditional foods in Tripura are plain and simple like any other states of the region but are colourful (yellow) because of the use of turmeric in their dishes. Pork, mithun (a strain of bison) and fish is their primary source of protein among its ethnic communities.

The Reangs, based in the eastern part of the state, near Bangladesh, prepare their food in bamboo hollows over charcoal fire. Few other ethnic communities also do the same during festival and special occasion but it is a common practice among the Reangs.

Being an insider and growing up in this region, it can be stressed that there are far more dishes and cuisines which are not mentioned by the food bloggers in their blogs. Again, many of the dishes are prepared cutting across in the other parts of the region. As for example the Phalap (smoked tea) also known as Singpho Phalap is a common drink in eastern part of Assam and Arunachal Pradesh. Kheer is a common dessert served in Assam and Tripura too. Pukhlein is prepared in the same way and is known as '*ghilla pitha*' in Assam. Fermented bamboo shoots and fermented and

[7] Dried and fermented fish.

dry fish is frequently mentioned by the blogger and is found to be common ingredients in the ethnic foods of the region, but there are variations in the process of their preparation according to the region.

2.6.3 Knowledge and Local Wisdom of Ethnic Food of Assam

The information provided by the food bloggers and the tourist website of food of Assam is very minimal. It is far more varied and diverse than mentioned in these sources. Knowledge about ethnic food prepared by different ethnic communities in different geographical regions of Assam are very rich across these groups. However, there are variations of their richness. Ethnic knowledge of food among the tribal communities living in the rural and semi-urban areas are more intact than those living in the urban area. Even the communities living in the urban region have their knowledge of the ethnic food but changes in their food habits are observed and seem to be evolving because of varied socio-cultural, economic and political factors.

2.6.3.1 Ethnic Knowledge on Food of Women of Assam

The nine women and the sole man, belonging to eight ethnic groups who were interviewed for this study had strong and intact knowledge of the ethnic food prepared in their respective communities.

Knowledge about Ethnic Food Is Intact Among the Communities of Assam across Different Age Groups

The participant shared the recipes and explained in details the preparation process of the ethnic foods prepared in their household on day-to-day basis.

> Sonowal Kachari: In our house a curry is made of rice flour and bamboo shoot in it. It is called '*Gura Aanja*'.

This is briefly explained by a female participant who is 34 years of age and belongs to Sonowal Kachari ethnic community of Assam.

Another participant, referred here as Assamese Christian, who is from the central part of the state, Nagaon, explained in details when asked about the ethnic food that is common in her household. She is 65years old. It is not from the tribal community. She narrated.

> Interviewer (I): If I ask about Assamese local cuisine or any dish what comes to your mind?
>
> Assamese Christian: *Khaar* and leafy vegetables (smiles).
>
> I: What is *khaar* cooked with?
>
> Assamese Christian: It can be cooked with Papaya. There are also certain leafy vegetables like '*morapat*' (jute leaves), leaves which are dried known as '*hukuta tita*', is used with *khaar*. Pumpkin leaves are also used to make *khaar*. There are various kinds of khaar dishes that can be made using leaves and vegetables.

Again, another participant referred here as Bodo 1 who belongs to Bodo community, which is a major tribe of the state and is 23 years of age very promptly responded to the same question in the following way.

Bodo 1: We eat *khaar* curry. We add *khaar* and *dal* (lentils) with papaya to make a curry.

I: So dal (lentils) is also used in *khaar*?

Bodo 1: Yes, *khaar* is added to dal as well. Another one we eat is *Meska Tenga*- (Roselle leaves), also called Tengamora. We don't use any oil. When the pan is hot, Roselle leaves are put in the pan. Once the leaves soften water is added to it and then we add salt, chili, fish, and *khaar*. Once it boils, it is done.

Enriched Knowledge about Specific Ethnic Food during Festivals and Special Occasions Is Intact

The participant also had clear knowledge of the ethnic food prepared during festivals or special occasions.

I: Are there any special dishes made during special occasions?

Bodo 1: '*sewa diya pitha*'. It is made in a bigger bamboo basket. It is like the '*Tekeli pitha*' (*pot pitah*) made with a cloth in one go. It is bigger in size, like a cake that can be cut into pieces.

The sole male participant is referred as *Chutia*, the ethnic community to which he belongs. He is a Vaishnavite now (follower of Vishu of Hinduism) who grew up in a village and now settled in Guwahati, the largest urban centre of the region. He is 54 years old. Chutia shared his knowledge about the special dish made in his community.

I: There weren't much sweet dishes, were there?

Chutia: Yes, there are which are made during special occasion. For sweets there are *pithas—til pitha, Ghila pitha, narikol pitha..*

Chutia: During *Bihu,* we do not eat rice for lunch, we eat '*proxad*'.

I: Tell me about *proxad*. What are mixed in *proxad*?

Mohan: In *proxad* we use '*mogu*' (green gram). In our area we use split grams and not whole grams. Their peels were completely removed. Along with '*mogu*' a little bit of raw rice grain is used. It is usually prepared nearly four hours before being serving. They are soaked in water with *Aroi*[8] rice. With a kilo of *mogu* and chickpea mix, a handful of rice is used. Along with it, grated coconut or small slices of coconut, banana, cucumber, orange, sugarcane or other fruits that are available are offered.

Preparing dishes with local ducks is a specialty in the eastern part of Assam. It is considered a respectable gesture to offer fish curry but more so duck curry to guest invited for lunch or dinner.

I: During special occasion does the food remain same? During festival or if a guest visits?

Chutia: Cooking duck (for guest) is a sort of a respectable gesture (by the host). So, duck and fish are mainly cooked as special dish for guests in the rural society.

[8] A common variety of local rice.

Diverse Food Culture and Practices Are Followed by the Ethnic Communities During Festivals

The cuisines are different for the tribal communities during their festive occasions. Intake of pork is common and their preparations are also distinctive. Following are the narrations of three women belonging to three different ethnic communities of Assam, Bodo, Rabha and Mising. They are students studying in university but are from rural places of Assam.

> I: What do you eat during festival?
>
> Bodo 2: During festival pork is main food. During the daytime, the people clean the intestines of the pig and cook it with its blood
>
> I: Do Rabha community also have this?
>
> Rabha: It is same. It is eaten as chutney. It is eaten during '*Magh*'[9] and is distributed among everyone in the community. Two or three pigs are killed during this festival.
>
> I: Is this dish made by Mising as well?
>
> Mising: No, we do not make it. *Ali-ai-Ligang* is a very important festival for us which is celebrated during the first Wednesday of *Phagun* (February). During this festival we cook pork with '*Dimoru*' leaves (a type of locally available leaf), pork with '*mati mah*' (a type of greenish black lentil). These are the main dishes along with *Saai Apong* (a type of Ash rice beer). During this festival we pray to the Sun and the Moon God for the better harvest of the crops.

Food Culture and Practices of the Ethnic Communities Differ According to Regions

Rabhas and Bodos are predominant ethnic communities in the western part of Assam. Whereas, Mising are concentrated in the north and south banks of river Brahmaputra of eastern part of Assam. Pork is their main food during their festivals. But the ingredients used and the process of their preparation are different. It also has some commonalities like using less oil, drinking home brewed beer, using locally available herbs etc. Another participant from the Kuki community living in the central part of Assam states that,

> Kuki: We have this traditional dish cooked with leaves. The leaves are called '*Sizo*', in our language and in English I think it's called '*Uriya*' leaf which is most important dish prepared during weddings and any festivals.

Down south of Assam in the Barak Valley, the food culture is totally different. They use variety and large number of spices, plenty of oil and involves intricate preparation process of their daily food as well as those prepared on special occasion. It is more distinctive among the Bengali speaking Muslim community.

> I: What about your community (Bengali Muslim)? What are the cuisines prepared in festivals or when there is guests or special occasion in your community?
>
> Bengali Muslim: Sweet is must. We have to have varieties.
>
> Bengali Muslim: During Eid Biryani has to be there, or pulao… and it has to be like… not veg pulao (laughs loudly). Then, … chicken. Chicken, beef, all have to be there. There has to be chicken dishes cooked separately, there has to be beef dish separately, there has to be mutton and then there also has to be fish (laughs).

[9] Mid of January to mid of February.

Religious practices have also changed the food habits in certain communities of Assam to a certain extent. The Vaishnavites turns vegetarians once they accept '*horon*' (a form of baptism). The Kukis forsake the ethnic locally prepared beer after they accept Christianity.

2.6.3.2 Local Wisdom and Ethnic Food of Assam

From the interaction with the participants, it is reflected that they could not clearly relate the values, be it nutrition, medicinal or others associated with the food they are consuming. However, their clear understanding of the process of preparing the ingredients of their food, the way they prepare their ethnic, the specific time chosen to prepare, harvest and process the locally available edible resources etc. gives a deep reflection of their wisdom associated with their food culture.

Local Wisdom Is Intrinsically Associated with the Food Culture of the Communities

Most of ethnic community have their own distinctive way of preparing and using ingredient in their cuisines. As for the example the preparation of a steer fried dish with 101 varieties of green leafy vegetables during Bohag Bihu.[10] These 101 varieties of leafy vegetable are collected from the open areas growing in the wild or those cultivated.

> It was my first Bihu celebration with my in-laws. My mother-in-law took me around our house, which had huge courtyard around, teaching me about the edible leaves that needs to be plucked for the 101 leafy vegetable dish. I could identify around 30 varieties on my own. Both of us collected more than 90 different types of leavy vegetable (laughing with amusement).

This is narrated by a Christian participant married to a Hindu who grew up in an urban area whose in-laws are from rural area. She is a 53-year-old women.

It is a common dish prepared during this festival in different parts of Assam. This instance clearly illuminates how knowledge is transferred orally from one generation to another by women.

Again, another participant narrated,

> Sonawal Kachari: It is said that no leaves are poisonous or harmful that day (Bohag Bihu). You can eat any type of leaves that day... like Teasle Gourd leaves. I've seen my grandmother putting leaves of '*Sokamora*' which we were afraid of eating otherwise.

These clearly reflects the deep local wisdom the women behold with regard to their rich knowledge of the locally available edible leafy vegetables in their vicinity.

Ethnic Cuisines Are Created by Different Locally Available Herbs, Seeds, Fruits, etc. by Different Communities

It is interesting to note here that almost all the communities use similar ingredients, like rice, leafy vegetable, bamboo shoots, fish, pork, chilies, etc. even though they

---

[10] It is a festival celebrated in mid of April in Assam, India. It is the Assamese new year.

are located in different corners of the state. As for example most of them use grinded rice (except the non-tribal Assamese communities), *khaar*, fermented bamboo shoots, dry fish etc. However, the variations of cuisines evolve from the use of locally available herbs

Mising: We eat *Dimoru* leaves (a type of local herb) with pork.

Sonowal Kachari: With rice flour, chicken is added and sometimes '*bashika*' flower is added. It is a red flower with long leaves and tastes bitter.

My mother also cooks chicken with herbs named '*matiphura*' or '*matigaldap*' which is found during the winter season in the areas covered with tall grasses. Another one my mother uses is '*Jaari*' leaves. It is a plant with yellow flowers and it tastes like schezwan pepper.

Mising: We eat pork with '*ou tenga*' (elephant apple).

Kuki: Yes, pork and '*sizo* leaf'. We cook pork and 'sizo' leaf together.

The knowledge of the herbs, flowers, fruits etc. and their uses in different cuisines again reflects the deep local wisdom of the women of the region.

Food and Ingratiates Are Processed, Harvested, etc. during Specific Time According to the Condition of Nature

The participants in this study had a very clear understanding of the time of harvesting the special ingredients with which they prepare special cuisines. Crabs are caught during summer season when the low-lying areas gets filled with water and prepared according to their unique ways.

Bodo 2: During rainy season crabs are found and is eaten with roselle leaves. Sometimes we also eat smoked crab chutney.

Weaver ant larvae is harvested during the outbreak of the winter season. The Bodo and the Mising communities have this unique method of extracting this ant larvae whose bites are very painful.

I: You eat '*Amroli Porua tup*' (weaver ant larvae). How do you catch them?

Bodo 2: The larvae are fried and eaten.

Mising: It is eaten during autumn and winter.

Mising: After the ant's nest is brought, we dip it in hot water in order to kill the ants. After that we separate the larvae and fry it with oil, onion and chili. No spices are used.

Pupa which are fed with mulberry leaves are also a part of exotic food culture of certain communities of Assam. The women have nurtured their wisdom to harvesting matured pupa by observing the change of their colour.

I: You eat pupa as well, right? Is the process similar with *Amroli Porua tup* (weaver ant larvae)?

Mising: No, we eat the pupa after the cocoon is fully made. We call it '*poki jua*' (ripen). We store the larvae until it turns brownish red in colour and forms cocoon which is called '*Kumari*'.

I: What is '*Kumari*'?

Mising: It is the stage before the worm turns into moth. In this stage the cocoon takes a reddish colour.

I: Does it come out of the cocoon?

Mising: No ma'am, we have to pull it out. After that we wash it properly with water and fry it.

Locally brew bear is the important drink among all the ethnic communities of the region. It is prepared with rice and millet. Different ethnic communities use different herbs, seeds, leaves, flowers etc. to mix with rice to brew it. *Sai Apong* (*Ash Apong*), is a specially brews rice beer with ashes of rice hasks. When asked how they know it is ready to brew? This is what was narrated which reflects the enriched wisdom of the participant from Mising community.

I: Does the beer come out on its own?
Mising: Yes and from the flavour of it we know it is ready. \
I: What does it smell like?
Mising: It smells sweet and has a strong unique smell.

Women fishing in groups is another common sight during the rainy summer days in the rural areas of Assam. One of the participants enlighten me as to when to get a good catch of fishes.

Bodo 1: During sunny day when it is very hot the water of the rivers gets heated up and then when there is a sudden heavy shower, we get good catches of fishes. Not when it is raining constantly. It should be very hot and sunny and then sudden rain.

Duck curry is served during specific seasons of the year in eastern part of Assam.

Chutia: Ducks are usually eaten during Kati (mid of October). The ducks get to eat a lot of fishes and worms during the monsoon season (June to September) as the fishes are available in plenty.

'Khaar', which is made from the ashes of the banana trunk and is prepared during the month of October just with the advent of the drought period in the region. To prepare '*khaar*' the lower bottom of the banana trunk is used which makes the '*khaar*' more alkalic. The ash of the peels of a special banana, 'bhim kaal', are used to prepare '*khaar*'.

2.6.3.3 Local Wisdom Is Intrinsically Related with Sustainable Use of Locally Available Resources

One of the distinctive features of food of this region that emerged in this study is the intake of fermented and dried food. Leaves, sprouts, fishes, seeds, etc. are fermented and dried for prolonged use. Since Assam lies in the monsoon region, it has a hot and humid climate during the summer season and it is during the summer season that rich vegetation flourishes. But at the same time, it is because of this climatic condition that fresh food items easily decomposes. To over this natural phenomena, people living in this region over a long period of time might have devised various mechanisms to preserve food for a longer period of time through fermentation and drying. These have led to sustainable uses of food.

I: What is '*hindol*'?
Rabha: Some fishes are dried and grinded with Colocasia. They are then stored in a bamboo tube. Brinjal and little bit of '*bhekuri tita*' (a type of seed which taste bitter) is added. They are then left there for fermentation. They can also be stored in a bottle but

bamboo tube is better as it can be stored for a longer period of time. Some people also put turmeric above it so that it doesn't get spoiled. It can be preserved for around 2 years.

Among the Mising community, '*namsing*' (a type of fermented fish) is a favourite delicacy. Here brinjal is also added. The dried fish is grinded with Colocasia and stored in a bamboo tube. They make sure that interior part of the bamboo is not contaminated by fungal growth. They have a sustainable way of drying fish (as well as pork) before it is put inside the bamboo tube.

Mising: The fish are kept in a bamboo colander (*saloni*) and then it is placed above the fire place of the kitchen to dry.

I: How is pork dried?

Mising: The pork is put in a bamboo cauldron which is also put above the fire place. S1: How long is the pork kept there?

S4: Not much long. If we put it in the morning, we can eat it for dinner. If it is kept for long, the pork becomes too hard and dry. It is then eaten with '*mati mah*' (black split bean).

The ethnic community make the best use of the kitchen's fireplace.

Mising: The fish is cleaned and is then threaded in the skewers. The skewers are put near the fire until it is cooked. They are not smoked. They get cooked from the heat.

The Sonowal Kachari participant shared an interesting herb used to prepare a dish when the fish becomes stale.

Sonowal Kachari: There is one more leaf called '*posotia paat*' (chaste tree leaf). At times when my father goes for fishing and brings fish, sometimes the fish starts to leave an odour. In such condition, my mother cooks the fish with these chaste tree leaves, because these leaves neutralize the odour from the fish and it also has medicinal properties.

Fermenting bamboo shoots is a common practice across all the ethnic communities in the region. Again, lemon is preserved in salt for years together and eater as pickle or when one has any digestion issue.

2.6.3.4 Women Are the Carriers and Distributors of Knowledge and Wisdom of Ethnic Food

One major finding that emerged from the interactions with the participants is that they have acquired all these information from the female members of the family. Be it their mother, grandmother, aunty and even from female members of their neighbour's houses. They did not receive any formal training or instructions but learned through observation, interaction and just by growing up in that environment. Acquiring ethnic knowledge and the wisdom about their food culture was through their socialization process too from which they got socially constructed knowledge (Vauquline, 2015) and wisdom.

Knowledge and Wisdom of Ethnic Food Was Transferred More by the Mothers, Grandmothers and Aunts

I: From where have you learned these?

Assamese Christian: From my grandmother and my aunties. To be honest I had a very keen interest, that's why I learned all of these. I spent my childhood with my grandmother. I used to keenly observe her... how she cooked, what she prepares?

Kuki: From childhood I have seen my grandmother and my mother extract ash from the banana trunks to prepare '*khaar*'.

Knowledge and Wisdom of Ethnic Food also Evolved by Growing Up in that Environment

A few of the participants learned from different sources as they grew up living in that environment. Acquiring this knowledge and wisdom was a natural process for them.

Chutia: I learned by myself because I grew up in a village. I'm telling you whatever I ate and saw my aunties, mother and grandmother preparing.

Bodo 1- We have seen this from childhood. Mother and grandmother used to cook like that. I learned by looking at them.

Mising: I learned from mother and grandmother by observing at how they cooked and how they cooked. I have also seen in the neighbourhood.

I: You?

Rabha: From my mother and from the neighbourhood.

Women as the Carriers of Knowledge and Wisdom Can Be Understood by their Mention in the Interaction

Throughout the interaction with the Kuki participant, she mentioned about her grandmother and her mother. Her grandmother has a huge influence on her knowledge about the ethnic food of her community. She learned from a very young age by assisting her old grandmother in cooking. It was also for the other participants.

Kuki: When I was small, my mother used to go for her work and I used to stay with my grandmother. I used to help her in cooking.

The Bengali Muslim participant from Barak valley region also mentioned about her female relatives from whom she acquired knowledge of the different spices used in the cuisines used in her community.

Bengali Muslim: Yes, I have seen my grandmother, my nani (maternal grandmother), my mother, all my maternal aunts and paternal aunts using spices to prepare our dishes.

Only one participant gathered the knowledge about the ethnic food of her community both from mother and father. But now she is more reliant on her mother.

Sonowal Kachari: Both of them knew it. My mother is not a Xoroniya Kachari. So my father taught my mother. But now when I prepare anything I call my mother for help and assistance.

2.7 Discussion

As mentioned, Northeast India is the site of two biodiversity hotspots of the world. In this study, it has emerged that a variety of ingredients are used to prepare the ethnic food for different occasions. The food culture of serving special cuisines to pregnant women during different months of her pregnancy exit across communities in Assam (as stated by an Assamese Christian). Again, the custom of serving the mother special dishes immediately after giving birth to a child is a common practice in Assam. This dish basically consists of '*tan mass*' (hard fish), black pepper and herbs. The locally available herbs and leafy vegetables are varied which again depicts the rich biodiversity of the region. It reflects the diversity of the food culture of the ethnic communities of the region. The diversity is seen more within than across the ingredients. As for example, fishes and leafy vegetables of diverse varieties are consumed and are part of the food culture of the region. Mentions made by the participants of 'big fish, small fish, fresh water snails, long snails, round snail' etc. reflects the diverse food acquired from the fresh water bodies of Assam. Participants mentioned about the use of different types of leaves, flowers, bulbs, shoots, ants, pupa, etc. in their own language in their cuisines. The diversity of the these naturally endowed edible food in the region is the reflection of blending of natural environment, the biodiversity hotspots with social culture, the food.

The cuisine prepared are dependent on the locally available herbs and vegetables. The influence of the multinational market as a form of globalization process, on the ingredients used in the rural and semi-urban areas of the study area, is not visible. The participants procured the ingredients from their own kitchen garden and locally available sources. The participants who belonged to the rural areas of Assam mentioned the existence of these ingredients for ethnic foods in abundance and plenty in their close vicinity. Their reliance on the other sources for food (market) is minimum. However, this may not be the same in the urban regions of Assam. However, there is strong impact of modernization in rice, the staple food of the region, through the use of high yield variety seeds. It has emerged in this study that many of the indigenous varieties of rice no longer exist now. As mentioned by the participant, Chutia, that, 'there were varieties of local rice when I was a child… Basmati, Manikimadhuri, Suagmoni… There were many more varieties. We usually ate Basmati. It's not the one with long grains, it's was a local variety'. In the recent years effective initiatives to conserve the indigenous seeds of the region in the form of seed banks are initiated (Deka et al., 2023). However, it is insignificant in related to the aggressive corporate marketing.

Deforestation for modernization, urbanization as well as for agriculture has intensified men animal conflicts in different parts of NEI. In many rural areas of the plains of the region people can hardly grow any crops or vegetables because of the menace created by monkeys. The buds, tender leaves, the flowers, etc. and at times the entire plant is consumed or destroyed by the monkeys leaving nothing for the cultivators. Invasion of poppy fields by elephants during the harvesting period is becoming a common site and increasing trend in the rice cultivated areas of the region. In such circumstances people rely of the food products which are easily available in the local markets. These are gradually changing the food habits and thereby the ethnic food and local wisdom.

There are traces of changes in the food culture because of transformation in the family structure in the region. Like elsewhere in India, more women in the region are getting educated and are involved in the labour market at paid workers. At the same time the families are shrinking from joint to nuclear family both in the rural and urban areas of the region. Women's role as provider of food of the family is still intact, however, time devoted in food preparation is shrinking. Due to paucity of time some of the home-made ingredients are getting replaced or substitute with easily available products of the market, as for example '*khaar*' by edible soda, home grinded rice flour by wheat flour, etc. Both of these are common ingredients used widely in the ethnic foods of the region. No massive changes of ethnic foods in the rural area have emerged in the study. Since food of this region is simpler and lacks lavish preparation, these changes are not remarkable but cautionary for the ethnic communities.

Women are playing prominent roles as carrier, practitioner and disseminator of the ethnic knowledge and local wisdom of food of the state of Assam and it is assumed that the scenario would be same in the context of Northeast India as it shares similar ethnic backgrounds. Role of men in this aspect also throws interesting insights. In the study it emerged that men are also involved in the food preparation process. In certain ethnic communities like Rabha, Bodo, Shantal (tea tribes) etc. it is the men in the community who are involved in preparation of food during festive occasions. Among the Sonowal Kachari it is the men who harvest the bamboo shoots. And in certain Assamese communities, it is the man in the family cooks when the woman is menstruating. Through these occasional involvement with food preparation, the men in these communities have nurtured ethnic knowledge and wisdom of ethnic food.

2.8 Conclusion

In this study it has emerged that globalization and modernization through multinational food outlets etc. are not able to impact the ethnic foods and their knowledge amongst the ethnic communities of the rural areas of the region in a massive way. However, these are subtle ways in which certain market made products are replacing home-made ingredients. This gradual beginning can be accelerated by the mushroom growth of multinational food outlets, both small and big, in the urban area. This the long run will influence the ethnic knowledge of food. Huge influx of rural to urban migration has influenced the foods in the urban area, Guwahati (the largest urban centre of the region). As stated by Chutia, 'I eat whatever I get here.'

Women play a distinctive role as the carrier and disseminator of the ethnic knowledge and wisdom of food. They have been the practice of sustainable use of food in the domestic space of home. The common practice of fermenting, sun drying, smoke drying, etc. illustrates minimal wastage of food and sustainable use of the natural resources. Again, rural women have created their own eco-space while traversing through multiple spaces (Dutta & Daulagupu, 2024) for food in their everyday life. On the other hand, there are changes in the position of women in the society through the processes and forces of modernization through employability, mobility, etc. which is transforming the role of women as food provider. This change in the social fabric of the society, which will accelerate in due course of time, is and will be a vital factor in influencing ethnic knowledge and wisdom of food in the region.

It is also a fact that ethnic knowledge and local wisdom of food are valued and practiced by the ethnic communities of the region in their day-to-day life as well as on special occasions. Even though the local wisdoms are diminishing, the special festivals of each community are able to pull them to their roots through their ethnic foods. In the entire North East India the essence of ethnic identity and identity as a community has huge significance. Innumerable social movements across the region revolving around ethnic identity is a witness to it. This study in multiple circumstances has pin-pointed food as a form of ethnic identity of the communities. Thereby ethnic food is valued, practiced and will continuity to exist.

This rich knowledge has emerged by interacting with just nine communities of Assam as a pilot study of an individually project. Further study of this type, taking into account the major ethnic communities of the entire NE region, will provide a pandora box of roles played by women on ethnic knowledge and local wisdom of ethnic food. It will be a valuable asset for Indian Knowledge System. Further study in this regard is very essential to document the role played by women as the custodian of the ethnic knowledge and wisdom of food in the mundane space of home, the kitchen, which requires scholarly research attention.

References

Abas, A., Aziz, A., & Awang, A. (2022). A systematic review on the local wisdom of indigenous people in nature conservation. *Sustainability, 14*(6), 3415.

Amri, Y. (2018). Mangupa; An oral tradition of Angkola community. *Budapest International Research and Critics Institute-Journal (BIRCI-Journal), 1*, 51–61.

Bowlby, S., Lewis, J., McDowell, L., & Foord, J. (2023). The geography of gender. In *New models in geography* (Vol. 2, pp. 157–175). Routledge.

Brickell, K. (2012). 'Mapping' and 'doing' critical geographies of home. *Progress in Human Geography, 36*(2), 225–244.

Census of India. (2011). Office of the Registrar General & Census Commissioner, India Ministry of Home Affairs, Government of India, New Delhi

D'Sylva, A., & Beagan, B. L. (2011). 'Food is culture, but it's also power': The role of food in ethnic and gender identity construction among Goan Canadian women. *Journal of Gender Studies, 20*(3), 279–289. https://doi.org/10.1080/09589236.2011.593326

Deka, D., Rodridgues, J., Kikon, D., Karlsson, B. G., Barbora, S., & Tulu, M. (2023). *Seeds and food sovereignty eastern himalaya experiences*. North East Social Research Centre. ISBN: 978-81-953024-9-9

Dutta, N., & Daulagupu, D. (2024). *Forests and fields of imagined futures, indigenous women and conservation practises in Northeast India* (NESRC Gender Series) (Vol. 7). North East Social Research Centre.

Ginting, J. S., Sumarno, E., Karina, N., & Lubis, M. A. R. (2020). Maintaining local wisdom-building a harmonious life: Inter-ethnic relations in Paya Itik Village, Galang District, Deli Serdang Regency. *Budapest International Research and Critics Institute (BIRCI-Journal): Humanities and Social Sciences, 3*(4), 3598–3607.

Gogoi, B., & Vauquline, P. (2023). A study of the relationship between mental illness and women in Assam from feminist perspective. *Utkal Historical Research Journal, XXXVI*(5 (III)).

Gogoi, B., Das, G., Patra, G., & Vauquline, P. (2023). Probing the socioeconomic and cultural position of women with mental illness in Assam. *Social Change and Development, XX*(2), 66–89.

Hartati, C. D., & Chandra, Y. N. (2020). Local wisdom in the form of traditional food, corn in Karo People tradition. In *2nd International Conference on Social Science and Character Educations (ICoSSCE 2019)* (pp. 212–216). Atlantis Press.

Holt, N. L., Pankow, K., Ormond, I., Jørgensen, H., Deal, C. J., & Pynn, S. R. (2022). Grounded theory. *International Review of Sport and Exercise Psychology, 15*(1), 199–225.

Kapoor, K., & Singh, A. K. (Eds.). (2005). *Indian knowledge systems.* Indian Institute of Advanced Study.

Kuasa, W. A., Rianse, U., Widayati, W., Sidu, D., Abdullah, W. G., Zulfikar, L., & Rianse, I. S. (2015). Local wisdom of farmers in meeting of local food. *International Journal of Sustainable Tropical Agricultural Sciences, 2*(1), 243–296.

Marchang, R. (2019). Economic, occupational and livelihood changes of Scheduled Tribes of North East India. *Institute for Social and Economic Change*, 1–22.

Mungmachon, M. R. (2012). Knowledge and local wisdom: Community treasure. *International Journal of Humanities and Social Science, 2*(13), 174–181.

Pollock, S. (2000). Indian knowledge systems on the eve of colonialism. *Intellectual History Newsletter, 22*, 1–16.

Prabha, J., & Jain, A. (2016). Northeast India: A unique biodiversity hotspot.

Samal, P. K., & Dhyani, P. P. (2006). Gender in the management of indigenous knowledge: Reflections from Indian Central Himalaya. *Current Science*, 104–108.

Shiva, V. (1992). Women's indigenous knowledge and biodiversity conservation. *India International Centre Quarterly, 19*(1/2), 205–214.

Shiva, V., & Mies, M. (2014). *Ecofeminism.* Bloomsbury Publishing.

Urquhart, C. (2022). *Grounded theory for qualitative research: A practical guide.*

Vauquline, P. (2015). Socialisation process, power relations and domestic violence: Marginal voices of Assamese Women. *Space and Culture, India, 3*(2), 54–71.

Wahlqvist, M. L. (2003). Regional food diversity and human health. *Asia Pacific Journal of Clinical Nutrition, 12*(3).

Warren, K. J. (1988). Toward an ecofeminist ethic. *Studies in the Humanities, 15*(2), 140.

Chapter 3
Comparative Ethnobotany between Northern Thailand and Northeastern India: Food Plants of Tibeto-Burman Groups in Lanna and Nagaland

Angkhana Inta, Jatupol Kampuansai, Wibhu Kutanan, Metawee Srikummool, Wittaya Pongamornkul, Prachaya Srisanga, Prateep Panyadee, and Neizo Puro

Abstract This chapter embarks on a comparative ethnobotanical study that aims to document and compare the food plants utilized by Tibeto-Burman groups at Lanna (northern Thailand) and Nagaland (northeast India). It also seeks to identify and discuss cultural similarities and differences in food plant usage, providing insights into their ethnobotanical heritage. Using a combination of free-listing, semi-structured interviews, group discussions, and the walk-to-the-wood method, a total of 59 key informants from nine ethnic groups—Akha, Bisu, Kachin, Karen, Lahu, Lisu, Angami, Ao, and Chakhesang—at 13 villages in northern Thailand and Nagaland were involved. The plant specimens were identified and categorized by food types. Use-reports were calculated for Use Value (UV) and percentage of overlap (% overlap). In relation to 394 species in 261 genera, 97 families and 8 unidentified species, a total of 1218 use-reports were compiled. Vegetables and fruits were the most common food types in both northern Thailand and Nagaland, with leaves and infructescences being the most frequently used parts. The percentage overlap

A. Inta (✉) · J. Kampuansai
Department of Biology, Faculty of Science, Chiang Mai University, Chiang Mai, Thailand
e-mail: angkhana.inta@cmu.ac.th; jatupol.k@cmu.ac.th

W. Kutanan
Department of Biology, Faculty of Science, Naresuan University, Phitsanulok, Thailand
e-mail: wibhuk@nu.ac.th

M. Srikummool
Department of Biochemistry, Faculty of Medical Science, Naresuan University, Phitsanulok, Thailand
e-mail: metaweesr@nu.ac.th

W. Pongamornkul · P. Srisanga · P. Panyadee
Queen Sirikit Botanic Garden, The Botanical Garden Organization, Chiang Mai, Thailand

N. Puro
Department of Botany, Nagaland University, Lumami, Nagaland, India

Y. Santasombat (ed.), *Bio-Cultural Diversity, Ethnicity, and Local Knowledge*,
https://doi.org/10.1007/978-981-95-4242-0_3

between the two regions was 24.5 percent for genera and 12.5 percent for species. The culturally significant food plants among the Tibeto-Burman groups in both regions include *Allium*, *Alpinia*, *Amaranthus*, *Clerodendrum*, *Colocasia*, *Dioscorea*, *Elsholtzia*, *Ficus*, *Musa*, *Senegalia*, *Solanum*, and *Zanthoxylum*. The diversity and overlap in plant usage reflect these communities' adaptation to their ecological settings and their ability to integrate a wide range of plant resources into their diets. Understanding these plants' roles offers valuable insights into the complex relationship between culture, ecology, and nutrition, emphasizing the importance of preserving traditional knowledge and biodiversity.

Keywords Biodiversity · Comparative ethnobotany · Plant specimens · Traditional knowledge · Nagaland and Lanna

3.1 Introduction

Ethnobotany, the study of the relationships between people and plants, offers unique insights into the cultural practices, traditions, and knowledge systems of diverse communities (Cotton, 1996). In regions such as Lanna in northern Thailand and Nagaland in northeastern India, Tibeto-Burman groups have long relied on their surrounding flora for sustenance, medicine, and cultural rituals. These groups, while geographically distant, share linguistic roots and certain cultural affinities that make their traditional knowledge of food plants particularly intriguing for comparative study.

The Tibeto-Burman groups in Lanna and Nagaland represent a fascinating tapestry of cultural and linguistic diversity. These communities are part of the larger Tibeto-Burman language family, which extends across several countries in the Himalayan region and Southeast Asia. In Lanna, these groups, including the Akha, Karen, Lahu, and Lisu, have historically lived in remote, mountainous areas, preserving their unique traditions, languages, and ways of life (Anderson, 1993). Their rich cultural heritage, including vibrant textiles, spiritual practices, and oral traditions, continues to thrive despite the challenges of modernization and integration into broader Thai society.

Similarly, in Nagaland, the Tibeto-Burman groups, including the Angami, Ao, Chakhesang, Chang, Chirr, Khiamniungan, Konyak, Liangmai, Lotha, Phom, Pochury, Poumai, Rengma, Sangtam, Sümi, Tikhir, Yimkhiung, and Zeme, form the "Naga tribe" or "Naga people", majority of the population (Chasie, 1999). The Naga tribes are known for their distinct languages, elaborate customs, and a strong sense of identity tied to their land and history. Each ethnic group has its own language, customs, and social structures, contributing to the intricate cultural mosaic of the region. Despite being part of modern India, the Naga people have retained their traditional ways of life, which include intricate wood carvings, traditional warfare practices, and vibrant festivals like the Hornbill Festival, which showcases the rich cultural diversity of the region (Department of Information Technology & Communication, 2024).

Both in Lanna and Nagaland, the Tibeto-Burman groups are characterized by their resilience, adaptability, and deep connection to their ancestral lands, making them integral to the cultural landscapes of their respective regions. The food plants of ethnic communities hold a vital place in their cultural identity, traditional knowledge, and daily sustenance (Anderson, 1993; Chasie, 1999; Department of Information Technology & Communication, 2024). Across diverse regions, from the highlands of the Himalayan region to the remote villages of Southeast Asia, these communities have developed a deep understanding of their local ecosystems, cultivating and gathering a wide variety of plants that are integral to their diets and traditional practices.

Ethnic groups often rely on a rich diversity of food plants, including wild edibles, cultivated crops, medicinal herbs, and spices, which have been passed down through generations. These plants are not just sources of nutrition; they embody the cultural wisdom, sustainable practices, and ecological knowledge that these communities have honed over centuries. Whether it is the use of rice and indigenous vegetables among the Karen people in Thailand (Khamfachuea et al., 2010; Junsongduang et al., 2014), the cultivation of millet and sorghum by the Maasai in East Africa (Bussmann et al., 2006), or the gathering of wild greens and roots by the Indigenous peoples of the Amazon (Turner et al., 2011), each group's relationship with food plants reflects its unique environment and history.

This work, "Comparative Ethnobotany between Northern Thailand and Northeastern India: Food Plants of Tibeto-Burman Groups in Lanna and Nagaland," delves into the rich botanical traditions of these two regions. Our research questions include: (1) What are the common and unique food plants used by Tibeto-Burman groups in Lanna and Nagaland? (2) How do the cultural practices and environmental factors in Lanna and Nagaland influence the selection and use of food plants among these groups? (3) How do the similarities and differences in food plant usage reflect broader cultural connections or divergences between the Tibeto-Burman groups in Lanna and Nagaland? To document and compare the food plants used by Tibeto-Burman groups in Lanna and Nagaland, as well as to identify and discuss the cultural similarities and differences in food plant usage between the Tibeto-Burman groups in Lanna and Nagaland, this study will look at the food plants used by both communities. This will provide insights into the ethnobotanical heritage of the two regions.

3.2 Materials and Methods

3.2.1 An Overview of the Cultural Background of Akha, Bisu, Kachin, Karen, Lahu, Lisu, Angami, Ao, and Chakhesang

The Akha people live in the highlands of Thailand, Myanmar, Laos, and the Yunnan Province in China. They maintain a strong connection to their animistic beliefs, which influence their social structure and daily life. The Akha are recognized for their intricate traditional headdresses and weaving skills.

The Bisu are a smaller ethnic group residing in Thailand and Myanmar, known for their unique language and cultural practices. Their traditional way of life revolves around agriculture, and they have rich oral traditions that preserve their history and beliefs.

The Kachin, mainly found in the Kachin State of Myanmar as well as a few villages in northern Thailand, are known for their warrior heritage and Christian beliefs, though many also practice animism. The Kachin have a rich tradition of festivals. For instance, the Manau festival is a significant cultural event that involves dance, music, and social gatherings.

The Karen are one of the largest ethnic groups in Myanmar and are also found in Thailand. Their society is traditionally agrarian, with a strong emphasis on community and spiritual practices that blend Buddhism and animism. Many Karen have also adopted Christianity over time.

The Lahu people, originally from Tibet, now predominantly reside in the mountainous regions of Myanmar, China, Thailand, and Laos (Anderson, 1993; Princess Maha Chakri Sirindhorn Anthropology Centre (Public Organization), 2024). They practice a mix of animism, Buddhism, and Christianity, with their traditional culture emphasizing the importance of community and shared agricultural work.

The Lisu are an ethnic group primarily found in Myanmar, Thailand, and China. Their culture is characterized by a deep connection to nature, reflected in their agricultural practices. The Lisu are known for their colorful traditional dress and their celebration of New Year with vibrant festivals.

The Angami people are an indigenous group from Nagaland in India (Chasie, 1999; Department of Information Technology & Communication, 2024). They have a rich tradition of storytelling, agriculture, and crafts, particularly known for their woodwork and shawls. Their society was traditionally structured around clan systems, and they have a history of resistance against external domination.

The Ao are one of the major Naga tribes in Nagaland, India. They are known for their distinct dialects and rich oral traditions. The Ao have a significant Christian population, and their festivals, like Moatsu, play a vital role in maintaining their cultural identity, focusing on agricultural cycles and community bonding.

The Chakhesang are a sub-group of the Naga people in India, residing mainly in Nagaland. They have a rich cultural heritage that includes unique festivals. As an example, Sukrunye celebrates community well-being. Their society places a strong emphasis on agriculture and traditional crafts.

The cultural backgrounds of the ethnic groups of Akha, Bisu, Kachin, Karen, Lahu, Lisu, Angami, Ao, and Chakhesang reflect a rich tapestry of traditions, beliefs, and lifestyles shaped by their unique histories and environments. While their languages share a common linguistic root in the Tibeto-Burman family, all groups maintain their respective distinct speech patterns and cultural practices. Despite these differences, a deep connection to nature and traditional knowledge binds them all, playing a crucial role in their way of life. This connection underscores the importance of preserving their heritage, as it holds invaluable insights into sustainable living and cultural diversity.

3.2.2 *Study Sites*

The study was carried out in 13 villages located in Chiang Mai (5), Chiang Rai (1), Mae Hong Son (4) provinces in northern Thailand, and Nagaland (3) in northeast India (see Fig. 3.1 and Table 3.1). The study sites were selected based on the presence of nine ethnic groups, including Akha, Bisu, Kachin, Karen, Lahu, Lisu, Angami, Ao, and Chakhesang. These groups were represented by 2, 1, 1, 2, 3, 4, 1, 1, and 1 village, respectively. The villages in the study had a range of household sizes from 8 to 600 and a range of population sizes from 68 to 3811 individuals. The vegetation at the study sites primarily consisted of a mixture of deciduous forest, dipterocarp forest, and dry evergreen forest. The three distinct forest types inhabit distinct but partly overlapping areas. In particular, the dry dipterocarp forest's tree species composition showed comparable dominant trees with a mixed stand of deciduous dipterocarps (Khamyong et al., 2018). The elevation of the study sites varied from 628 to 1846 meters above sea level.

There were 59 key informants in total including Akha (7), Bisu (4), Kachin (4), Karen (12), Lahu (8), Lisu (13), Angami (3), Ao (1), and Chakhesang (7) of which 31 were males and 28 were females. Most of them were farmers. The key informants were aged between 25 and 84 years, and they primarily followed Animism, Buddhism, and Christianity as their religion.

In the village of Pa Ya Sai, located in Chiang Mai, all four ethnic groups lived together. In this village, each ethnic group was separately interviewed to gather information on their traditional knowledge of food plants.

3.2.3 *Ethnobotanical Investigation*

The ethnobotanical investigation in this study was carried out with a focus on understanding the local knowledge and practices related to food plants. The process began with the explanation of the study objective to the village head who then suggested key informants for the research. The informants were experienced villagers. The gathering of information was conducted using a combination of free-listing,

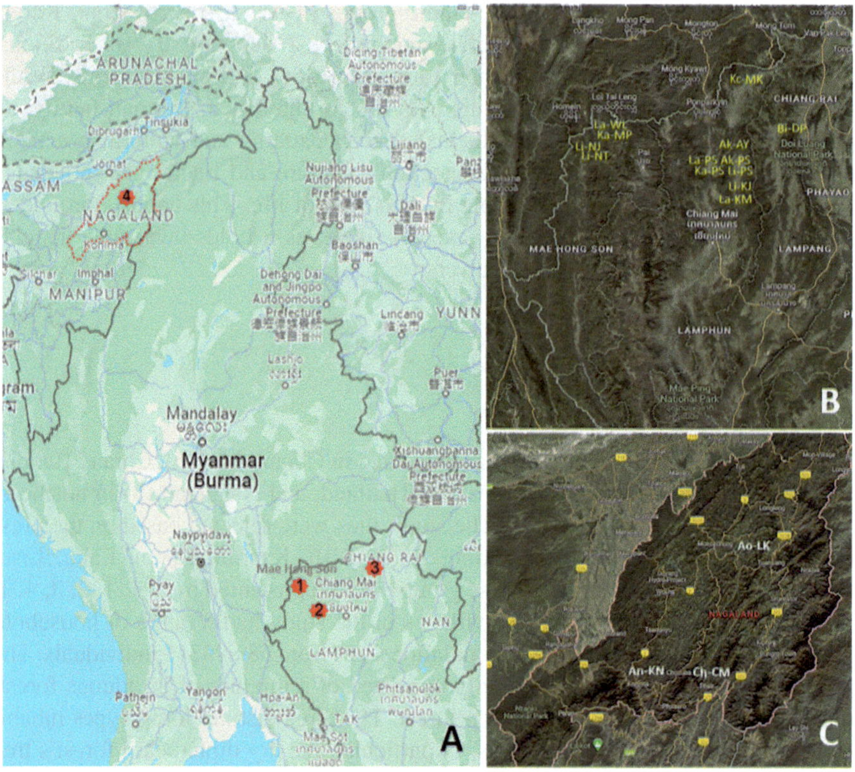

Fig. 3.1 The location of 13 studied villages where the ethnobotanical survey of food plants was conducted; (**a**) Mae Hong Son (1), Chiang Mai (2), Chiang Rai (3) in northern Thailand and Nagaland (4) in India; (**b**) The 10 studied villages in northern Thailand including A Ya (AY), Khon Muang (KM), Khun Che (KC), Mueang Pam (MP), Nong Tong (NT), Nong Pha Chum (NC), Pa Ya Sai (PS), Wana Luang (WL), Doi Chom Phu (DP), Mai Samakkee (MK); (**c**) The 3 studied villages in Nagaland including Khonoma (KN), Chizami (CM), Longkhüm (LK); the nine ethnic groups including Akha (Ak), Bisu (Bi), Kachin (Kc), Karen (Ka), Lahu (La), Lisu (Li), Angami (An), Ao, and Chakhesang (Ch)

semi-structured interviews, group discussion and the walk-to-the-wood method. The questions were designed to collect data on the (i) vernacular name of the plants, (ii) parts used, (iii) methods for preparation, and (iv) characteristics of the plant material.

A total of 59 informants were recruited in this study. The free-listing method was used to gather an initial list of plants used by the informants during group discussion. This was followed by semi-structured interviews, which allowed for a more in-depth exploration of the plants, including the parts used, preparation methods, and purposes of use. The walk-to-the-wood method involved accompanying the informants as they visited their home gardens, nearby forests, and other areas to collect specimens of the plants discussed. This hands-on approach provided an opportunity to observe the plants in their natural habitats and gather additional information from the informants.

Table 3.1 Demographic and ecological characteristics of study villages in northern Thailand and Nagaland

Village	District	Province/ state	Religion	Elevation	Vegetation	Ethnicity	No. of informant (male/ female)	Range of informant age	No. of household	No. of population
A Ya (AY)	Phrao	Chiang Mai	Animism, Buddhism, Christianity	1137	Dry evergreen forest	Akha (Ak)	5 (2/3)	35–75	8	68
Chizami (CM)	Phek	Nagaland	Christianity	981	Dry evergreen forest	Chakhesang (Ch)	7 (7/0)	43–84	600	2592
Doi Chom Phu (DP)	Mae Lao	Chiang Rai	Animism, Buddhism	546	Dipterocarp forest	Bisu (Bi)	4 (3/1)	45–70	48	200
Khon Muang (KM)	Phrao	Chiang Mai	Animism, Christianity	628	Mixed deciduous forest and dipterocarp forest	Lahu (La)	3 (0/3)	40–65	35	151
Khun Che (KC)	Phrao	Chiang Mai	Animism, Buddhism, Christianity	1409	Dry evergreen forest	Lisu (Li)	3 (0/3)	50–65	120	750
Khonoma (KN)	Kohima	Nagaland	Christianity	1621	Dry evergreen forest	Angami (An)	3 (3/0)	26–54	424	1943
Longkhüm (LK)	Mokokchung	Nagaland	Christianity	1846	Dry evergreen forest	Ao	1 (1/0)	50	546	3811
Mai Samakkee (MK)	Chiang Dao	Chiang Mai	Christianity	769	Dipterocarp forest	Kachin (Kc)	4 (2/2)	30–65	103	200
Mueang Pam (MP)	Pang Ma Pha	Mae Hong Son	Animism, Buddhism, Christianity	749	Mixed deciduous forest & dipterocarp forest	Karen (Ka)	10 (5/5)	45–70	60	369

(continued)

Table 3.1 (continued)

Village	District	Province/ state	Religion	Elevation	Vegetation	Ethnicity	No. of informant (male/ female)	Range of informant age	No. of household	No. of population
Nong Pha Chum (NC)	Pang Ma Pha	Mae Hong Son	Animism, Buddhism	982	Mixed deciduous forest & dipterocarp forest	Lisu (Li)	3 (0/3)	35–60	180	464
Nong Tong (NT)	Pang Ma Pha	Mae Hong Son	Animism, Buddhism	680	Mixed deciduous forest & dipterocarp forest	Lisu (Li)	4 (1/3)	25–70	105	400
Pa Ya Sai (PS)	Phrao	Chiang Mai	Animism, Buddhism, Christianity	687	Mixed deciduous forest & dipterocarp forest	Akha (Ak)	2 (1/1)	60–70	8	48
						Karen (Ka)	2 (1/1)	40–50	9	39
						Lahu (La)	3 (1/2)	35–50	7	39
						Lisu (Li)	3 (1/2)	45–55	2	8
Wana Luang (WL)	Pang Ma Pha	Mae Hong Son	Animism, Christianity	630	Dipterocarp forest	Lahu (La)	2 (0/2)	45–60	131	387

All participants provided informed consent prior to their participation in the study. The participants were fully informed about the purpose and procedures of the study and were free to withdraw at any time. Those who were literate provided written consent, while those who were illiterate provided verbal consent with a witness present.

3.2.4 *Plant Identification and Categorization*

Photographs obtained during the field survey were included with the plant specimens that were placed as voucher specimens at the Herbarium of Queen Sirikit Botanic Garden (QBG). The accepted scientific names and families of the identified plant species were determined based on the Plants of the World Online Database (powo.science.kew.org), which is a comprehensive database that provides information on the current accepted name and family for plant species. In some cases, living specimens were collected to ensure future accurate identification and facilitate propagation efforts.

The classification of food types in this study abides by the Economic Botany Data Collection Standard (Cook, 1995), including beverages, cereals, coloring agents, condiments, fruits, snacks, spices, and vegetables.

3.2.5 *Data Analysis*

3.2.5.1 Use-Reports & Group-Informant

In this study, use-reports (URs) were gathered as a primary source of information for ethnobotanical and ethnomedicinal research. These URs provide a comprehensive insight into the traditional knowledge and practices of a specific community regarding the use of various plant species. The URs were collected through discussions with group informants who were members of the same ethnic group from a particular village. The term "group informant" was chosen because the ethnobotanical research was conducted through group discussions. A total of 15 group informants were interviewed to obtain the URs.

The decision to use group discussions in this study was made due to the several advantages it has over individual interviews. First, group discussions helped reduce the informants' fear and apprehension towards the researchers. This was crucial in areas where building trust with the community is challenging. Second, they created a space for informants to share knowledge with one another, thereby facilitating the transmission of ethnobotanical knowledge within the community. Lastly, group discussions were more time-efficient given the study's time constraints. These benefits collectively influenced the choice to use group discussions for this ethnobotanical investigation.

3.2.5.2 Use Value

The Use Value (UV) is an ethnobotanical index that assesses the relative importance of plant species within a community by considering the number of uses reported by informants. It measures how often a plant is mentioned and used for food. The UV index provides insight into which food plants are most important to the Tibeto-Burman groups in Lanna and Nagaland, reflecting their traditional knowledge and practices and identifying the species most valuable in their daily lives and cultural heritage. This index is represented as follows:

$$UV = \left(\sum Ui\right) / N$$

where Ui is the number of use-reports mentioned by each informant and N refers to the total number of informants (Zenderland et al., 2019). Use Values are high when there are many use-reports for a plant, implying that the plant is important, and in contrast, UV approaches zero when there are few reports related to its use.

3.2.5.3 The Percentage of Genera and Species Overlap between Nagaland and Northern Thailand (% Overlap)

To determine the similarities and differences in food plant species between the Tibeto-Burman groups in Nagaland and northern Thailand, an analysis was conducted on the genus and species of food plants from both regions to calculate the percentage of overlap (% overlap). Food plant genus and species with a Use Value equal to or greater than the median was selected for this analysis.

To calculate the median of a set of Use Value, we need to first arrange the numbers in ascending order. If the set has an odd number of values, the median is the middle number. If the set has an even number of values, the median is the average of the two middle numbers. For example, in the set 3, 7, 9 the median is 7. In the set 2, 4, 6, 8 the median is (4 + 6)/2 = 5.

3.3 Results

3.3.1 Food Plants Diversity

A total of 1218 use-reports were compiled from 59 informants, covering 13 villages in Chiang Mai, Chiang Rai, and Mae Hong Son provinces in northern Thailand, and in Nagaland in northeastern India. These use-reports are related to 394 species in 261 genera, 97 families and 8 unidentified species (see Table 3.2 and Table S1). The families with the most food plant species in northern Thailand and in Nagaland were as follows: Fabaceae (25 species, 74 use-reports; 8 species, 24 use-reports), Zingiberaceae (16 species, 34 use-reports; 5 species, 11 use-reports), Poaceae (13

Table 3.2 Food plant species of the Tibeto-Burman groups in northern Thailand and Nagaland ([a]ba = barks, bu = bulbs, co = corms, infl = inflorescences, infr = infructescences, l = leaves, r = roots, rh = rhizomes, s = stems, se = seeds, tu = tubers, wh = whole plants)

Botanical name	Family	Plant part[a]	Type of food	Ethnic group	Country/state
Abelmoschus moschatus Medik.	Malvaceae	infr	Vegetables	Angami	Nagaland
Acmella paniculata (Wall. ex DC.) R.K. Jansen	Asteraceae	infl, l	Vegetables	Akha, Angami, Karen, Lahu, Lisu	Nagaland, Thailand
Aconogonon molle (D.Don) H.Hara	Polygonaceae	l	Vegetables	Angami	Nagaland
Acorus gramineus Aiton	Acoraceae	infl	Vegetables	Lahu	Thailand
Acronychia sp.	Rutaceae	l	Vegetables	Karen	Thailand
Actinidia callosa Lindl.	Actinidiaceae	infr	Fruits	Angami	Nagaland
Actinidia sp.	Actinidiaceae	infr	Fruits	Angami	Nagaland
Adenanthera pavonina L.	Fabaceae	l	Vegetables	Lahu	Thailand
Aegle marmelos (L.) Corrêa	Rutaceae	infr	Beverages	Bisu	Thailand
Agapetes sp.	Ericaceae	infl	Vegetables	Ao	Nagaland
Ageratum conyzoides L.	Asteraceae	infl, l	Vegetables	Ao	Nagaland
Albizia chinensis (Osbeck) Merr.	Fabaceae	infl, l	Vegetables	Ao	Nagaland
Allium ascalonicum L.	Amaryllidaceae	bu	Spices, vegetables	Akha, Ao, Bisu, Kachin, Karen, Lahu	Nagaland, Thailand
Allium chinense G.Don	Amaryllidaceae	l	Spices, vegetables	Akha, Ao, Karen, Lahu	Nagaland, Thailand
Allium hookeri Thwaites	Amaryllidaceae	wh	Spices, vegetables	Ao	Nagaland
Allium sativum L.	Amaryllidaceae	bu	Spices, vegetables	Ao, Kachin, Bisu	Nagaland, Thailand
Allium sp.	Amaryllidaceae	wh	Vegetables	Ao	Nagaland
Allium tuberosum Rottler ex Spreng.	Amaryllidaceae	bu, l	Spices, vegetables	Angami, Karen	Nagaland, Thailand
Allospondias laxiflora (Kurz) Lace	Anacardiaceae	l	Vegetables	Lisu	Thailand
Alocasia macrorrhizos (L.) G.Don	Araceae	l	Vegetables	Ao	Nagaland

(continued)

Table 3.2 (continued)

Botanical name	Family	Plant part[a]	Type of food	Ethnic group	Country/state
Alpinia galanga (L.) Willd.	Zingiberaceae	infl, l, rh	Spices, vegetables	Akha, Bisu, Karen, Lahu	Thailand
Alpinia malaccensis (Burm.f.) Roscoe	Zingiberaceae	l	Vegetables	Lisu	Thailand
Alpinia nigra (Gaertn.) Burtt	Zingiberaceae	l	Vegetables	Angami	Nagaland
Alpinia zerumbet (Pers.) B.L.Burtt & R.M.Sm.	Zingiberaceae	infl, infr, l	Vegetables	Akha, Karen, Lisu	Thailand
Amaranthus blitum L. subsp. *oleraceus* (L.) Costea	Amaranthaceae	l	Vegetables	Lisu	Thailand
Amaranthus cruentus L.	Amaranthaceae	l	Snacks, vegetables	Angami	Nagaland
Amaranthus sp.	Amaranthaceae	l	Vegetables	Ao, Bisu, Kachin	Thailand
Amaranthus spinosus L.	Amaranthaceae	l	Vegetables	Akha, Angami, Ao, Karen, Lahu	Nagaland, Thailand
Amaranthus tricolor L.	Amaranthaceae	l	Vegetables	Angami	Nagaland
Amaranthus viridis L.	Amaranthaceae	l	Vegetables	Angami	Nagaland
Amomum dealbatum Roxb.	Zingiberaceae	l	Vegetables	Angami, Lisu	Nagaland, Thailand
Amomum sp.	Zingiberaceae	l	Vegetables	Ao	Nagaland
Amomum subulatum Roxb.	Zingiberaceae	infr	Spices	Akha	Thailand
Amorphophallus napalensis (Wall.) Bogner & Mayo	Araceae	l	Vegetables	Angami	Nagaland
Amorphophallus sp.	Araceae	l, tu	Vegetables	Akha, Karen, Lahu, Lisu	Thailand
Ananas comosus (L.) Merr.	Bromeliaceae	infr	Fruits	Ao	Nagaland
Annona muricata L.	Annonaceae	infr	Fruits	Bisu	Thailand
Annona reticulata L.	Annonaceae	infr	Fruits	Akha, Karen, Lahu	Thailand
Annona squamosa L.	Annonaceae	infr	Fruits	Bisu	Thailand
Antidesma acidum Retz.	Phyllanthaceae	infr, l	Condiments, fruits, vegetables	Akha, Bisu, Karen, Lahu, Lisu	Thailand
Antidesma bunius (L.) Spreng.	Phyllanthaceae	l	Vegetables	Angami	Nagaland

Aphananthe cuspidata (Blume) Planch.	Cannabaceae	infr	Fruits	Angami	Nagaland
Aporosa villosa (Lindl.) Baill.	Phyllanthaceae	infr	Fruits	Karen, Lahu, Lisu	Thailand
Arachis hypogaea L.	Fabaceae	se	Vegetables	Karen, Lahu, Lisu	Thailand
Ardisia corymbifera Mez	Primulaceae	l	Beverages	Lisu	Thailand
Ardisia crispa (Thunb.) A.DC.	Primulaceae	infr	Fruits, vegetables	Angami	Nagaland
Ardisia polycephala Wall. ex A.DC.	Primulaceae	l	Vegetables	Bisu	Thailand
Ardisia sp.	Primulaceae	infr, l	Fruits, vegetables	Bisu, Karen	Thailand
Arenga westerhoutii Griff.	Arecaceae	l, se	Fruits, vegetables	Bisu, Lahu	Thailand
Artemisia lactiflora Wall. ex DC.	Asteraceae	l	Vegetables	Kachin	Thailand
Artemisia vulgaris L.	Asteraceae	l	Vegetables	Ao	Nagaland
Artocarpus heterophyllus Lam.	Moraceae	infr	Fruits, vegetables	Bisu, Lahu, Lisu	Thailand
Artocarpus lacucha Buch.-Ham.	Moraceae	infr	Fruits	Lisu	Thailand
Artocarpus nitidus Trécul	Moraceae	infr	Fruits	Lisu	Thailand
Asparagus officinalis L.	Asparagaceae	l	Vegetables	Angami	Nagaland
Averrhoa carambola L.	Oxalidaceae	infr, l	Fruits, vegetables	Ao, Bisu	Nagaland, Thailand
Azadirachta indica A.Juss.	Meliaceae	infr, l	Vegetables	Bisu	Thailand
Baccaurea ramiflora Lour.	Phyllanthaceae	infr	Fruits	Akha, Angami, Ao, Karen, Lahu, Lisu	Nagaland, Thailand
Baccaurea sp.	Phyllanthaceae	infr	Fruits	Ao	Nagaland
Basella alba L.	Basellaceae	infr, infl, l	Vegetables	Bisu	Thailand
Bauhinia purpurea L.	Fabaceae	infr, l	Vegetables	Angami, Karen, Lahu, Lisu	Nagaland, Thailand
Bauhinia sp.	Fabaceae	l	Vegetables	Bisu	Thailand
Bauhinia variegata L.	Fabaceae	infr, infl, l, se	Vegetables	Akha, Angami, Karen, Lisu	Nagaland, Thailand
Begonia palmata D.Don	Begoniaceae	l	Vegetables	Angami, Ao	Nagaland
Begonia roxburghii (Miq.) A.DC.	Begoniaceae	l	Vegetables	Angami	Nagaland

(continued)

Table 3.2 (continued)

Botanical name	Family	Plant part[a]	Type of food	Ethnic group	Country/state
Begonia sp.	Begoniaceae	l	Vegetables	Ao	Nagaland
Bidens pilosa L.	Asteraceae	l	Beverages	Angami, Lisu	Nagaland, Thailand
Bischofia javanica Blume	Phyllanthaceae	infr, l	Fruits, vegetables	Karen, Lahu, Lisu	Thailand
Boesenbergia rotunda (L.) Mansf.	Zingiberaceae	r, rh	Spices	Bisu	Thailand
Bombax ceiba L.	Malvaceae	infr, se	Vegetables	Bisu	Thailand
Brassica juncea (L.) Czern.	Brassicaceae	infr, l	Vegetables	Ao	Nagaland
Brassica sp.	Brassicaceae	l	Vegetables	Kachin	Thailand
Breynia androgyna (L.) Chakrab. & N.P.Balakr.	Phyllanthaceae	l	Vegetables	Bisu, Lahu, Lisu	Thailand
Cajanus cajan (L.) Huth	Fabaceae	infr, se	Vegetables	Akha, Karen, Lahu, Lisu	Thailand
Calamus sp.	Arecaceae	s	Vegetables	Akha, Karen, Lahu, Lisu	Thailand
Calamus tenuis Roxb.	Arecaceae	infr	Vegetables	Angami	Nagaland
Camellia sinensis (L.) Kuntze	Theaceae	l	Beverages	Kachin	Thailand
Canarium subulatum Guillaumin	Burseraceae	se	Snacks	Bisu	Thailand
Canavalia gladiata (Jacq.) DC.	Fabaceae	infr	Vegetables	Angami, Ao	Nagaland
Canna indica L.	Cannaceae	rh	Vegetables	Akha, Lahu	Thailand
Capparis sp.	Capparaceae	l	Vegetables	Akha	Thailand
Capsicum annuum L.	Solanaceae	infr	Spices, vegetables	Ao, Bisu	Nagaland, Thailand
Capsicum chinense Jacq.	Solanaceae	infr	Spices	Ao	Nagaland
Capsicum frutescens L.	Solanaceae	infr	Spices	Akha, Bisu, Kachin, Karen, Lahu	Thailand
Carallia brachiata (Lour.) Merr.	Rhizophoraceae	infr	Vegetables	Angami	Nagaland
Cardamine occulta Hornem.	Brassicaceae	l	Vegetables	Angami	Nagaland
Careya arborea Roxb.	Lecythidaceae	infr, l	Vegetables	Akha, Karen	Thailand
Carica papaya L.	Caricaceae	infr	Fruits	Bisu, Kachin, Lahu	Thailand

Caryota maxima Blume	Arecaceae	l	Vegetables	Lisu	Thailand
Caryota sp.	Arecaceae	l	Vegetables	Bisu, Lahu	Thailand
Caryota urens L.	Arecaceae	infr	Vegetables	Angami, Ao	Nagaland
Castanea sativa Mill.	Fagaceae	infr	Vegetables	Ao	Nagaland
Castanopsis acuminatissima (Blume) A.DC.	Fagaceae	infr	Snacks	Akha, Karen, Lahu, Lisu	Thailand
Castanopsis diversifolia (Kurz) King ex Hook.f.	Fagaceae	infr	Snacks	Akha, Karen, Lahu, Lisu	Thailand
Castanopsis indica (Roxb. ex Lindl.) A.DC.	Fagaceae	infr	Snacks	Angami	Nagaland
Cayratia sp.	Vitaceae	infr, l	Condiments, fruits	Karen, Lisu	Thailand
Celastrus paniculatus Willd.	Celastraceae	l	Vegetables	Karen, Lisu	Thailand
Centella asiatica (L.) Urb.	Apiaceae	l	Vegetables	Akha, Ao, Karen, Lahu, Lisu	Nagaland, Thailand
Chenopodium album L.	Amaranthaceae	l	Vegetables	Angami	Nagaland
Chenopodium giganteum D.Don	Amaranthaceae	l	Vegetables	Angami	Nagaland
Chimonobambusa callosa (Munro) Nakai	Poaceae	infr	Vegetables	Angami	Nagaland
Chloranthus elatior Link	Chloranthaceae	l	Vegetables	Bisu	Thailand
Chloranthus nervosus Collett & Hemsl.	Chloranthaceae	wh	Beverages	Lisu	Thailand
Choerospondias axillaris (Roxb.) B.L.Burtt & A.W.Hill	Anacardiaceae	infr	Vegetables	Angami	Nagaland
Cinnamomum sp.	Lauraceae	ba	Spices	Kachin	Thailand
Cinnamomum verum J.Presl	Lauraceae	ba	Spices	Angami	Nagaland
Citrus × aurantiifolia (Christm.) Swingle	Rutaceae	infr	Condiments	Bisu, Kachin Lisu,	Thailand
Citrus maxima (Burm.) Merr.	Rutaceae	infr	Fruits	Karen	Thailand
Clausena excavata Burm.f.	Rutaceae	l	Vegetables	Karen	Thailand
Cleidion javanicum Blume	Euphorbiaceae	l	Vegetables	Bisu	Thailand
Clematis subumbellata Kurz	Ranunculaceae	r	Beverages	Lisu	Thailand

(continued)

Table 3.2 (continued)

Botanical name	Family	Plant part[a]	Type of food	Ethnic group	Country/state
Clerodendrum colebrookianum Walp.	Lamiaceae	infr, l	Snacks, vegetables	Akha, Angami, Ao, Karen, Lahu, Lisu	Nagaland, Thailand
Clerodendrum infortunatum L.	Lamiaceae	l	Vegetables	Bisu	Thailand
Clerodendrum villosum Blume	Lamiaceae	infr, l	Vegetables	Akha, Karen, Lahu	Thailand
Clinacanthus nutans (Burm.f.) Lindau	Acanthaceae	l	Vegetables	Bisu	Thailand
Clitoria ternatea L.	Fabaceae	infl	Coloring agents	Bisu	Thailand
Cnidoscolus aconitifolius (Mill.) I.M.Johnst.	Euphorbiaceae	l	Vegetables	Bisu	Thailand
Coccinia grandis (L.) Voigt	Cucurbitaceae	l	Vegetables	Bisu	Thailand
Cocos nucifera L.	Arecaceae	infr	Fruits	Akha, Bisu, Karen	Thailand
Codiaeum variegatum (L.) Rumph. ex A.Juss.	Euphorbiaceae	l	Vegetables	Bisu, Karen	Thailand
Coffea arabica L.	Rubiaceae	infr	Beverages	Ao	Nagaland
Coix lacryma-jobi L.	Poaceae	infr	Cereals, snacks	Angami, Bisu	Nagaland, Thailand
Coleus amboinicus Lour.	Lamiaceae	l	Vegetables	Bisu	Thailand
Colocasia esculenta (L.) Schott	Araceae	co, l	Vegetables	Akha, Angami, Ao, Kachin, Karen, Lahu, Lisu	Nagaland, Thailand
Colocasia sp.	Araceae	co, l	Vegetables	Ao	Nagaland
Colubrina asiatica (L.) Brongn.	Rhamnaceae	l	Vegetables	Bisu, Kachin	Thailand
Conyza sp.	Asteraceae	l	Vegetables	Lisu	Thailand
Cornus capitata Wall.	Cornaceae	infr	Vegetables	Angami	Nagaland
Crassocephalum crepidioides (Benth.) S.Moore	Asteraceae	l	Vegetables	Akha, Angami, Bisu, Karen, Lahu, Lisu,	Nagaland, Thailand
Crateva religiosa G.Forst.	Capparaceae	l	Vegetables	Bisu, Kachin	Thailand
Crateva sp.	Capparaceae	l	Vegetables	Kachin	Thailand

Cratoxylum cochinchinense (Lour.) Blume	Hypericaceae	l	Vegetables	Akha, Karen, Lahu	Thailand
Cucumis hystrix Chakrav.	Cucurbitaceae	infr	Vegetables	Angami	Nagaland
Cucumis sativus L.	Cucurbitaceae	infr	Vegetables	Akha, Karen, Lahu, Lisu	Thailand
Cucurbita moschata Duchesne	Cucurbitaceae	infr, l	Vegetables	Akha, Karen, Lahu, Lisu	Thailand
Curcuma angustifolia Roxb.	Zingiberaceae	infl	Vegetables	Angami	Nagaland
Curcuma longa L.	Zingiberaceae	rh	Spices	Bisu, Kachin	Thailand
Curcuma mangga Valeton & Zijp	Zingiberaceae	rh	Vegetables	Bisu	Thailand
Curcuma sp.	Zingiberaceae	infl	Vegetables	Karen, Lisu	Thailand
Cycas simplicipinna (Smitinand) K.D.Hill	Cycadaceae	l	Vegetables	Lisu	Thailand
Cycas sp.	Cycadaceae	l	Vegetables	Lisu	Thailand
Cymbopogon citratus (DC.) Stapf	Poaceae	l	Spices	Bisu, Kachin, Lahu	Thailand
Dalbergia sp.	Fabaceae	ba	Snacks	Lisu	Thailand
Dalrympelea pomifera Roxb	Staphyleaceae	infr, l	Fruits, vegetables	Akha, Karen	Thailand
Debregeasia longifolia (Burm.f.) Wedd.	Urticaceae	infr	Fruits	Angami	Nagaland
Dendrocalamus asper (Schult.f.) Backer	Poaceae	s	Vegetables	Bisu, Karen, Lisu	Thailand
Dendrocalamus giganteus Munro	Poaceae	s	Vegetables	Angami	Nagaland
Dendrocalamus hamiltonii Nees & Arn. ex Munro	Poaceae	s	Vegetables	Lahu	Thailand
Dendrocalamus membranaceus Munro	Poaceae	s	Vegetables	Lisu	Thailand
Dendrocalamus strictus (Roxb.) Nees	Poaceae	s	Vegetables	Kachin, Lahu	Thailand
Dichrocephala integrifolia (L.f.) Kuntze	Asteraceae	l	Vegetables	Lisu	Thailand
Dillenia parviflora Griff.	Dilleniaceae	infr	Fruits	Akha, Karen, Lahu, Lisu	Thailand
Dioscorea bulbifera L.	Dioscoreaceae	bu	Vegetables	Angami, Ao	Nagaland
Dioscorea sp.	Dioscoreaceae	tu	Vegetables	Ao, Akha, Karen, Lahu, Lisu,	Thailand
Diospyros kaki L.f.	Ebenaceae	infr	Fruits	Angami	Nagaland
Diplazium esculentum (Retz.) Sw.	Aspleniaceae	l	Vegetables	Angami, Ao	Nagaland

(continued)

Table 3.2 (continued)

Botanical name	Family	Plant part[a]	Type of food	Ethnic group	Country/state
Dischidia nummularia R.Br.	Apocynaceae	l	Snacks	Akha	Thailand
Docynia indica (Colebr. ex Wall.) Decne.	Rosaceae	infr	Fruits, condiments	Angami	Nagaland
Dracaena angustifolia (Medik.) Roxb.	Asparagaceae	infl	Vegetables	Bisu	Thailand
Dracontomelon dao (Blanco) Merr. & Rolfe	Anacardiaceae	se	Vegetables	Bisu	Thailand
Dregea volubilis (L.f.) Benth. ex Hook.f.	Apocynaceae	infl, l	Vegetables	Bisu, Karen	Thailand
Duabanga grandiflora (Roxb. ex DC.) Walp.	Lythraceae	infl	Vegetables	Karen	Thailand
Elaeagnus conferta Roxb.	Elaeagnaceae	infr	Fruits	Ao	Nagaland
Elaeagnus latifolia L.	Elaeagnaceae	infr	Fruits	Angami, Bisu, Karen, Lahu, Lisu	Nagaland, Thailand
Elaeagnus umbellata Thunb.	Elaeagnaceae	infr	Fruits	Angami	Nagaland
Elaeocarpus floribundus Blume	Elaeocarpaceae	infr	Fruits	Angami	Nagaland
Elatostema lineolatum Wight	Urticaceae	l	Vegetables	Angami	Nagaland
Elatostema platyphyllum Wedd.	Urticaceae	l	Vegetables	Angami	Nagaland
Elatostema sessile J.R.Forst. & G.Forst.	Urticaceae	l	Vegetables	Angami	Nagaland
Eleutherococcus trifoliatus (L.) S.Y.Hu	Araliaceae	l	Vegetables	Karen, Lisu	Thailand
Elsholtzia blanda (Benth.) Benth.	Lamiaceae	l	Spices, vegetables	Ao	Nagaland
Elsholtzia griffithii Hook.f.	Lamiaceae	infl, l	Spices, vegetables	Akha, Karen, Lahu	Thailand
Elsholtzia kachinensis Prain	Lamiaceae	l, s	Spices, vegetables	Akha, Lahu	Thailand
Elsholtzia sp.	Lamiaceae	l	Spices	Angami, Chakhesang	Nagaland
Elsholtzia stachyodes (Link) Raizada & H.O.Saxena	Lamiaceae	l	Spices, vegetables	Akha, Karen	Thailand
Embelia sessiliflora Kurz	Primulaceae	infr, l	Fruits, vegetables	Akha, Karen, Lahu, Lisu	Thailand
Embelia tsjeriam-cottam (Roem. & Schult.) A.DC.	Primulaceae	l	Vegetables	Karen, Lahu	Thailand

Entada rheedei Spreng.	Fabaceae	l, se	Cereals, vegetables	Karen, Lahu	Thailand
Eryngium foetidum L.	Apiaceae	l	Condiments, vegetables	Akha, Kachin, Karen, Lahu	Thailand
Erythrina subumbrans (Hassk.) Merr.	Fabaceae	l	Vegetables	Karen	Thailand
Etlingera araneosa (Baker) R.M.Sm.	Zingiberaceae	infl, infr	Vegetables	Lahu	Thailand
Etlingera elatior (Jack) R.M.Sm.	Zingiberaceae	infl	Vegetables	Lahu	Thailand
Euodia sp.	Rutaceae	l	Vegetables	Lahu	Thailand
Eupatorium capillifolium (Lam.) Small ex Porter & Britton	Asteraceae	l	Vegetables	Bisu	Thailand
Fagopyrum dibotrys (D.Don) H.Hara	Polygonaceae	s	Vegetables	Angami, Lisu	Nagaland, Thailand
Fagopyrum esculentum Moench	Polygonaceae	l	Vegetables	Angami	Nagaland
Ficus auriculata Lour.	Moraceae	infr, l	Fruits, vegetables	Akha, Angami, Ao, Karen, Lahu, Lisu	Nagaland, Thailand
Ficus fistulosa Reinw. ex Blume	Moraceae	l	Vegetables	Angami	Nagaland
Ficus geniculata Kurz	Moraceae	l	Vegetables	Angami	Nagaland
Ficus hispida L.f.	Moraceae	infr	Fruits, vegetables	Angami, Bisu, Akha, Lahu, Lisu	Nagaland, Thailand
Ficus racemosa L.	Moraceae	infr, l	Fruits, vegetables	Akha, Karen	Thailand
Ficus semicordata Buch.-Ham. ex Sm.	Moraceae	infr	Fruits	Angami, Bisu, Karen, Lisu	Nagaland, Thailand
Ficus sp.	Moraceae	l	Vegetables	Akha, Karen, Lahu	Thailand
Ficus virens Aiton	Moraceae	l	Vegetables	Akha, Kachin, Karen, Lisu	Thailand
Flacourtia indica (Burm.f.) Merr.	Salicaceae	infr, l	Fruits, vegetables	Karen	Thailand
Garcinia xanthochymus Hook.f. ex T. Anderson	Clusiaceae	infr	Fruits	Angami, Ao	Nagaland
Gardenia sootepensis Hutch.	Rubiaceae	infr	Fruits	Akha, Karen, Lahu	Thailand
Garuga pinnata Roxb.	Burseraceae	ba	Condiments	Lisu	Thailand
Gigantochloa albociliata (Munro) Kurz	Poaceae	s	Vegetables	Bisu, Karen	Thailand

(continued)

Table 3.2 (continued)

Botanical name	Family	Plant part[a]	Type of food	Ethnic group	Country/state
Glinus herniarioides (Gagnep.) Tardieu	Molluginaceae	l	Vegetables	Karen	Thailand
Glochidion sp.	Phyllanthaceae	l	Vegetables	Karen, Lahu, Lisu	Thailand
Glycine max (L.) Merr.	Fabaceae	se	Condiments	Kachin	Thailand
Gmelina arborea Roxb. ex Sm.	Lamiaceae	infl	Condiments	Akha, Karen	Thailand
Gnaphalium unionis Sch.Bip. ex Hochst.	Asteraceae	l	Vegetables	Angami	Nagaland
Grewia eriocarpa Juss.	Malvaceae	infr	Fruits	Lisu	Thailand
Gymnema inodorum (Lour.) Decne.	Apocynaceae	l	Vegetables	Bisu	Thailand
Gynura bicolor (Roxb. ex Willd.) DC.	Asteraceae	l	Vegetables	Angami	Nagaland
Gynura nepalensis DC.	Asteraceae	l	Vegetables	Angami	Nagaland
Hellenia speciosa (J.Koenig) S.R.Dutta	Costaceae	l	Vegetables	Angami, Karen	Nagaland, Thailand
Heptapleurum leucanthum (R.Vig.) Y.F.Deng	Araliaceae	l	Vegetables	Bisu	Thailand
Herpetospermum pedunculosum (Ser.) C.B.Clarke	Cucurbitaceae	l	Vegetables	Angami	Nagaland
Hibiscus sabdariffa L.	Malvaceae	infl	Beverages	Ao	Nagaland
Hodgsonia macrocarpa (Blume) Cogn.	Cucurbitaceae	infr	Snacks, vegetables	Angami	Nagaland
Houttuynia cordata Thunb.	Saururaceae	l,rh	Vegetables	Angami, Bisu, Kachin	Nagaland, Thailand
Hovenia dulcis Thunb.	Rhamnaceae	infr	Fruits	Angami	Nagaland
Hoya sp.	Apocynaceae	infl	Vegetables	Karen	Thailand
Hultholia mimosoides (Lam.) Gagnon & G.P.Lewis	Fabaceae	infl, l	Vegetables	Bisu	Thailand
Illicium verum Hook.f.	Schisandraceae	infr	Spices	Kachin	Thailand
Impatiens latifolia L.	Balsaminaceae	l	Vegetables	Angami	Nagaland
Ipomoea batatas (L.) Lam.	Convolvulaceae	r	Vegetables	Ao	Nagaland

Irvingia malayana Oliv. ex A.W.Benn.	Irvingiaceae	se	Snacks	Bisu	Thailand
Jasminum sambac (L.) Aiton	Oleaceae	infl	Beverages	Bisu	Thailand
Juglans regia L.	Juglandaceae	infr	Snacks, vegetables	Angami	Nagaland
Justicia santapaui Bennet	Acanthaceae	l	Vegetables	Angami	Nagaland
Kaempferia sp.	Zingiberaceae	l	Condiments	Kachin	Thailand
Lablab purpureus (L.) Sweet	Fabaceae	infr	Vegetables	Akha, Angami, Ao, Kachin, Karen, Lahu, Lisu	Nagaland, Thailand
Lagenaria siceraria (Molina) Standl.	Cucurbitaceae	infr	Vegetables	Akha, Ao, Bisu, Karen, Lahu, Lisu	Nagaland, Thailand
Lasia spinosa (L.) Thwaites	Araceae	l	Vegetables	Akha, Angami, Bisu, Lahu	Nagaland, Thailand
Lecanthus peduncularis (Royle) Wedd.	Urticaceae	l	Vegetables	Angami	Nagaland
Leea indica (Burm.f.) Merr.	Vitaceae	infr	Fruits	Karen, Lisu	Thailand
Leea sp.	Vitaceae	infr	Fruits	Karen, Lisu	Thailand
Lepisanthes rubiginosa (Roxb.) Leenh.	Sapindaceae	infr	Fruits	Bisu	Thailand
Leptochilus sp.	Polypodiaceae	l	Vegetables	Karen	Thailand
Leucaena leucocephala (Lam.) de Wit	Fabaceae	l	Vegetables	Bisu, Lisu	Thailand
Leucocasia gigantea (Blume) Schott	Araceae	l	Vegetables	Lisu	Thailand
Litsea cubeba (Lour.) Pers.	Lauraceae	infr	Spices	Angami, Lisu	Nagaland, Thailand
Livistona jenkinsiana Griff.	Arecaceae	infr	Fruits, vegetables	Angami	Nagaland
Livistona sp.	Arecaceae	infr	Fruits, vegetables	Ao	Nagaland
Livistona speciosa Kurz	Arecaceae	infr	Fruits, vegetables	Akha, Bisu, Karen, Lahu	Thailand
Luffa aegyptiaca Mill.	Cucurbitaceae	infr	Vegetables	Akha, Ao, Bisu, Karen, Lisu, Lahu	Nagaland, Thailand
Lygodium flexuosum (L.) Sw.	Lygodiaceae	l	Vegetables	Akha, Karen	Thailand
Mahonia napaulensis DC.	Berberidaceae	infr	Fruits	Angami	Nagaland
Mangifera indica L.	Anacardiaceae	infr, l	Fruits, vegetables	Akha, Karen, Lahu, Lisu	Thailand

(continued)

Table 3.2 (continued)

Botanical name	Family	Plant part[a]	Type of food	Ethnic group	Country/state
Mangifera sp.	Anacardiaceae	infr,l	Fruits, vegetables	Karen	Thailand
Manihot esculenta Crantz	Euphorbiaceae	l, r	Cereals, vegetables	Bisu, Lisu	Thailand
Maranta arundinacea L.	Marantaceae	rh	Vegetables	Bisu	Thailand
Markhamia stipulata (Wall.) Seem.	Bignoniaceae	infl, infr	Vegetables	Akha, Karen	Thailand
Mayodendron igneum (Kurz) Kurz	Bignoniaceae	infl	Vegetables	Akha, Karen	Thailand
Melastoma malabathricum L.	Melastomataceae	infr	Fruits	Angami, Lisu	Nagaland, Thailand
Melastoma sp.	Melastomataceae	infr	Fruits	Akha, Karen, Lahu	Thailand
Melicope pteleifolia (Champ. ex Benth.) T.G.Hartley	Rutaceae	l	Condiments	Karen	Thailand
Mentha × villosa Huds.	Lamiaceae	l	Spices, vegetables	Angami	Nagaland
Mentha sp.	Lamiaceae	l	Spices	Kachin	Thailand
Merremia sp.	Convolvulaceae	infr	Vegetables	Karen	Thailand
Microcos sp.	Malvaceae	infr	Fruits	Lisu	Thailand
Millettia sp.	Fabaceae	infl	Vegetables	Karen	Thailand
Momordica charantia L.	Cucurbitaceae	infr, l	Vegetables	Akha, Bisu, Karen, Lahu	Thailand
Momordica cochinchinensis (Lour.) Spreng.	Cucurbitaceae	infr, l	Vegetables	Bisu, Lisu	Thailand
Morinda citrifolia L.	Rubiaceae	l	Vegetables	Bisu	Thailand
Moringa oleifera Lam.	Moringaceae	infl, infr, l	Vegetables	Bisu, Karen	Thailand
Morus alba L.	Moraceae	infr	Fruits	Angami, Bisu	Nagaland, Thailand
Morus macroura Miq.	Moraceae	l	Vegetables	Lisu	Thailand
Morus nigra L.	Moraceae	infr	Fruits, vegetables	Angami	Nagaland
Musa acuminata Colla	Musaceae	infl, l	Vegetables	Ao, Lahu	Nagaland, Thailand
Musa sikkimensis Kurz	Musaceae	infr	Fruits	Angami	Nagaland

Musa thomsonii (King) A.M.Cowan & Cowan	Musaceae	infr	Fruits	Angami	Nagaland
Myrica esculenta Buch.-Ham. ex D.Don	Myricaceae	infr	Fruits	Angami	Nagaland
Nephelium lappaceum L.	Sapindaceae	infr	Fruits	Angami	Nagaland
Nicotiana tabacum L.	Solanaceae	l	Vegetables	Angami	Nagaland
Ocimum americanum L.	Lamiaceae	l	Spices, vegetables	Lahu	Thailand
Ocimum basilicum L.	Lamiaceae	l	Spices, vegetables	Kachin	Thailand
Ocimum tenuiflorum L.	Lamiaceae	l	Spices, vegetables	Bisu, Lahu	Thailand
Oenanthe javanica (Blume) DC.	Apiaceae	l	Vegetables	Akha, Angami, Karen, Lahu	Nagaland, Thailand
Olax imbricata Roxb.	Olacaceae	infr	Fruits	Angami	Nagaland
Oroxylum indicum (L.) Kurz	Bignoniaceae	infl, infr, l	Vegetables	Akha, Bisu, Karen, Lahu, Lisu	Thailand
Oryza sativa L.	Poaceae	se	Cereals	Angami, Kachin	Nagaland
Oxalis corniculata L.	Oxalidaceae	infr	Fruits	Angami	Nagaland
Paederia foetida L.	Rubiaceae	l	Vegetables	Angami	Nagaland
Paederia pilifera Hook.f.	Rubiaceae	l	Vegetables	Karen, Lahu, Lisu	Thailand
Parkia sp.	Fabaceae	se	Vegetables	Angami	Nagaland
Passiflora edulis Sims	Passifloraceae	infr, l	Fruits, vegetables	Akha, Karen, Lisu	Thailand
Passiflora subpeltata Ortega	Passifloraceae	infr	Fruits	Angami	Nagaland
Peliosanthes teta Andrews	Asparagaceae	infl	Vegetables	Bisu	Thailand
Perilla frutescens (L.) Britton	Lamiaceae	se	Condiments	Ao, Kachin, Lahu, Lisu	Nagaland, Thailand
Persicaria chinensis (L.) H.Gross	Polygonaceae	infr, l, s	Vegetables	Angami, Akha, Lisu	Nagaland, Thailand
Persicaria nepalensis (Meisn.) H.Gross	Polygonaceae	l	Vegetables	Angami	Nagaland
Persicaria odorata (Lour.) Soják	Polygonaceae	l	Condiments	Lahu	Thailand

Table 3.2 (continued)

Botanical name	Family	Plant part[a]	Type of food	Ethnic group	Country/state
Phaseolus sp.	Fabaceae	infr	Vegetables	Chakhesang	Nagaland
Phlogacanthus curviflorus (Nees) Nees	Acanthaceae	infl, l, r	Beverages, vegetables	Karen, Kachin, Lisu	Thailand
Phlogacanthus pulcherrimus T.Anderson	Acanthaceae	l	Vegetables	Kachin	Thailand
Phyllanthus acidus (L.) Skeels	Phyllanthaceae	ba, infr	Condiments, fruits	Akha, Angami, Ao, Bisu, Kachin, Karen, Lahu, Lisu	Nagaland, Thailand
Phyllanthus emblica L.	Phyllanthaceae	ba, infr	Condiments, fruits	Lisu	Thailand
Physalis angulata L.	Solanaceae	infr	Fruits	Angami	Nagaland
Physalis peruviana L.	Solanaceae	infr	Fruits	Angami	Nagaland
Pilea scripta (Buch.-Ham. ex D.Don) Wedd.	Urticaceae	l	Vegetables	Angami	Nagaland
Pimpinella cambodgiana H.Boissieu	Apiaceae	wh	Beverages	Lisu	Thailand
Piper nigrum L.	Piperaceae	infr, se	Spices	Bisu, Kachin	Thailand
Piper sarmentosum Roxb.	Piperaceae	l	Vegetables	Bisu	Thailand
Piper sp.	Piperaceae	l	Vegetables	Akha, Lahu, Lisu	Thailand
Piper sp.	Piperaceae	s	Condiments	Kachin, Lahu	Thailand
Piper wallichii (Miq.) Hand.-Mazz.	Piperaceae	s	Spices	Bisu	Thailand
Plantago asiatica L. subsp. *erosa* (Wall.) Z.Yu Li	Plantaginaceae	l	Vegetables	Angami	Nagaland
Plantago major L.	Plantaginaceae	l	Vegetables	Angami	Nagaland
Polygonum sp.	Polygonaceae	l	Condiments	Kachin	Thailand
Portulaca sp.	Portulacaceae	l	Condiments	Bisu	Thailand
Potentilla indica (Andrews) Th.Wolf	Rosaceae	infr	Fruits	Angami	Nagaland
Protium serratum (Wall. ex Colebr.) Engl.	Burseraceae	infr	Fruits	Akha, Karen, Lahu, Lisu	Thailand
Prunus cerasoides Buch.-Ham. ex D.Don	Rosaceae	infr	Fruits	Angami	Nagaland
Prunus napaulensis (Ser.) Steud.	Rosaceae	infr	Fruits	Angami	Nagaland

Pseudognaphalium affine (D.Don) Anderb.	Asteraceae	l	Vegetables	Angami	Nagaland
Psidium guajava L.	Myrtaceae	infr	Fruits	Akha, Angami, Bisu, Karen, Lahu, Lisu	Nagaland, Thailand
Psophocarpus tetragonolobus (L.) DC.	Fabaceae	infr	Vegetables	Akha, Ao, Karen, Lahu, Lisu	Nagaland, Thailand
Punica granatum L.	Lythraceae	infr	Fruits	Bisu	Thailand
Pyrus pashia Buch.-Ham. ex D.Don	Rosaceae	infr	Fruits	Angami	Nagaland
Rhododendron arboreum Sm.	Ericaceae	infl	Vegetables	Angami	Nagaland
Rhus chinensis Mill.	Anacardiaceae	infr	Fruits	Akha, Angami, Karen, Lahu, Lisu	Nagaland, Thailand
Rubus blepharoneurus Cardot	Rosaceae	infr	Fruits	Lisu	Thailand
Rubus dielsianus Focke	Rosaceae	infr	Fruits	Akha, Karen	Thailand
Rubus ellipticus Sm.	Rosaceae	infr	Fruits	Akha, Angami	Nagaland, Thailand
Rubus ellipticus Sm. var. *obcordatus* (Franch.) Focke	Rosaceae	infr	Fruits	Lisu	Thailand
Rubus leucanthus Hance	Rosaceae	infr	Fruits	Lisu	Thailand
Rubus niveus Thunb.	Rosaceae	infr	Fruits	Angami	Nagaland
Rubus sp.	Rosaceae	infr	Fruits	Akha	Thailand
Rubus sumatranus Miq.	Rosaceae	infr	Fruits	Angami	Nagaland
Rumex crispus L.	Polygonaceae	l	Vegetables	Angami	Nagaland
Rumex patientia L.	Polygonaceae	l	Vegetables	Angami	Nagaland
Saccharum officinarum L.	Poaceae	s	Beverages	Akha, Karen, Lahu	Thailand
Saurauia macrotricha Kurz	Actinidiaceae	infr	Fruits	Angami	Nagaland
Schefflera sp.	Araliaceae	l	Vegetables	Karen	Thailand
Schima wallichii (DC.) Korth.	Theaceae	l	Vegetables	Karen, Lahu	Thailand
Schleichera oleosa (Lour.) Oken	Sapindaceae	infr	Fruits	Karen, Lisu	Thailand

(continued)

Table 3.2 (continued)

Botanical name	Family	Plant part[a]	Type of food	Ethnic group	Country/state
Scoparia dulcis L.	Plantaginaceae	l	Vegetables	Bisu, Lisu	Thailand
Sechium edule (Jacq.) Sw.	Cucurbitaceae	infr, l	Vegetables	Akha, Angami, Karen, Lahu, Lisu	Nagaland, Thailand
Senegalia pennata (L.) Maslin	Fabaceae	l	Vegetables	Akha, Karen, Lahu, Lisu	Thailand
Senegalia rugata (Lam.) Britton & Rose	Fabaceae	l	Vegetables,condiments	Akha, Bisu, Karen, Lahu, Lisu	Thailand
Senna occidentalis (L.) Link	Fabaceae	l	Vegetables	Bisu	Thailand
Senna siamea (Lam.) H.S.Irwin & Barneby	Fabaceae	l	Vegetables	Bisu	Thailand
Senna tora (L.) Roxb.	Fabaceae	se	Vegetables	Karen	Thailand
Sesamum indicum L.	Pedaliaceae	se	Condiments	Akha, Karen, Lahu	Thailand
Setaria italica (L.) P.Beauv.	Poaceae	se	Cereals	Angami	Nagaland
Smallanthus sonchifolius (Poepp.) H. Rob.	Asteraceae	r	Fruits	Ao	Nagaland
Smilax ovalifolia Roxb. ex D.Don	Smilacaceae	l	Vegetables	Karen	Thailand
Solanum americanum Mill.	Solanaceae	l	Vegetables	Akha, Angami, Bisu, Karen, Lahu, Lisu	Nagaland, Thailand
Solanum erianthum D.Don	Solanaceae	r	Beverages	Lisu	Thailand
Solanum lycopersicum L.	Solanaceae	infr	Vegetables	Angami, Akha, Karen, Lahu, Lisu	Nagaland
Solanum sp.	Solanaceae	infr	Beverages, vegetables	Karen, Lahu, Lisu	Thailand
Solanum spirale Roxb.	Solanaceae	l	Vegetables	Angami, Bisu	Nagaland, Thailand
Solanum torvum Sw.	Solanaceae	infr	Vegetables	Akha, Bisu, Kachin, Lahu, Lisu	Thailand
Solanum tuberosum L.	Solanaceae	tu	Vegetables	Kachin	Thailand
Solanum violaceum Ortega	Solanaceae	infr	Vegetables	Akha, Karen, Lahu	Thailand
Sonchus arvensis L.	Asteraceae	l	Vegetables	Lisu	Thailand
Sorghum bicolor (L.) Moench	Poaceae	se	Beverages, cereals	Akha, Angami, Karen, Lahu, Lisu	Nagaland

Spondias pinnata (L.f.) Kurz	Anacardiaceae	l,infr,ba	Condiments, fruits, vegetables	Akha, Angami, Bisu, Karen, Lahu	Nagaland, Thailand
Sterculia lanceolata Cav.	Malvaceae	se	Snacks	Bisu	Thailand
Sterculia sp.	Malvaceae	se	Snacks	Akha, Lahu	Thailand
Stixis suaveolens (Roxb.) Baill.	Resedaceae	infr	Fruits	Angami	Nagaland
Streblus asper Lour.	Moraceae	infr	Fruits	Bisu	Thailand
Syzygium cumini (L.) Skeels	Myrtaceae	infr	Fruits	Akha, Angami, Karen, Lahu, Lisu	Nagaland, Thailand
Syzygium megacarpum (Craib) Rathakr. & N.C.Nair	Myrtaceae	l	Vegetables	Karen	Thailand
Tacca chantrieri André	Dioscoreaceae	l	Vegetables	Lahu	Thailand
Tamarindus indica L.	Fabaceae	ba, infl, infr, l	Condiments, fruits, vegetables	Akha, Bisu, Karen, Lahu, Lisu	Thailand
Terminalia bellirica (Gaertn.) Roxb.	Combretaceae	infr	Fruits	Angami, Karen, Lahu	Nagaland, Thailand
Terminalia chebula Retz.	Combretaceae	infr	Fruits	Akha, Angami, Bisu, Karen, Lahu, Lisu	Nagaland, Thailand
Tetradium fraxinifolium (Hook.) T.G.Hartley	Rutaceae	infr	Spices	Angami	Nagaland
Tetrastigma leucostaphylum (Dennst.) Alston	Vitaceae	infr	Fruits	Angami	Nagaland
Tetrastigma sp.	Vitaceae	infr, l	Condiments, fruits	Akha	Thailand
Thunbergia laurifolia Lindl.	Acanthaceae	infl	Vegetables	Bisu	Thailand
Thyrsostachys siamensis Gamble	Poaceae	s	Vegetables	Lisu	Thailand
Thysanolaena latifolia (Roxb. ex Hornem.) Honda	Poaceae	s	Vegetables	Karen	Thailand
Tiliacora triandra (Colebr.) Diels	Menispermaceae	f	L	Bisu	Thailand
Trevesia palmata (Roxb. ex Lindl.) Vis.	Araliaceae	infl	Vegetables	Akha, Bisu, Karen, Lahu, Lisu	Thailand
Trevesia sundaica Miq.	Araliaceae	l	Vegetables	Angami	Nagaland

(continued)

Table 3.2 (continued)

Botanical name	Family	Plant part[a]	Type of food	Ethnic group	Country/state
Trichodesma khasianum C.B.Clarke	Boraginaceae	l	Vegetables	Angami	Nagaland
Trichosanthes dunniana H.Lév.	Cucurbitaceae	infr	Vegetables	Angami	Nagaland
Trichosanthes pubera Blume	Cucurbitaceae	l	Vegetables	Lisu	Thailand
Urtica ardens Link	Urticaceae	l	Vegetables	Angami	Nagaland
Urtica fissa E.Pritz.	Urticaceae	l	Vegetables	Angami	Nagaland
Vaccinium sprengelii (G.Don) Sleumer ex Rehder	Ericaceae	l	Vegetables	Karen, Lahu	Thailand
Viburnum mullaha Buch.-Ham. ex D.Don	Adoxaceae	infr	Fruits	Angami	Nagaland
Vigna unguiculata (L.) Walp.	Fabaceae	infr	Vegetables	Kachin, Lisu	Thailand
Wendlandia tinctoria (Roxb.) DC.	Rubiaceae	infl, l	Vegetables	Bisu, Lahu, Lisu	Thailand
Wurfbainia schmidtii (K.Schum.) Škorničk. & A.D.Poulsen	Zingiberaceae	infr	Vegetables	Bisu	Thailand
Xylia xylocarpa (Roxb.) W.Theob.	Fabaceae	ba, l, se	Condiments, snacks, vegetables	Karen, Lisu	Thailand
Youngia japonica (L.) DC.	Asteraceae	l	Vegetables	Lisu	Thailand
Zanthoxylum acanthopodium DC.	Rutaceae	infr	Spices	Angami	Nagaland
Zanthoxylum armatum DC.	Rutaceae	infr	Spices	Angami	Nagaland
Zanthoxylum asiaticum (L.) Appelhans, Groppo & J.Wen	Rutaceae	infr, l	Fruits, vegetables	Akha, Angami, Karen, Lisu	Nagaland, Thailand
Zanthoxylum myriacanthum Wall. ex Hook.f.	Rutaceae	l	Spices	Kachin	Thailand
Zanthoxylum oxyphyllum Edgew.	Rutaceae	l	Vegetables	Angami	Nagaland
Zanthoxylum sp.	Rutaceae	infr, l	Spices	Angami	Nagaland
Zea mays L.	Poaceae	infr	Beverages, cereals	Bisu, Lisu	Thailand
Zingiber officinale Roscoe	Zingiberaceae	l, rh	Spices	Angami, Ao, Bisu, Kachin	Nagaland, Thailand
Zingiber sp.	Zingiberaceae	l	Vegetables	Lisu	Thailand

Ziziphus rugosa Lam.	Rhamnaceae	ba, infr	Condiments, fruits	Lisu	Thailand
Ziziphus sp.	Rhamnaceae	infr	Fruits	Lahu, Lisu	Thailand
unknown-Araliaceae	Araliaceae	l	Vegetables	Lisu	Thailand
unknown-Ericaceae	Ericaceae	l	Vegetables	Lahu	Thailand
unknown-Fabaceae	Fabaceae	l	Vegetables	Lisu	Thailand
unknown-Poaceae	Poaceae	s	Vegetables	Kachin	Thailand
unknown-Poaceae	Poaceae	infl	Vegetables	Lisu	Thailand
unknown-Sapindaceae	Sapindaceae	infr	Fruits	Lisu	Thailand
unknown-Urticaceae	Urticaceae	l,infr	Vegetables	Lahu	Thailand
unknown-Zingiberaceae	Zingiberaceae	rh	Spices	Lahu	Thailand

species, 33 use-reports; 6 species, 11 use-reports) Lamiaceae (13 species, 39 use-reports; 5 species, 18 use-reports), Moraceae (12 species, 44 use-reports; 7 species, 24 use-reports), Asteraceae (9 species, 22 use-reports; 10 species, 19 use-reports), Solanaceae (9 species, 49 use-reports; 8 species, 24 use-reports), Cucurbitaceae (9 species, 50 use-reports; 7 species, 14 use-reports), Rutaceae (9 species, 17 use-reports; 6 species, 13 use-reports), and Rosaceae (6 species, 9 use-reports; 8 species, 15 use-reports), respectively (see Fig. 3.2).

3.3.2 Type of Food Plants

A total of 1218 use-reports were classified into eight types according to their usage (see Fig. 3.3). The types of food plants with the most species in northern Thailand and Nagaland were vegetables, followed by fruits.

3.3.3 Plant Parts

In terms of plant parts used, the infructescences were the most commonly used part for food in Nagaland, mentioned in 46.8 percent of all use-reports, followed by leaves (37.7 percent), as shown in Fig. 3.4. Meanwhile, in northern Thailand, leaves and infructescences were the most used parts for food, both mentioned in about 38 percent of all use-reports.

3.3.4 Ethnobotanical Indices: Use Value (UV)

The total number of food plant species used by the Tibeto-Burman groups in northern Thailand and Nagaland was 394 species across 261 genera. The food plant genera and species with a Use Value (UV) equal to or greater than the median was shown in Table 3.3 and 3.4.

The genera used by the Tibeto-Burman groups in Nagaland with the greatest UV values were *Solanum* (1.45), *Ficus* (1.45), *Zanthoxylum* (0.82), *Colocasia* (0.55), *Allium* (0.55), *Amaranthus* (0.55), and *Musa* (0.55). The genera used by the Tibeto-Burman groups in Thailand with the greatest UV values were *Solanum* (0.83), *Ficus* (0.71), *Colocasia* (0.31), *Clerodendrum* (0.31), *Dioscorea* (0.29), *Alpinia* (0.27), and *Senegalia* (0.27) (see Table 3.3).

The species used by the Tibeto-Burman groups in Nagaland with the greatest UV, each having a value of 0.45, were *Ficus auriculata* Lour., *Clerodendrum colebrookianum* Walp., *Phyllanthus emblica* L., and *Houttuynia cordata* Thunb. The species used by the Tibeto-Burman groups in Thailand with the greatest UV was *Colocasia esculenta* (L.) Schott, with an UV of 0.31 (see Table 3.4).

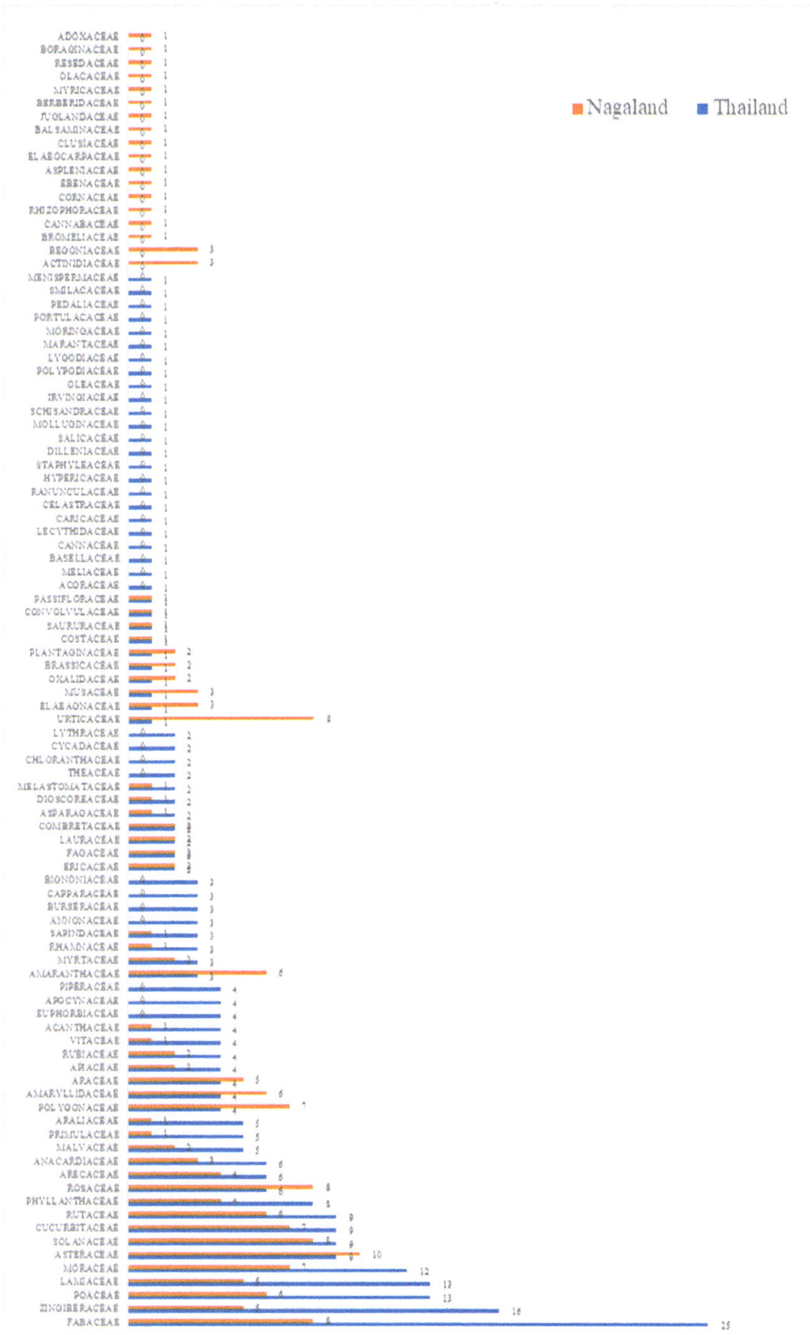

Fig. 3.2 Number of food plant species in each family of the Tibeto-Burman groups in northern Thailand and Nagaland

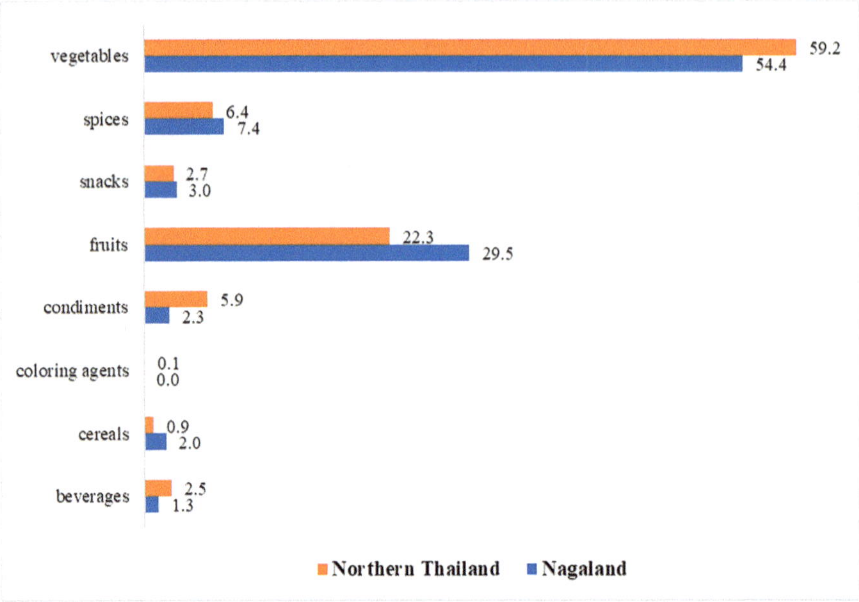

Fig. 3.3 Percentage of food plant species in each food type used by the Tibeto-Burman groups in northern Thailand and Nagaland

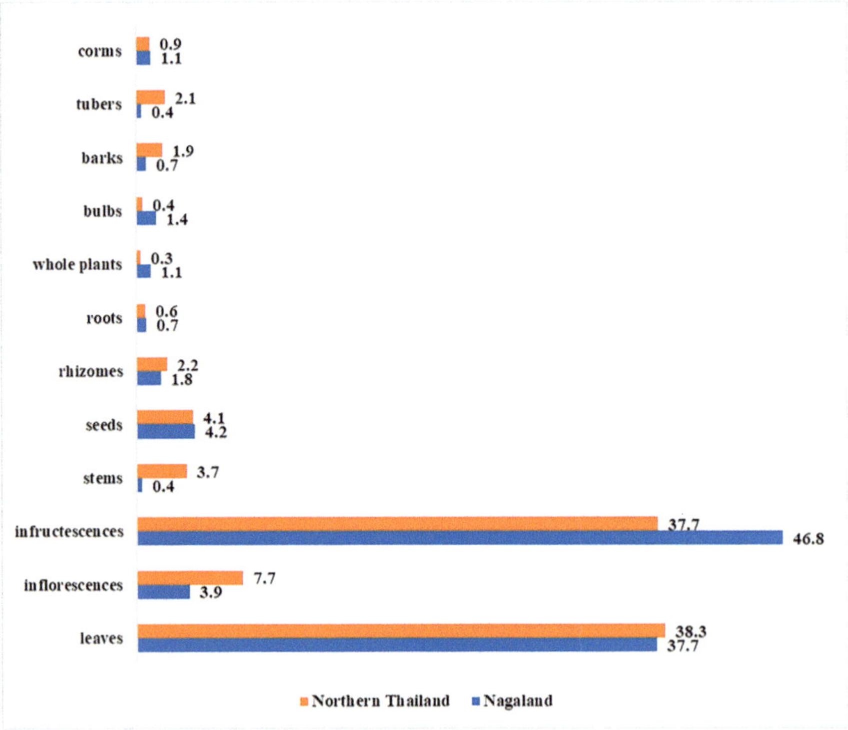

Fig. 3.4 Percentage of food plant species in each plant part used by the Tibeto-Burman groups in northern Thailand and Nagaland

Table 3.3 The food plant genera with a use value equal to or greater than the median used by the Tibeto-Burman groups in northern Thailand and Nagaland

Nagaland (gN) (24 genera)	UVgN median = 0.36	Northern Thailand (gT) (26 genera)	UVgT median = 0.18	Overall (gTo) (29 genera)	UVgTo median = 0.19
Solanum	1.45	*Solanum*	0.83	*Solanum*	0.95
Ficus	1.45	*Ficus*	0.71	*Ficus*	0.85
Zanthoxylum	0.82	*Colocasia*	0.31	*Colocasia*	0.36
Colocasia	0.55	*Clerodendrum*	0.31	*Clerodendrum*	0.34
Allium	0.55	*Dioscorea*	0.29	*Dioscorea*	0.29
Amaranthus	0.55	*Alpinia*	0.27	*Elsholtzia*	0.27
Musa	0.55	*Senegalia*	0.27	*Zanthoxylum*	0.27
Clerodendrum	0.45	*Elsholtzia*	0.25	*Phyllanthus*	0.25
Phyllanthus	0.45	*Terminalia*	0.23	*Allium*	0.25
Bauhinia	0.45	*Castanopsis*	0.23	*Alpinia*	0.24
Rubus	0.45	*Momordica*	0.23	*Terminalia*	0.24
Houttuynia	0.45	*Phyllanthus*	0.21	*Bauhinia*	0.24
Begonia	0.45	*Crassocephalum*	0.21	*Rubus*	0.24
Prunus	0.45	*Psidium*	0.21	*Senegalia*	0.22
Elsholtzia	0.36	*Lagenaria*	0.21	*Syzygium*	0.22
Syzygium	0.36	*Oroxylum*	0.21	*Amaranthus*	0.22
Artocarpus	0.36	*Piper*	0.21	*Castanopsis*	0.20
Livistona	0.36	*Allium*	0.19	*Momordica*	0.20
Morus	0.36	*Bauhinia*	0.19	*Crassocephalum*	0.20
Canarium	0.36	*Rubus*	0.19	*Psidium*	0.20
Garcinia	0.36	*Syzygium*	0.19	*Calamus*	0.20
Gynura	0.36	*Calamus*	0.19	*Centella*	0.20

(continued)

Table 3.3 (continued)

Nagaland (gN) (24 genera)	UVgN median = 0.36	Northern Thailand (gT) (26 genera)	UVgT median = 0.18	Overall (gTo) (29 genera)	UVgTo median = 0.19
Plantago	0.36	*Centella*	0.19	*Mangifera*	0.20
Stixis	0.36	*Mangifera*	0.19	*Lagenaria*	0.19
		Capsicum	0.19	*Oroxylum*	0.19
		Tamarindus	0.19	*Piper*	0.19
				Capsicum	0.19
				Rhus	0.19
				Artocarpus	0.19

Table 3.4 The food plant species with a use value equal to or greater than the median used by the Tibeto-Burman groups in northern Thailand and Nagaland

Nagaland (sN) (16 species)	UVsN median = 0.23	Northern Thailand (sT) (48 species)	UVsT median = 0.13	Overall (sTo) (31 species)	UVsTo median = 0.14
Ficus auriculata Lour.	0.45	*Colocasia esculenta* (L.) Schott	0.31	*Colocasia esculenta* (L.) Schott	0.32
Clerodendrum colebrookianum Walp.	0.45	*Dioscorea* sp.	0.29	*Dioscorea* sp.	0.25
Phyllanthus emblica L.	0.45	*Solanum torvum* Sw.	0.25	*Solanum torvum* Sw.	0.25
Houttuynia cordata Thunb.	0.45	*Crassocephalum crepidioides* (Benth.) S.Moore	0.21	*Ficus auriculata* Lour.	0.22
Colocasia esculenta (L.) Schott	0.36	*Psidium guajava* L.	0.21	*Crassocephalum crepidioides* (Benth.) S.Moore	0.20
Bauhinia variegata L.	0.36	*Lagenaria siceraria* (Molina) Standl.	0.21	*Psidium guajava* L.	0.20
Stixis suaveolens (Roxb.) Baill.	0.36	*Oroxylum indicum* (L.) Kurz	0.21	*Centella asiatica* (L.) Urb.	0.20
Solanum torvum Sw.	0.27	*Centella asiatica* (L.) Urb.	0.19	*Solanum americanum* Mill.	0.20
Centella asiatica (L.) Urb.	0.27	*Solanum americanum* Mill.	0.19	*Clerodendrum colebrookianum* Walp.	0.20
Solanum americanum Mill.	0.27	*Ficus hispida* L.f.	0.19	*Lagenaria siceraria* (Molina) Standl.	0.19
Rhus chinensis Mill.	0.27	*Momordica charantia* L.	0.19	*Oroxylum indicum* (L.) Kurz	0.19
Zingiber officinale Roscoe	0.27	*Tamarindus indica* L.	0.19	*Ficus hispida* L.f.	0.19
Litsea cubeba (Lour.) Pers.	0.27	*Calamus* sp.	0.19	*Rhus chinensis* Mill.	0.19
Juglans regia L.	0.27	*Ficus auriculata* Lour.	0.17	*Phyllanthus emblica* L.	0.19
Livistona jenkinsiana Griff.	0.27	*Rhus chinensis* Mill.	0.17	*Momordica charantia* L.	0.17
Oxalis corniculata L.	0.27	*Luffa aegyptiaca* Mill.	0.17	*Tamarindus indica* L.	0.17
		Solanum lycopersicum L.	0.17	*Luffa aegyptiaca* Mill.	0.17
		Syzygium cumini (L.) Skeels	0.17	*Solanum lycopersicum* L.	0.17
		Lablab purpureus (L.) Sweet	0.17	*Syzygium cumini* (L.) Skeels	0.17
		Mangifera indica L.	0.17	*Calamus* sp.	0.15

(continued)

Table 3.4 (continued)

Nagaland (sN) (16 species)	UVsN median = 0.23	Northern Thailand (sT) (48 species)	UVsT median = 0.13	Overall (sTo) (31 species)	UVsTo median = 0.14
		Spondias pinnata (L.f.) Kurz	0.17	*Lablab purpureus* (L.) Sweet	0.15
		Antidesma acidum Retz.	0.17	*Mangifera indica* L.	0.15
		Capsicum frutescens L.	0.17	*Spondias pinnata* (L.f.) Kurz	0.15
		Protium serratum (Wall. ex Colebr.) Engl.	0.17	*Baccaurea ramiflora* Lour.	0.15
		Clerodendrum colebrookianum Walp.	0.15	*Eryngium foetidum* L.	0.15
		Baccaurea ramiflora Lour.	0.15	*Terminalia chebula* Retz.	0.15
		Eryngium foetidum L.	0.15	*Bauhinia variegata* L.	0.15
		Terminalia chebula Retz.	0.15	*Antidesma acidum* Retz.	0.14
		Cucurbita moschata Duchesne	0.15	*Capsicum frutescens* L.	0.14
		Senegalia rugata (Lam.) Britton & Rose	0.15	*Protium serratum* (Wall. ex Colebr.) Engl.	0.14
		Trevesia palmata (Roxb. ex Lindl.) Vis.	0.15	*Psophocarpus tetragonolobus* (L.) DC.	0.14
		Phyllanthus emblica L.	0.13		
		Psophocarpus tetragonolobus (L.) DC.	0.13		
		Elaeagnus latifolia L.	0.13		
		Ficus virens Aiton	0.13		
		Oenanthe javanica (Blume) DC.	0.13		
		Saccharum officinarum L.	0.13		
		Sechium edule (Jacq.) Sw.	0.13		
		Sorghum bicolor (L.) Moench	0.13		
		Alpinia galanga (L.) Willd.	0.13		

(continued)

Table 3.4 (continued)

Nagaland (sN) (16 species)	UVsN median = 0.23	Northern Thailand (sT) (48 species)	UVsT median = 0.13	Overall (sTo) (31 species)	UVsTo median = 0.14
		Alpinia zerumbet (Pers.) B.L.Burtt & R.M.Sm.	0.13		
		Amorphophallus sp.	0.13		
		Castanopsis diversifolia (Kurz) King ex Hook.f.	0.13		
		Dillenia parviflora Griff.	0.13		
		Elsholtzia griffithii Hook.f.	0.13		
		Livistona speciosa Kurz	0.13		
		Piper sp.	0.13		
		Senegalia pennata (L.) Maslin	0.13		

3.3.5 The Percentage of Genera and Species Overlap between Northern Thailand and Nagaland (% Overlap)

Food plant genera and species with a Use Value equal to or greater than the median were selected to analyze the percentage of overlap (% overlap) to determine the similarities and differences in food plant species between the Tibeto-Burman groups in Nagaland and northern Thailand. The percentage overlap of genera and species used for food between Nagaland and northern Thailand was 24.5 percent and 12.5 percent, respectively (Fig. 3.5).

3.4 Discussion

3.4.1 Diversity of Food Plant of Tibeto-Burman Groups in Lanna and Nagaland

The Tibeto-Burman groups in Lanna and Nagaland demonstrate a significant diversity of food plants, utilizing 394 different species across both regions compared with other research (Aryal et al., 2018; Balemie & Kebebew, 2006; Boesi, 2014; Deb et al., 2019; Punchay et al., 2020; Uprety et al., 2012). This rich variety reflects their deep knowledge of local ecosystems and their ability to adapt to different environmental conditions. In both areas, communities rely on a broad spectrum of families, including common ones like Fabaceae, Solanaceae, Moraceae, Asteraceae,

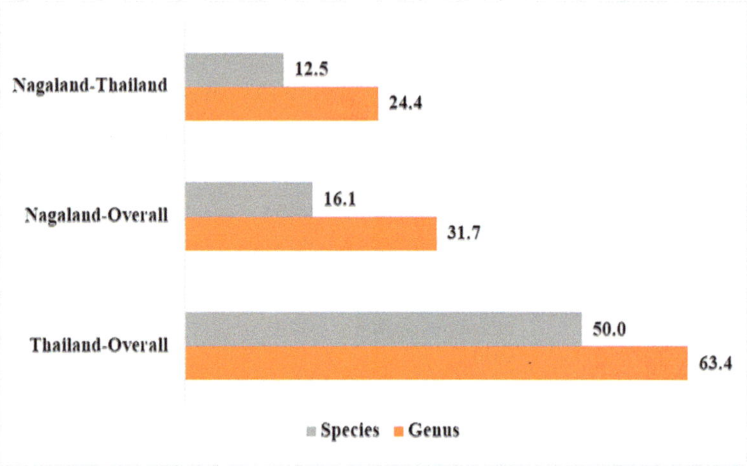

Fig. 3.5 The percentage of genera and species used for food that overlap (% overlap) between Nagaland and northern Thailand

Cucurbitaceae, Rosaceae, Lamiaceae, Rutaceae, Poaceae, and Zingiberaceae, which serve as key food sources (Sutjaritjai et al., 2019; Deshmukh & Waghmode, 2011; POWO, 2024). The use of such a wide array of species underscores the adaptability of the Tibeto-Burman groups who have developed extensive ethnobotanical knowledge to sustain themselves in diverse landscapes. This diversity is not only crucial for providing nutritional security but also for preserving cultural practices and fostering a sustainable relationship with their natural surroundings.

These families are essential plant families that play a crucial role in human nutrition and food security. These families encompass a wide variety of species that are integral to diets worldwide. Rich in protein and essential for vegetarian and vegan diets, the Fabaceae family includes Arachis hypogaea, Cajanus cajan, Glycine max, and Lablab purpureus. Solanaceae comprises staple vegetables like tomatoes, potatoes, and peppers, providing essential vitamins and minerals. Fruits such as wild figs (*Ficus auriculata, F. hispida, F. semicordata*) and wild vegetables (*Acmella paniculata, Bidens pilosa, and Crassocephalum crepidioides*) belong to the Moraceae and Asteraceae families, respectively, and they provide nutrients and dietary fiber. *Cucumis sativus, Cucurbita moschata, Lagenaria siceraria, Luffa aegyptiaca, Momordica charantia*, and *Sechium edule* are all members of the Cucurbitaceae family, which is essential for vitamins and hydration. Rosaceae encompasses many wild berries, such as *Rubus dielsianus*, *R. ellipticus*, *R. leucanthus*, and *Docynia indica*, which are key sources of antioxidants. Herbs with therapeutic and flavor-enhancing qualities, such as *Elsholtzia blanda, E. griffithii, E. kachinensis*, and *E. stachyodes*, belong to the Lamiaceae family. Rutaceae, is used for condiment and spices, such as *Citrus* × *aurantiifolia*, *Zanthoxylum acanthopodium*, *Z. armatum*, *Z. asiaticum*, and *Z. myriacanthum*. For the populations of the Tibeto-Burman group, rice, wheat, maize, and bamboo are basic foods and vegetables. These plants belong to the Poaceae family, also known as the grass family. Zingiberaceae, including *Alpinia galanga*, *Curcuma longa*, and *Zingiber officinale*, not only adds flavor but also offers health benefits. Together, these plant families provide a wide range of nutrients, flavors, and health benefits, highlighting their importance in diverse culinary traditions and food systems of the Tibeto-Burman Groups.

3.4.2 Food Types and Parts

The analysis of food types and plant parts used in Lanna and Nagaland provides valuable insights into the dietary practices and cultural preferences of these regions. The data from the 1218 use-reports, categorized into eight types of food plants, reveal that vegetables are the most commonly used food type in both Lanna and Nagaland, followed closely by fruits. This suggests that both regions rely heavily on plant-based foods, with a particular emphasis on vegetables and fruits as key components of their diets.

In Nagaland, the predominance of infructescences as the most commonly used plant part for food, comprising 46.8 percent of all use-reports, indicates a strong preference for fruits and edible fruiting structures. This preference may reflect the local availability of such plants and their importance in traditional diets. The substantial use of infructescences might also suggest that fruiting structures are highly valued for their taste, texture, and nutritional content, which includes vitamins, minerals, and fibers that are essential for a balanced diet. Leaves, which account for 37.7 percent of the use-reports in Nagaland, are also a significant part of the diet, highlighting their role as a versatile and nutritious food source.

In Lanna, the use of leaves and infructescences is almost equal, with both parts being mentioned in about 38 percent of all use-reports. Given Lanna's great biodiversity and cultural traditions that value a range of flavors and textures, this balanced use suggests a varied dietary practice that involves a wide range of plant parts. The equal emphasis on leaves and infructescences in Lanna suggests that both are integral to the diet, contributing to both the nutritional quality and culinary diversity of meals. Leaves are often used in fresh vegetable dishes alongside chili paste, in spicy soups, and in stir-fries, while infructescences may be consumed fresh or cooked, highlighting their versatility.

The differences and similarities in the use of plant parts between Nagaland and Lanna underscore the complex interplay between ecological availability, cultural preferences, and nutritional needs. While both regions prioritize vegetables and fruits, the specific plant parts use varies, reflecting localized adaptations and traditional knowledge systems. Understanding these patterns not only enhances our appreciation of regional food cultures but also provides insights into sustainable food practices and the importance of maintaining biodiversity for food security and nutrition.

3.4.3 Important and Overlapping Food Plants of Tibeto-Burman Groups in Lanna and Nagaland

The analysis of Use Value (UV) and percentage overlap between the food plant species used by the Tibeto-Burman groups in Lanna and Nagaland provides insights into the cultural and botanical connections between these two regions. The total number of food plant species utilized by these groups was 394, distributed across 261 genera. A better understanding of the most important plants in their diets is made possible by concentrating on genera and species with UV values equal to or higher than the median.

In Nagaland, the genera with the highest UV values were *Solanum* and *Ficus* (both at 1.45), followed by *Zanthoxylum* (0.82), and a cluster of genera—*Colocasia*, *Allium*, *Amaranthus*, and *Musa*—each with a UV of 0.55. This indicates that these genera are highly valued in Nagaland, possibly due to their nutritional benefits, cultural significance, or versatility in traditional cooking methods. On the other hand, like in Nagaland, the most highly prized genera in Lanna were *Solanum* (0.83) and *Ficus* (0.71), but there were also *Clerodendrum* (0.31), *Dioscorea* (0.29), *Alpinia* (0.27), and *Senegalia* (0.27), demonstrating a distinct variety of plant genera that are important for culture and nourishment. This variation reflects differences in local ecosystems, dietary needs, and cultural practices.

At the species level, Nagaland's Tibeto-Burman groups highly valued *Ficus auriculata*, *Clerodendrum colebrookianum*, *Phyllanthus emblica*, and *Houttuynia cordata*, each with a UV of 0.45. These species are likely central to local cuisines and traditions, possibly used in specific dishes or for their medicinal properties. In Lanna, *Colocasia esculenta* was the most valued species, with a UV of 0.31, suggesting a preference for this species due to its versatility and nutritional value.

The percentage overlap between the two regions was 24.5 percent for genera and 12.5 percent for species. This suggests that even while there are some similarities across the plant species utilized, there are also notable disparities, most likely as a result of ecological and cultural differences. The lower species overlap suggests a broader range of specific plants being used in each region, despite some shared genera. This could imply that while both regions might recognize the value of certain genera, the specific species chosen from these genera differ based on local availability and preference.

These findings highlight the complex interplay of ecological, cultural, and nutritional factors that shape the food plant usage of the Tibeto-Burman groups in Lanna and Nagaland. The shared use of certain genera, such as *Solanum* and *Ficus*, suggests a common recognition of their value, possibly related to their adaptability and nutritional benefits. However, the differences in species and UV values underline the importance of local ecological knowledge and cultural practices in shaping dietary preferences. Understanding these overlaps and differences can provide valuable insights into regional biodiversity, conservation priorities, and the preservation of traditional knowledge and practices related to food plant use.

3.4.4 Cultural Significant Food Plants of Tibeto-Burman Groups

The cultural significance of food plants among the Tibeto-Burman groups in Lanna and Nagaland is deeply intertwined with their traditional practices, dietary preferences, and ecological environments. Each of the recognized genera—*Solanum, Ficus, Zanthoxylum, Colocasia, Allium, Amaranthus, Musa, Clerodendrum, Dioscorea, Alpinia, Elsholtzia, and Senegalia*—has a distinct function in these communities, which is indicative of their extensive botanical expertise and rich cultural legacy.

Solanum and *Ficus* are prominent genera in both regions, underscoring their importance in traditional diets. *Solanum* species, such as eggplants and certain wild relatives, are integral to various dishes, providing essential nutrients and flavors. Similarly, *Ficus* species, including fruits like wild figs, are valued for their nutritional content. The dual significance of these genera highlights a shared appreciation for their culinary and cultural value.

Zanthoxylum, known for its aromatic fruits and spices, adds a distinctive flavor to traditional dishes. Its use in both Lanna and Nagaland indicates a common preference for spicy and aromatic ingredients, reflecting a broader cultural trend towards incorporating bold flavors in cuisine. This genus illustrates how specific plant characteristics, such as pungency and aroma, can shape dietary traditions.

Colocasia, with its edible petioles and tubers, and *Allium*, which includes onions and garlic, are critical for their culinary versatility and flavor enhancement. *Colocasia* is often used in stews and curries, while *Allium* species are fundamental in seasoning and adding depth to dishes. Their widespread use highlights their essential role in enriching the flavor profiles of traditional meals.

Amaranthus, a leafy green, and *Musa*, which includes bananas, are significant for their nutritional benefits and versatility. *Amaranthus* is valued for its high protein content and vitamins, while *Musa* is a staple fruit and vegetable, providing important energy and nutrients. The incorporation of these plants into daily diets underscores their importance in maintaining nutritional health and diversity.

Clerodendrum, *Dioscorea*, and *Alpinia* are notable for their unique contributions to traditional food systems. *Clerodendrum* species are often used in herbal remedies and culinary dishes, *Dioscorea* provides essential tubers for energy-rich meals, and *Alpinia* adds spice and medicinal benefits. These genera reflect the intersection of dietary needs and traditional knowledge, where plants are valued for both their practical uses and their roles in cultural practices.

Elsholtzia and *Senegalia* further enrich the traditional diets of these communities. Known for its fragrant leaves, *Elsholtzia* is utilized for medicinal and flavoring purposes, while *Senegalia* offers edible components for regional cuisines. *Senegalia rugata* is also frequently used in Thai ceremonies. These genera highlight the diverse applications of plants in traditional food systems, from flavoring and seasoning to providing essential nutrients.

Overall, the cultural significance of these food plants among the Tibeto-Burman groups is a testament to their deep connection with the natural environment and traditional practices. The diversity in plant usage reflects the adaptation of these communities to their ecological settings and their ability to integrate a wide range of plant resources into their diets. Understanding these plants' roles offers valuable insights into the complex relationship between culture, ecology, and nutrition, emphasizing the importance of preserving traditional knowledge and biodiversity.

However, there is a notable gap in understanding the specific preparation methods that make these plants part of daily use. Future data collections should focus on documenting these techniques to provide a more comprehensive view of how traditional knowledge and practices influence the culinary use of these plants. This deeper understanding will enhance our appreciation of their cultural and nutritional roles and contribute to preserving and promoting traditional food practices among the Tibeto-Burman groups.

3.5 Conclusion

This comparative ethnobotanical study of food plants used by Tibeto-Burman groups in Lanna and Nagaland provides valuable insights into the cultural and nutritional significance of these plants across diverse environments. The analysis reveals that both regions share a deep reliance on a variety of plant genera and species, reflecting their rich botanical knowledge and traditional practices. Key genera such as *Solanum*, *Ficus*, *Zanthoxylum*, and *Colocasia* emerge as central to both diets, illustrating their widespread importance in regional cuisines.

Despite these similarities, distinct differences in specific species and preparation methods highlight the unique adaptations and preferences of each community. While the variations highlight the various methods in which these plants are incorporated into regional diets, the significant percentage overlaps in genera and food-use species suggest that information is shared.

A significant finding of this study is the gap in detailed data regarding the preparation methods of these food plants. Future research should prioritize documenting these techniques to gain a more comprehensive understanding of how traditional knowledge influences food practices. Such insights will not only enhance our appreciation of the cultural and nutritional roles of these plants but also contribute to the preservation and promotion of traditional food practices. By bridging these knowledge gaps, we can better support the conservation of biodiversity and the cultural heritage of the Tibeto-Burman groups in both Lanna and Nagaland.

Acknowledgement We would like to extend our sincerest gratitude to the villagers in both Nagaland and northern Thailand for their hospitality and warm welcome during our fieldwork. This study would not have been possible without their support and cooperation. Additionally, we would like to acknowledge the Fundamental Fund 2022–2024 (FF65–67) provided by the Ministry of Higher Education, Science, Research, and Innovation, Thailand.

References

Anderson, E. F. (1993). *Plants and people of the golden triangle*. Dioscorides Press. 279p.

Aryal, K. P., Poudel, S., Chaudhary, R. P., Chettri, N., Chaudhary, P., Ning, W., & Kotru, R. (2018). Diversity and use of wild and non-cultivated edible plants in the Western Himalaya. *Journal of Ethnobiology and Ethnomedicine, 14*, 10. https://doi.org/10.1186/s13002-018-0211-1

Balemie, K., & Kebebew, F. (2006). Ethnobotanical study of wild edible plants in Derashe and Kucha districts, South Ethiopia. *Journal of Ethnobiology and Ethnomedicine, 2*, 53. https://doi.org/10.1186/1746-4269-2-53

Boesi, A. (2014). Traditional knowledge of wild food plants in a few Tibetan communities. *Journal of Ethnobiology and Ethnomedicine, 10*, 75. https://doi.org/10.1186/1746-4269-10-75

Bussmann, R. W., Gilbreath, G. G., Solio, J., Lutura, M., Lutuluo, R., Kunguru, K., Wood, N., & Mathenge, S. G. (2006). Plant use of the Maasai of Sekenani Valley, Maasai Mara, Kenya. *Journal of Ethnobiology and Ethnomedicine, 2*, 22. https://doi.org/10.1186/1746-4269-2-22

Chasie, C. (1999). *The Naga imbroglio: A personal perspective*. Standard Printers and Publishers. 230p.

Cook, F. E. M. (1995). *Economic botany data collection standard*. Whitstable Litho. 146p.

Cotton, C. M. (1996). *Ethnobotany: Principles and applications*. Wiley. 424p.

Deb, C. R., Khru-omo, N., & Paul, A. (2019). Underutilized edible plants of Nagaland: A survey and documentation from Kohima, Phek and Tuensang District of Nagaland, India. *American Journal of Plant Sciences, 10*, 162–178. https://doi.org/10.4236/ajps.2019.101014

Department of Information Technology & Communication. (2024). *People & culture*. Official State Portal, Government of Nagaland. Retrieved July 3, 2024, from https://nagaland.gov.in/pages/people-culture

Deshmukh, B., & Waghmode, A. (2011). Role of wild edible fruits as a food resource: Traditional knowledge. *IJPLS, 2*.

Junsongduang, A., Balslev, H., Inta, A., Jampeetong, A., & Wangpakapattanawong, P. (2014). Karen and Lawa medicinal plant use: Uniformity or ethnic divergence? *Journal of Ethnopharmacology, 151*, 517–527. https://doi.org/10.1016/j.jep.2013.11.009

Khamfachuea, K., Trisonthi, P., & Trisonthi, C. (2010). Ethnobotany of the Karen at Ban Chan and Chaem Luang subdistricts, Mae Chaem district, Chiang Mai province. *Thai Journal of Botany, 2*, 275–297.

Khamyong, N., Wangpakapattanawong, P., Chairuangsri, S., Inta, A., & Tiansawat, P. (2018). Tree species composition and height-diameter allometry of three forest types in northern Thailand. *Chiang Mai University Journal of Natural Sciences, 17*(4), 289–305. https://doi.org/10.12982/CMUJNS.2018.0021

POWO. (2024). *Plant of the world online*. Retrieved June 20, 2024, from http://www.plantsoftheworldonline.org/

Princess Maha Chakri Sirindhorn Anthropology Centre (Public Organization). (2024). Retrieved July 20, 2024, from https://ethnicity.sac.or.th/database-ethnic

Punchay, K., Inta, A., Tiansawat, P., Balslev, H., & Wangpakapattanawong, P. (2020). Traditional knowledge of wild food plants of Thai Karen and Lawa (Thailand). *Genetic Resources and Crop Evolution, 67*, 1277–1299. https://doi.org/10.1007/s10722-020-00910-x

Sutjaritjai, N., Wangpakapattanawong, P., Balslev, H., & Inta, A. (2019). Traditional uses of Leguminosae among the Karen in Thailand. *Plants (Basel, Switzerland), 2019*(8), 600. https://doi.org/10.3390/plants8120600

Turner, N. J., Łuczaj, L. J., Migliorini, P., Pieroni, A., Dreon, A. L., Sacchetti, L. E., & Paoletti, M. G. (2011). Edible and tended wild plants, traditional ecological knowledge and agroecology. *Critical Reviews in Plant Sciences, 30*, 198–225. https://doi.org/10.1080/07352689.2011.554492

Uprety, Y., Poudel, R. C., Shrestha, K. K., Rajbhandary, S., Tiwari, N. N., Shrestha, U. B., & Asselin, H. (2012). Diversity of use and local knowledge of wild edible plant resources in Nepal. *Journal of Ethnobiology and Ethnomedicine, 8*, 16. https://doi.org/10.1186/1746-4269-8-16

Zenderland, J., Hart, R., Bussmann, R. W., Paniagua Zambrana, N. Y., Sikharulidze, S., Kikvidze, Z., Kikodze, D., Tchelidze, D., Khutsishvili, M., & Batsatsashvili, K. (2019). The use of "use value": Quantifying importance in ethnobotany. *Economic Botany, 73*, 293–303. https://doi.org/10.1007/s12231-019-09480-1

Chapter 4
The Forest Guardians: A Comparative Analysis of Indigenous Knowledge and Community Forest Governance in Northern Thailand and Nagaland, India

Arratee Ayuttacorn, Lungsang Zeliang, and Khotole Khieya

Abstract This comparative study critically examines community forest management practices in northern Thailand and Nagaland, Northeast India. It investigates the processes through which forest-dwelling populations establish social networks with non-governmental organizations (NGOs), government agencies, and the private sector. The research further explores the resilience of indigenous knowledge systems and their integration with contemporary practices to enhance the effectiveness of forest management. The research was conducted in five Karen communities in Northern Thailand, recognized as exemplary models of community forest management, as well as in four villages located in the Kohima and Peren Districts of Nagaland, where ecotourism and development projects are actively being implemented. Data collection methods included in-depth interviews, focus group discussions, and participant observation from January 2022 to November 2023. The findings reveal that both Karen communities in Thailand and communities in Nagaland have developed community-based Payment for Environmental Services (PES) schemes, albeit with distinct approaches. The Karen communities use PES to finance not only conservation activities but also social welfare initiatives, leveraging these schemes to position themselves as forest protectors and strengthen their claims to land rights. In contrast, the communities in Nagaland mostly utilize community-based PES to support conservation efforts and finance development projects because they gain from the increased revenue from ecotourism. Their land ownership reduces their dependence on external support; however, the sustainability of these initiatives is constrained by limited resource-sharing across villages. Both indigenous communities in Thailand and Nagaland effectively leverage their social

A. Ayuttacorn (✉)
Faculty of Social Sciences, Chiang Mai University, Chiang Mai, Thailand
e-mail: arratee.a@cmu.ac.th

L. Zeliang · K. Khieya
Department of Education, School of Humanities and Education, Nagaland University, Meriema, Nagaland, India
e-mail: lu_zeliang@yahoo.in; khotole@nagalanduniversity.ac.in

Y. Santasombat (ed.), *Bio-Cultural Diversity, Ethnicity, and Local Knowledge*,
https://doi.org/10.1007/978-981-95-4242-0_4

networks to secure support for livelihood diversification, emphasizing the significance of collaboration with external institutions to sustain their forest management initiatives. Additionally, Indigenous Peoples in both regions utilize traditional practices and actively construct new indigenous identities to negotiate state policies, foster community cohesion, and promote sustainable forest conservation.

Keywords Community forest management · Indigenous knowledge · Social networks · Payment for Environmental Services (PES) · Northern Thailand · Nagaland

4.1 Introduction

Forests in Thailand have been governed by the state, so this inadvertently causes significant challenges for forest dwellers due to certain policies regulating the access to and use of forest resources. The establishment of national parks and the implementation of restrictive laws have adversely impacted the livelihoods of residents in 4192 villages, covering 687,200 hectares within national parks and wildlife sanctuaries (Sueb Nakhasathien Foundation, 2020, August 27). These groups have lived in these forest areas for generations, but because of territorial control regulations, they are now considered unlawful residents (Vandergeest, 1996a: 259). Forest governance in Thailand is further complicated by several factors, including ineffective policy implementation (Hirsch, 1990), land tenure insecurity (Walker, 2004), socio-economic transformations (Delang, 2002), and the inadequate integration of indigenous knowledge into forest management practices (Forsyth & Walker, 2014: 415; Salam et al., 2006).

The Thai government has employed strategies such as territorialization through mapping technology (Vandergeest, 1996b; Leblond & Pham, 2014), racialization discourse (Vandergeest, 2003), and narratives around swidden farming (Laungaramsri, 2005) to assert control over forest governance and increase forest cover. These strategies have led to conflicts between state agencies and local communities (Roth, 2008; Hares, 2009). For instance, the Community Forest Act of 2019 adopts a decentralized approach, empowering communities to establish community forests within designated reserved forests and government lands. However, this framework excludes communities residing within national parks and wildlife sanctuaries (Royal Forest Department (RFD), Community Forest Management Office, 2022, November 21). Consequently, indigenous communities living within protected forest areas are not granted the authority to manage or access forest resources.

In contrast to Thailand's forest governance, forests in Nagaland are managed by families and communities. Article 371(A) of the Indian Constitution Act of 1962 provides special provisions for Nagaland, ensuring the protection of customary law, religious beliefs, and social practices of the Naga people. This constitutional provision requires the consent of the Nagaland Legislative Assembly before any central law can be enforced within the state (Thong, 2023). As a result, customary ownership of land and forests by communities and individuals in Nagaland is protected, and they are exempt from the Forest Conservation Act of 1980.

However, due to illegal logging activities, community-based forest management in Nagaland faces significant challenges related to hunting and deforestation. The state's forest cover has declined notably, from 82.75 percent in 2005 to 75.33 percent in 2017 (Sarma et al., 2019), and further to 73.90 percent in 2021. Between 2019 and 2021, Nagaland reportedly lost 235 square kilometers of forest cover (Morung Express, 2022). Shifting cultivation and high rates of firewood consumption are primary drivers of deforestation (Rawat et al., 2018). Additionally, some communities struggle to regulate logging activities, particularly charcoal production, on private or clan-owned lands (George & Yhome, 2008). In response, the Nagaland state government has initiated efforts to introduce Joint Forest Management (JFM) and support communities in diversifying their livelihoods through tourism, cash crops, and tree plantations.

Joint Forest Management (JFM) in India extends institutional arrangements to Community Conservation Areas (CCAs). The forest department is repositioning itself to control over community initiatives, while self-initiated forest protection groups have gained formal recognition through JFM. Grassroots responses, often promoted by NGOs, aim to build local institutions to strengthen bargaining power and amplify community voices. However, the changes from JFM sometimes disrupt internal power balances, diminishing local decision-making. Village leaders may struggle to mobilize entire communities for forest protection, and some members living outside official village boundaries face disenfranchisement or loss of access to forest areas, leading to conflicts between villages (Sarin, 1997: 23).

Community-based forest management is often presented as a way to empower communities and minimize state interference. The Community-Based Natural Resource Management (CBNRM) model operates on the assumption that communities are effective resource managers owing to their traditional knowledge and practices. CBNRM is promoted to policymakers as a powerful tool to address environmental and social justice issues as a single package (Li, 2002: 267). However, it can unwittingly reinforce state control by aligning communities with state conservation agendas, often oversimplifying internal complexities and conflicts. Indigenous Peoples may not prioritize forest preservation, engaging instead in activities like small-scale logging, resource collection, and cash crop cultivation. Community forestry practices, such as tree ordination ceremony in northern Thailand, are viewed as social movements to reclaim resources that have been distributed unfairly (Hirsch, 1997: 15).

Indigenous Peoples in Thailand, such as the Karen, have portrayed themselves as forest protectors, leveraging this image to transform indigenous knowledge into symbolic power. This creates a space of resistance against the hegemonic discourse imposed by state forestry agencies. Karen leaders combine oral traditions with modern scientific tools, such as maps, to communicate effectively with academics,

NGOs, the Royal Forest Department (RFD), and international movements (Trakansuphakorn, 2008).

Similarly, in Nagaland, indigenous communities have strategically navigated colonial and postcolonial representations to assert their agency. Historically, the region has been depicted as a place of headhunters, perpetuating a narrative of Naga primitivism since British colonial rule (Theunuo, 2016: 90). Through performances at the Hornbill Festival, the Nagas have appropriated these exotic images to engage with audiences around the world. This revitalization of traditional culture and indigenous identities functions as a form of national branding, effectively promoting tourism in the region (Longkumer, 2015).

Thailand and Nagaland present contrasting models of forest governance, each defined by distinct land ownership, legal frameworks, and regulations. However, indigenous communities in both regions share similar approaches to community forest management, including swidden cultivation (Erni, 2015), which is deeply rooted in indigenous knowledge and belief systems.

A paradox arises when examining the state's perspective on swidden agriculture. Both Thai and Nagaland governments classify swidden farming—known as *rai mun wien* in Thailand and *jhum* cultivation in Nagaland—as wasteful, unsustainable, and environmentally destructive practices (Das, 2018: 84; Laungaramsri, 2005). Consequently, government agencies often promote wet-rice cultivation, cash crops (Cramb et al., 2009: 342), and tree plantations (Hirsch & Lohmann, 1989) as alternatives to shifting agriculture. This policy creates a contradictory approach: on one hand, the state implements exclusionary measures to restrict swidden practices for forest conservation; on the other hand, it promotes cash crops like corn to drive economic development, even though these practices can be unsustainable. Moreover, the government has struggled to control the forest encroachment resulting from cash crop cultivation.

While shifting cultivation faces criticism from the state, NGOs and academics often depict Karen agricultural practices as ecologically friendly and subsistence-oriented (Walker, 2001). However, market forces have penetrated Karen villages, leading to the cultivation of cash crops and the use of chemicals to boost productivity (Kanazawa et al., 2006).

This research project aims to identify an effective community forest management mechanism through a comparative study of forest governance practices in Thailand and Nagaland. By examining these two regions, the study seeks to shed light on the specific challenges of forest governance, including issues related to land ownership rights, power dynamics among stakeholders, and the need for alternative livelihood opportunities within indigenous communities. The study examines how communities in both regions develop mechanisms to finance conservation activities, social welfare, and infrastructure development. These communities establish social networks with various institutions, including NGOs, the private sector, and government agencies, to secure support for livelihood diversification and enhance their resilience. The research also explores the integration of indigenous knowledge with contemporary practices, aiming to empower communities, maintain cultural heritage, and improve forest management.

This paper is organized into five sections. The first section outlines the methodology and describes the study areas. The second section provides an overview of forest governance in Thailand and Nagaland. The third section examines community forest management practices and livelihood diversification strategies in both regions. The fourth section explores how communities establish social networks with various institutions to secure financial support. The final section focuses on the resilience of indigenous knowledge, emphasizing its integration with scientific knowledge to enhance conservation efforts and negotiation with state policies.

4.2 Method and Study Areas

This comparative study focused on five Karen villages in three provinces in northern Thailand: Chiang Mai, Chiang Rai, and Lampang. It examined communities within reserve forests, demonstrating the best practices in community forest management. These included the Hin Lad Nai Community (Chiang Rai), the Baan Klang and Baan Mae Saan Communities (Lampang), and the Mae Lan Kham and Huay E-Khang Communities (Chiang Mai) (Fig. 4.1). All these villages are struggling with the expansion of national parks and wildlife sanctuaries and the communal land titling implementation.

Huay E-Khang, with 125 households, is located at Mae Win Subdistrict, Mae Wang District, Chiang Mai. It lies within a Class 1 watershed zone and the Mae Khan and Mae Wang Reserved Forests. The community faced discrepancies between a new map created by the Royal Forest Department (RFD) and the one they developed with Mae Win Subdistrict authorities.

Mae Lan Kham, comprising 144 households, is in Samoeng-Tai Subdistrict, Samoeng District, Chiang Mai. This area, also within a Class 1 watershed zone, falls under the Samoeng Reserve Forest, with parts slated for inclusion in Ob-Khan National Park, where the park's boundary overlaps with a community forest.

Hin Lad Nai, with 24 households, is located at Baan Pong Subdistrict, Wiang Papao District, Chiang Rai. Positioned within a Class 1 watershed zone, it encompasses the Mae Poon-noi, Mae Poon-loung, and Huay Pongment Reserve Forests and is being prepared for designation as a wildlife sanctuary.

Baan Klang, with 78 households, and Mae Saan, with 128 households, are both situated at Baan Dong Subdistrict, Mae Moh District, Lampang. Categorized under the Mae Tah and Mae Mai Reserve Forests, these villages are within Class 1 and Class 2 watershed areas. While facing significant challenges related to the sustainable utilization of forest resources, both were slated for inclusion in Tham Pha-thai National Park at the time of data collection.

The research sites in Nagaland include Khonoma and Dzuleke in Kohima District, as well as Benreu and Ngwalwa in Peren District (Fig. 4.2). Kohima District is home to the Angami tribe, while Peren District is inhabited by the Zeliang and Kuki tribes. These villages were selected because their communities had established Community Conservation Areas (CCAs) and diversified their livelihoods into

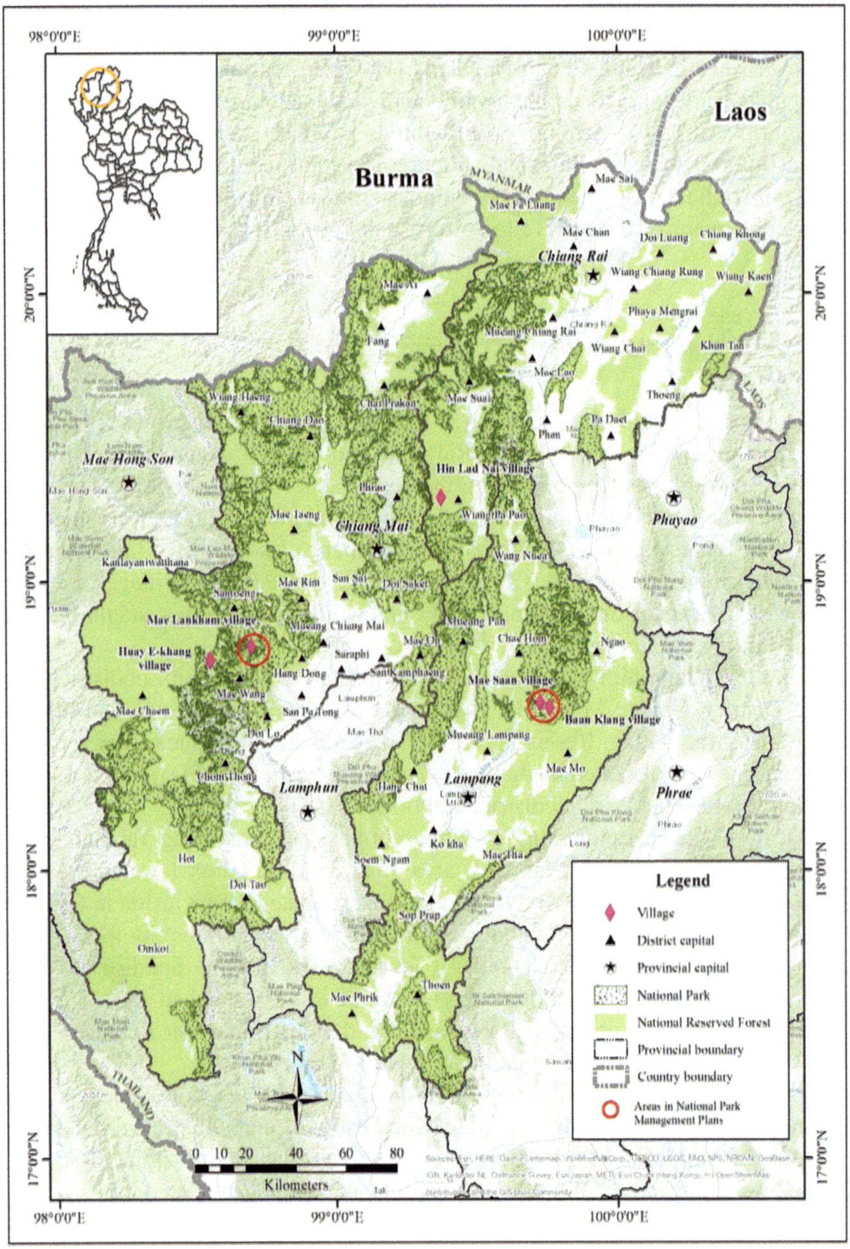

Fig. 4.1 Map of research areas in northern Thailand

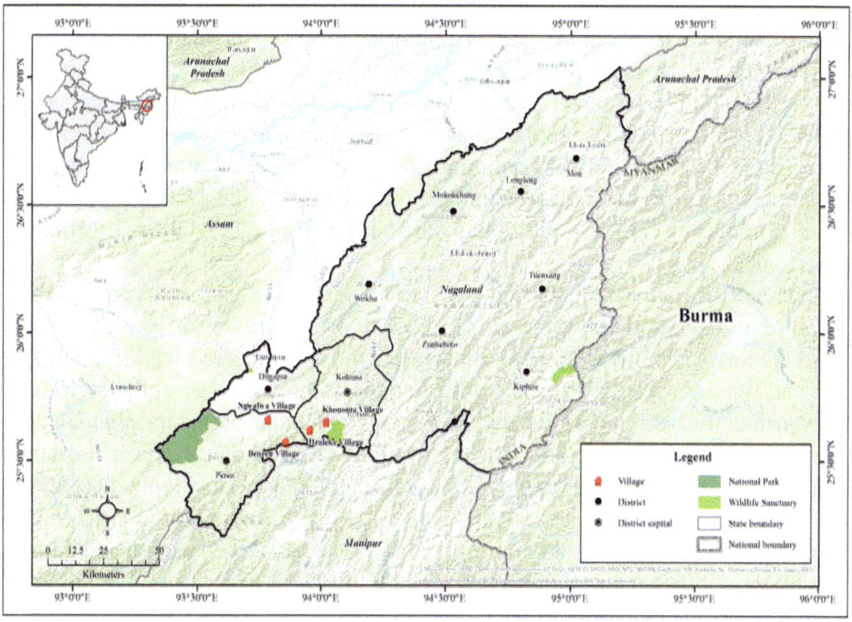

Fig. 4.2 Maps of research areas in Nagaland

eco-tourism, handicrafts, cash crops, and livestock, thereby reducing their reliance on forest resources.

Located at the Dzukou Valley, which is about 20 kilometers from Kohima, Khonoma Village, spans 123 square kilometers at an elevation of 5320 feet above sea level (Papu & Nathani, 2020). Recognized as an Important Bird Area (IBA), Khonoma is situated at the foothills of the Barail Mountains, part of the Eastern Himalayan and Indo-Burma Biodiversity Hotspot. Forests cover 75 percent of the village's land, with 20 percent allocated to paddy fields and wasteland. The residential area occupies just two square kilometers. Inhabited by the Angami tribe, Khonoma has a population of approximately 2370 people in 424 families as of 2023.[1] Most villagers have practiced Christianity since 1890. On December 11, 1998, the community established the Khonoma Nature Conservation and Tragopan Sanctuary (KNCTS), spanning 70 square kilometers. The primary concern was the rapid decline of the Blyth's Tragopan bird population. In response, the community implemented a hunting ban and designated the forest as a wildlife sanctuary (Chase & Singh, 2012).

Dzuleke Village originated as an extension of Khonoma Village. Historically, individuals who committed murder were exiled to Dzuleke for seven years before returning to Khonoma. Over time, as Khonoma villagers and newly married couples sought to expand their farmland, they moved to Dzuleke, drawn by its natural

[1] https://www.indiagrowing.com/Nagaland/Kohima/Sechu_Zubza/Khonoma

beauty. Eventually, the exiled individuals were relocated, and Dzuleke was fully established as a settlement. As of 2024, Dzuleke had an estimated population of 197 people, comprising 49 families.[2] The primary occupation is agriculture, with local rice varieties cultivated on terraces for subsistence. However, many residents prefer growing cash crops such as potatoes and cabbage in their gardens and jhum fields due to higher productivity. The community has established Community Conservation Areas, managed by the village council board. In 2014, the Dzuleke community began a homestay business with financial support from the Tata Trust to improve their homes.

Benreu Village, located in Peren district of Nagaland, is part of the Mount Pauna Community Conservation Area, an 843-hectare forest known for its rich biodiversity. This forest is home to some of the world's most endangered wildlife, including rare and unique species endemic to the region. Benreu Village collaborates on a long-term conservation plan for the CCA, which includes boundary mapping, biodiversity assessments, and habitat conservation. The community receives support from the Nagaland Empowerment of People through Energy Development (NEPED) and the Foundation for Ecological Security (FES). They enforce strict norms, including bans on logging and hunting within the CCA, as well as seasonal hunting regulations and penalties (Sumi, 2025). These efforts have confirmed the presence of four globally endangered species: the Chinese Pangolin, Western Hoolock Gibbon, Blyth's Tragopan, and Great Indian Hornbill. As of 2023, Benreu Village has a population of 1037 people across 180 families.[3] Traditionally, most residents are farmers who cultivates rice in field terraces, taking advantage of fertile soil and abundant water from a nearby river. In addition to farming, villagers engage in secondary occupations, such as carpentry, stone carving, basketry, weaving, and selling vegetables and fruits, as well as earning income from homestays.

Ngwalwa Village, in Peren district, has 649 residents across 117 households, divided into six colonies.[4] Farmers here earn more from cash crops than traditional ones, with year-round banana sales further encouraging this shift. Government support is crucial, especially for cultivating betel nuts, spices, vegetables, and fruits, with seedlings provided through the NEPED organization. Most collective projects in Ngwalwa focus on land use for securing government aid for cash crops rather than conservation. For example, the village youth group has requested community land for tree planting, including rubber, palm, and bamboo, to generate income for the group.

Data collection in northern Thailand occurred from January to September 2022, while in Nagaland, it was conducted between March and November 2023. The methods used included interviews, focus group discussions, participant observation, and the collection of secondary data from documents. Interviews and focus groups

[2] https://www.indiagrowing.com/Nagaland/Kohima/Sechu_Zubza/Dzuleke

[3] https://www.indiagrowing.com/Nagaland/Peren/Pedi/Benreu

[4] https://www.indiagrowing.com/Nagaland/Peren/Pedi/Ngwalwa

were guided by a semi-structured interview format and continued until data saturation was reached.

In terms of language, Thai and English were used in Thailand and Nagaland, respectively. One research team member is fluent in Nagamese, Zeme, and English. For interviews in Khonoma and Dzuleke where respondents could not speak English, an interpreter for the Angami dialect was employed. In northern Thailand, five focus groups were conducted with a total of 20 participants, along with 17 in-depth individual interviews. In Nagaland, 28 in-depth interviews and nine focus groups involving 36 participants were arranged.

4.3 Forest Governance in Nagaland and Thailand

Thailand and Nagaland offer distinct models of forest governance, reflecting their unique legal frameworks, cultural practices, and historical contexts. In Thailand, forest management has traditionally been centralized, with state agencies exerting significant control over forest lands, often leading to tensions with local communities. In contrast, Nagaland's forest governance is deeply rooted in customary laws and community-based management, allowing indigenous practices to guide conservation efforts.

State Forest Governance in Thailand

Thailand's forest resources are regulated by the Royal Forest Department (RFD) and the Department of National Parks, Wildlife, and Plant Conservation. For decades, forest communities have pushed for legal recognition of their land rights. The RFD proposed the first draft of the Community Forest Act in 1991 to encourage local participation in forest protection, but it was criticized for maintaining state control. Various versions, including those by the People's version, National Economic and Social Development Board (NESDB), and conservationist groups, were debated in parliament (Forsyth & Walker, 2014). Although a revised version was approved by the National Legislative Assembly in 2007, it was rejected by the Constitutional Court. The first Community Forest Law was finally enacted in 2019, after a 28-year struggle (Thai PBS, 2019).

Through the Community Forest Management (CFM) model, the Community Forest Act of 2019 seeks to foster cooperation between communities and the government in the sustainable conservation, management, and use of forest resources and biodiversity. The legislation, however, restricts the establishment of community forests to National Reserved Forests and other government properties—national parks and animal sanctuaries are not included. Consequently, as of 2020, it is prohibited for different communities in 4192 villages, which encompass 687,200 hectares of national parks and wildlife sanctuaries (Seub Nakhasathien Foundation, 2020, August 27), to establish community forests and utilize forest resources.

The northern region of Thailand has the highest forest cover, encompassing 63.53 percent of its land area, compared to the national forest cover of 31.57 percent in 2022 (approximately 16 million hectares) (Forest Land Management Office, 2023). Community forest initiatives have made significant progress in this region, with 4766 villages establishing community forests by 2021, covering 684,800 hectares. However, this accounts for only 4.3 percent of Thailand's total forest area (Community Development Knowledge, 2021).

Before the 2019 Community Forest Act, the military government established the National Land Policy Board (NLPB) in 2014 to address landlessness. The NLPB allocated plots in reserved forests and government lands for collective use by underprivileged groups, following proposals from civil society. This approach aimed to avoid the challenges of individual ownership and promote watershed conservation. By 2021, the NLPB had issued permits for 267 areas (112,000 hectares), benefiting 60,000 people with access to over 74,000 plots (RECOFTC, 2021, September 17).

Despite its aim to empower communities, the NLPB program faces significant limitations. Land use activities are controlled by provincial governors rather than local organizations, restricting income generation opportunities. For instance, in Mae Tha village, villagers are unable to harvest the teak trees planted on collective land (Green News, 2021). Furthermore, the NLPB criteria exclude most Class 1 and 2 watershed areas (arable land) and national parks, limiting participation for many forest-dwelling communities. Although communal land titles provide some security and benefits (Wittayapak & Baird, 2018), they raise concerns about long-term uncertainty and potential restrictions.

Conflicts have emerged due to discrepancies between NLPB mapping and the RFD. In Huay E-Khang village, Chiang Mai Province, differences between RFD-created official land use maps and community maps certified by Subdistrict Administrative Organizations (SAOs) have resulted in the reduction or disappearance of arable land on official maps, possibly due to the inclusion of rotational farming areas within designated forest zones. Concerns have also been raised about the accuracy of watershed quality distribution on RFD maps.

Furthermore, the NLPB project has exacerbated tensions not only between villagers and the RFD but also within the communities themselves. While some villagers fear losing their land, others seek to engage with the NLPB project for its offer of 20-year guaranteed land use rights, seeing it as a way to legitimize their land claims. However, those who view the land as ancestral inheritance resist participation, perceiving the arrangement as a rental system.

The Thai state continues to employ exclusionary practices to assert control over forest resources, as evidenced by the following cases. First, in 2023, the Department of National Parks, Wildlife, and Plant Conservation initiated plans to designate several communities in the Lampang province, including Baan Klang and Baan Mae San, as part of the Tham Pha Thai National Park. This forest management policy has effectively prohibited residents from accessing and utilizing resources that their communities have long managed and preserved. In response, these communities, in collaboration with the Northern Peasant Federation (NPF) and P-Move, have advocated for the exclusion of community areas from the proposed national park boundaries to mitigate conflicts between local communities and state authorities (Green News, 31 August 2023).

Besides, in November 2024, the Cabinet approved a draft royal decree aimed at the conservation and management of natural resources within national parks and wildlife sanctuaries. However, this legislation has presented significant challenges for forest-dependent communities. The law stipulates that only individuals holding Thai nationality are eligible to reside within the designated areas—a criterion that some individuals are in the process of acquiring. Furthermore, the decree imposes restrictions on land use, limiting it to a 20-year duration and a maximum allocation of 20 rai (3.2 hectares) per household. Affected communities have criticized the legislation as being incompatible with their traditional practices and livelihoods, calling on the government to revise the policy to better align with their circumstances (Green News, 12 November 2024).

This situation highlights the complex challenges of land use management, particularly in areas where state policies intersect with traditional practices and community livelihoods. It underscores the necessity of collaborative approaches that integrate official regulations with community-based management, ensuring that local knowledge and cultural contexts are respected.

Community-Based Forest Governance in Nagaland

Northeast India is notable for its substantial forest cover, representing 25 percent of the country's total forest area (George & Yhome, 2008). According to the 2021 Indian Forest Survey, forests cover 52 percent of Nagaland's geographical area, amounting to 862,900 hectares. Interestingly, community and private ownership account for 46 percent of this forested area, while government forests, including reserved areas and wildlife sanctuaries, constitute only 6 percent (Department of Environment, Forest & Climate Change, Government of Nagaland, 2021). Nagaland is a model of community-based forest management, with 88.3 percent of its forest area under indigenous community control, in stark contrast to the 90 percent managed by government departments in other parts of India. Jhum cultivation, a traditional form of shifting agriculture, is a significant land-use practice in Nagaland, occupying about 45 percent of the land. In Kohima District, for example, swidden cultivation covers a notable 70.2 percent, underscoring the persistence of this traditional practice (Cairns, 2007: 16).

Nagaland's land and forest ownership operates under a customary law framework, which is pivotal in regulating access to forest resources and ensuring both individual rights and collective ownership. This system categorizes land into three types: community, clan (khel), and individual/family. Community land, managed by village committees, includes reserved forests, woodlands, and communal areas. The extent of community land varies by village and tribe; for example, in Chakhro Angami village, community land accounts for 75 percent of the total area, while in other Angami villages, individual lands may dominate (Martemjen, 2017: 54). Community members can request permission from the village committee to use community land or access forest resources for personal or family needs.

Clan land includes areas designated for housing, cultivation, and forested areas, and it is managed and allocated by senior clan members. Individual/family land comprises arable land, rice fields, and private forests, typically following a patriarchal inheritance system with some modern adaptations, such as daughters receiving land if there are no sons. Land ownership also has a significant social role, with larger holdings often associated with higher status (Kikhi & Ghosh, 2016: 58).

Nagaland's legal framework poses significant challenges for forest governance. Article 371(A) of the Indian Constitution (1962) protects customary practices, limiting the application of central forest laws, such as the Forest Rights Act of 1980, on private and community lands, which often have unclear ownership boundaries. Additionally, the government imposes symbolic taxes on shifting cultivation, collected through local administrators. These taxes, typically just a few rupees per household, serve more to assert state authority and foster loyalty than to generate revenue (Das, 2018: 103).

New land laws have promoted a shift from customary practices like *jhum* toward individual ownership and sedentary agriculture. The Nagaland Jhum Act of 1970 recognized the right to transfer *jhum* land within community boundaries, acknowledging both customary land rights and establishing private authority over the land. *Jhum* cultivation often generates limited income and is perceived by the state as a poverty trap for farmers. To address this, governments have implemented development initiatives promoting cash crop plantations as alternatives to *jhum* practices. This strategy includes encouraging the cultivation of rubber, coffee, and other commercially viable crops in upland areas, with farmers receiving subsidies to extend terrace cultivation for cash crops, while *jhuming* is discouraged (Sarma, 2022).

Local communities in Nagaland have developed their own customary institutions and laws for managing natural resources. Community Conservation Areas (CCAs) have been established based on traditional beliefs and community rules to promote sustainable conservation. Each community has its own enforcement system, with fines ranging from 10,000 to 30,000 rupees (135–410 USD) for violations,[5] particularly to protect endangered species like Blyth's tragopan. However, the effectiveness of these village councils varies. Discrepancies in property rights can undermine conservation efforts; some councils lack the authority to regulate private land use, leading to deforestation for activities like commercial charcoal production. In contrast, other villages successfully impose restrictions on private land, prohibiting logging and even controlling land use for cultivation (George & Yhome, 2008). Additionally, enforcing hunting bans remains difficult due to the deep cultural significance of meat consumption in Naga traditions.[6]

While community regulations, such as logging and hunting bans with fines, serve as deterrents, their effectiveness depends on consistent enforcement across village boundaries. Isolated conservation efforts can lead to a "self-exclusion" scenario, where hunting practices shift to neighboring villages, causing conflict within and between communities. A promising solution is the expansion of CCAs into interconnected networks, creating larger protected areas for wildlife. However, the successful implementation of such networks requires substantial resources. Budgetary

[5] Interviewed with villagers from Khonoma, Dzuleke, and Benreu.

[6] Interviewed with Khonoma villager on 11 May 2023.

constraints within the Forest Department limit support for all conservation projects, often leading to selective funding based on specific criteria. Furthermore, it is uncommon for communities and the state to work together on Joint Forest Management (JFM) projects, which makes comprehensive conservation efforts even more difficult.[7]

The Nagaland Forest Management Project (NFMP), launched in 2017 with support from the Japan International Cooperation Agency (JICA), aims to enhance forest ecosystems and boost income generation. The project primarily focuses on rehabilitating *jhum* cultivation areas and providing livelihood support. Implementation began in 2019 across 31 villages, with tree plantation being a key activity (Department of Economics and Statistics, 2024: 51).

Nagaland's forest governance integrates traditional practices with state initiatives, with the Nagaland Forest Department prioritizing JFM in collaboration with communities that have established over 1000 CCAs. However, formalizing these CCAs requires meeting International Union for Conservation of Nature (IUCN) criteria, which can restrict activities such as hunting, fishing, and *jhum* cultivation. Currently, only 407 CCAs, managed by clans or colonies, meet these standards.

The Nagaland Forest Department also encourages the establishment of new formal CCAs and the expansion of existing ones to create larger conservation corridors, which would improve connectivity and the monitoring of illegal activities. While financial support was provided to 277 villages in the past, it ceased during 2014–2015. Now, CCAs seeking funding for conservation projects must submit proposals and sign Memorandums of Understanding (MOU) with the Forest Department, detailing conservation activities like building firebreaks and watershed management. Currently, only 128 CCAs participate in this conservation program, covering 88,900 hectares.[8]

The establishment of CCAs by the Forest Department aims to formalize community organizations and secure state recognition. Communities voluntarily create CCAs to receive support from international organizations such as Japan's JICA, Germany's Kreditanstalt für Wiederaufbau (KfW), and the Food and Agriculture Organization (FAO). However, a major obstacle in establishing these conservation areas arises from the state requesting or receiving land donations from the community, clans, and families.

Despite these efforts, the Nagaland Forest Department faces significant challenges. *Jhum* cultivation is seen as a threat to forest preservation, and human-wildlife conflicts occur when animals damage crops. Additionally, the "island syndrome" effect, where conservation areas become isolated pockets surrounded by unprotected land, further complicates broader conservation efforts.

In summary, Thailand's forest governance is primarily state-driven and centralized, with little room for community involvement or control. By prioritizing

---

[7] Interviewed with the administrative personnel from the Nagaland Forest Department on 23 May 2023.

[8] Interviewed with the administrative personnel from the Nagaland Forest Department on 8 May 2023.

conservation efforts through top-down laws, the Thai government implements exclusionary tactics that limit community access to forest resources. On the other hand, Nagaland has a community-based strategy in which local governance and traditional rules are important. The state aims to establish its power by formalizing Community-Conserved Areas (CCAs) and Joint Forest Management (JFM) programs, even if the community owns the land and forest. Notwithstanding these disparate strategies, both areas struggle to strike a balance between conservation goals with the needs and the rights of local communities. This emphasizes the critical importance of integrating traditional knowledge and practices with modern forest management strategies to achieve more equitable and sustainable outcomes.

4.4 Community Forest Management Practices and Livelihood Diversification

A comparative analysis of community forest management practices in northern Thailand and Nagaland reveals shared challenges arising from increasing reliance on forest resources. As resource scarcity intensified, competition emerged among both intra- and inter-community members. In response, these communities devised innovative mechanisms to regulate and equitably distribute forest resources. Notably, the implementation of community-based Payment for Environmental Services (PES) emerged as a common strategy. By collecting fees from resource users and reinvesting in conservation and community development, these communities have effectively mitigated conflicts and fostered resource sharing.

Community Forest Management (CFM) in Northern Thailand

The Karen communities in the research areas have invented their own systems for managing the forests. These include setting rules and regulations, defining agricultural and community forest boundaries, and designating conservation zones. Notwithstanding their long history of harmonious coexistence with the forest, these communities have faced a lack of government assistance, necessitating self-reliant strategies. To address this, they have established bamboo shoot and honey funds, allocating a portion of sales proceeds to support conservation projects, including infrastructure maintenance, watershed management, fire prevention, and land rights advocacy.

Establishing community funds with proceeds from the selling of bamboo shoots and honey is consistent with the Payment for Environmental Services (PES) principles. By giving monetary value to ecosystem benefits, this mechanism encourages environmental stewardship (Engel et al., 2008). In this situation, the community serves as both the service provider and the service recipient. The money raised is then used to support community welfare programs, like public utilities and healthcare, illustrating a system that incentivizes environmental conservation.

In 1999, the Hin Lad Nai community in Chiang Rai Province established the Bamboo Shoot Fund, followed by the Honey Fund in 2013 when beekeeping gained popularity. Additional contributions to the fund came from study tours and donations from various organizations, such as academics, famous monks, the private sector, and NGOs. The community collects 20,000 baht (US$555) a year. This fund was utilized by the community to maintain the forest and create firebreaks, addressing the insufficient support from government agencies. The budget provided by the Subdistrict Administrative Organization for firebreaks was limited to covering food expenses only. The community also allocated the funds to maintain the mountain water supply system, roads, and welfare services.[9]

The Baan Klang community in Mae Moh District, Lampang Province, initiated the Bamboo Shoot Fund in 2004, inspired by the Hin Lad Nai community. Villagers collected bamboo shoots three times annually during the rainy season, contributing all proceeds to the fund, which generated up to 60,000 baht (US$1666) per year. After nearly 20 years, the fund has accumulated a balance of approximately 100,000 baht (US$2777). The community uses this fund for forest conservation activities, maintaining firebreaks, extinguishing forest fires, and repairing roads and water pipes. Additionally, the community received financial support from the Mae Moh Electricity Generating Authority of Thailand as compensation for the Ban Dong sub-district, where coal mine soil was dumped. These funds were used to repair the church, purchase saplings, and establish a cattle bank to promote animal husbandry as an alternative to corn cultivation. Community leaders advocate animal husbandry, emphasizing the harmful effects of corn farming on soil and forest resources due to its high chemical requirements.[10]

In 2016, the Mae Saan community in Lampang Province, neighboring Baan Klang Village, established a community fund. Contributions are collected from each family that earns income from bamboo shoots and *ma-kwaen* (local herb), accumulating approximately 20,000 baht annually. The community also receives compensation from the Electricity Generating Authority of Thailand and the National Park Office, which brought the total fund to around 400,000 baht (US$11,111) in 2021. The village uses this fund for building firebreaks and conducting forest conservation activities; however, the budget is sometimes insufficient due to the frequent occurrence of forest fires. To ensure future financial stability, the village committee plans to increase household contributions to the fund, recognizing that the annual income from selling *ma-kwaen* in the village amounts to several million baht.[11]

The Huay E-Khang community in Mae Wang District, Chiang Mai Province, began collecting taxes from community members in 2019, with each person contributing 38 baht (US$1.05) annually. This amount is based on the cost of maintaining a 1600-hectare area per person. The community also receives contributions from visitors and donors. The funds collected are used for watershed management, creating firebreaks, patrolling the forest, and covering legal expenses for community members prosecuted by the state. This self-reliant approach allows the community

[9] Interviewed with Hin Lad Nai leader on 28 May 2022.

[10] Interviewed with Baan Klang village head on 14 May 2022.

[11] Interviewed with Mae Saan village head on 12 May 2022.

to manage and care for its own resources independently, without relying on the insufficient state budget for forest resource management.[12]

The Mae Lan Kham community established the Rice Merit Fund in 2004 as an extension of the Rice Bank project, which was initiated by the Thai-Norway project in 1991 to support highland communities. The project originally allowed community members to borrow rice during shortages, with repayment made after the harvest. Although temporarily suspended due to a sufficient rice supply, the initiative was later resumed, with each of the 144 families contributing three sacks of unhusked rice. The rice was sold to raise money for the fund, which was used to finance the purchase of GPS devices for plot mapping, conducted with assistance from the Northern Farmers Network and the IMPECT organization. The community also receives annual financial support from the municipality, the Watershed Management Unit, and a well-known monk for forest fire management.

All Karen communities have utilized their village funds for advocacy purposes. This includes covering travel expenses to Bangkok for land rights advocacy during the resettlement policy, as well as attending seminars on forest fires and rotational farming in Chiang Mai. Additionally, farmers have collaborated with the Northern Peasant Federation (NPF) to negotiate with the state for community-based forest management.

Community Conservation Areas (CCAs) in Nagaland and Ecotourism

In Nagaland, a study on community conservation areas, community networking, and the economy in four villages in Kohima and Peren District found that different methods of managing community forests were used. Khonoma and Dzuleke utilize the community-based PES principle in managing their forest resources, integrating this approach into their conservation strategies. In contrast, Benreu Village does not have community funds but receives support from government agencies and NGOs for its conservation efforts, which include both plant and animal conservation activities. Despite having a community conservation area, Ngwalwa Village lacks a formal management framework to properly manage it.

The Khonoma Nature Conservation and Tragopan Sanctuary (KNCTS), established by the Khonoma village in December 1998, exemplifies the effectiveness of community-driven forest conservation initiatives. Spanning 7000 hectares, KNCTS was created to protect the endangered Blyth's tragopan birds. The villagers implemented a hunting ban after witnessing the detrimental effects of uncontrolled hunting, and since the forest's designation as a wildlife sanctuary, it has become a significant destination for bird watching. KNCTS also enforces strict regulations prohibiting the commercial collection of forest products and promotes eco-friendly fishing practices, with penalties for violators (Chase & Singh, 2012). Although KNCTS operates independently of traditional government frameworks, the Nagaland Village Councils Act empowers village councils to establish mechanisms for forest management, further supporting such community-led efforts.

[12] Interviewed with Huay E-Khang village head on 2 March 2022.

In 2005, the Nagaland Forest Department played a crucial role in securing official status for the Khonoma Nature Conservation and Tragopan Sanctuary (KNCTS) by designating it under the India Wildlife Protection Act. The KNCTS committee comprises representatives from three village clans: *Merhuma, Thevoma,* and *Semoma*. Their responsibilities include developing conservation strategies, patrolling the area to prevent hunting, and documenting their activities. Additionally, they conduct annual seminars for specific groups, such as women and youth, to raise awareness and foster a sense of ownership within the community.

However, KNCTS lacks its own income source, prompting the committee to actively seek funding from various stakeholders. This includes the Nagaland Forest Department, NGOs, and international conservation organizations such as the International Union for Conservation of Nature (IUCN) and the World Wildlife Fund (WWF), as well as other organizations from neighboring Assam. The funds acquired support alternative livelihood programs for villagers who previously relied on the forest. These projects encourage villagers to move away from unsustainable practices by distributing livestock, such as pigs and chickens, as well as providing rice to support their transition.

KNCTS has implemented a robust system to prohibit illegal hunting, with the village council establishing clear rules and guidelines, including outlined penalties. The Khonoma Youth Organization (KYO) plays a crucial role in patrolling these protected areas. Violators are subject to community fines, which get harsher for repeated infractions. First-time offenders face a fine of 20,000 rupees (US$238), while subsequent offenses incur a higher fine of 30,000 rupees (US$357). Additionally, fines are even higher for individuals from outside Khonoma. KYO previously operated a reward system where informants reporting hunting activities could receive one-third of the collected fine, with the remainder split between KYO and the informant's clan. KYO's patrolling and monitoring activities require financial resources, and they rely on the funding from the Eco-Tourism Management Committee (ETMC), which generates income from tourism.[13]

The villages in the study areas of Nagaland which are Khonoma, Dzuleke and Benreu have focused on eco-tourism as an alternative livelihood to generate household income. Communities have established village funds, which they use for forest conservation activities and monitoring costs, given that the budget from the Forest Department is insufficient for all communities.

The villages of Khonoma, Dzuleke, and Benreu in the Nagaland study areas have prioritized ecotourism as a substitute source of income for their households. Since the budget from the Nagaland Forest Department is insufficient, the communities have established village funds, which they can utilize for forest conservation activities and monitoring costs.

On October 25, 2005, Khonoma Village received support from the Nagaland Tourism Department and the Union Ministry of Tourism and Culture, amounting to 30 million rupees (US$357,142) for ecotourism development. Government

[13] Interviewed with the head of KNCTS on 8 May 2023.

agencies established the Khonoma Green Village Project (KGVP) to implement the country's first green village initiative. The project includes infrastructure development, sanitation improvements, solar power systems, and landscape enhancements to attract tourists. The goal of this project is to create an ecotourism village that serves as a model for other villages in Nagaland and beyond (Papu & Nathani, 2020).

The Ecotourism Management Committee (ETMC) was established in 2003, with directors selected from three clans to oversee tourism and community welfare. The community views tourism as a community enterprise that provides services; therefore, tourists are expected to pay for these services. At the same time, the community hopes to gain attention from tourists for the high quality of its products. When Khonoma Village began its tourism business, the goal was to ensure that visitors felt safe and welcome. To achieve this, the community organized training for homestay owners to improve service quality, encouraging tourists to willingly pay entrance fees: 10 rupees (US$0.1) for students, 50 rupees (US$0.6) for adults, and 100 rupees (US$1.2) for foreigners.

While tourism generates income for the community, the ETMC also recognizes that it can pose challenges. Due to the presence of endangered birds, the community is a delicate source of biodiversity that is susceptible to biological asset theft. Therefore, efficient tourism management is crucial to limiting the impact of tourists on the environment. All visitors must register before visiting the village and hire a guide to guarantee this.[14]

Income from guest houses has become a primary source of livelihood for many families in Khonoma, largely due to the high unemployment rate in Nagaland. Finding work in the city is challenging, even for those with a bachelor's degree, as government jobs, which are stable, often require connections. Those unable to find work in Nagaland often migrate to other parts of India. However, the village's status as a tourist destination has allowed many who previously worked in cities to return to their community, opening restaurants and guest houses. Many of the local youth work as guides and participate in organizing traditional performances, such as dances, to welcome tourists.

Khonoma villagers also produce processed food products like dried fruits, pickled vegetables, chilies, and bamboo shoots for sale to tourists. However, production remains limited due to a lack of marketing channels and financial support. Women's groups generate additional income through weaving. When not working on farms or in gardens, they can produce up to 20 pieces of cloth per year, providing an important supplementary income for their households.[15]

For families with the resources to renovate or expand their homes into homestays, tourism provides a significant source of income through accommodation and food services. Women typically take on the primary caregiving roles, including cooking and cleaning. Room rates in Khonoma Village range from 1500 to 2000 rupees (US$18–24) per day, excluding meals. Currently, there are only 12 registered

[14] Interviewed with the ETMC secretary on 10 May 2023.

[15] Interviewed with a group of guesthouse owners on 13 May 2023.

homestays in the village. Homestay owners are required to pay an annual registration fee to the ETMC based on the number of rooms in their guesthouse. For two to four rooms, the fee is 500 rupees (US$6) per room per year; for five rooms or more, the fee is 3000 rupees (US$36) per year. Guides also pay a registration fee of 3000 rupees (US$36) per year. A portion of the income collected by the ETMC is allocated to the Khonoma Youth Organization (KYO) to support the monitoring of illegal activities related to logging and hunting.

After 2005, Khonoma Village implemented the Green Project, funded by government agencies and supported by various organizations, to promote alternative livelihoods such as growing cash crops, fish farming, and tourism. Community organizations, including youth and women's groups, established Self Help Groups (SHGs) to receive funding from the government in the form of loans or subsidized projects. These funds were used to purchase capital, tools, machines, and weaving equipment. Additionally, since many of the men were skilled carpenters, they formed construction companies and raised funds to support their ventures.

Dzuleke Village, an extension of Khonoma Village, was established for married young people seeking to expand their arable land. The main occupation of Dzuleke villagers is agriculture; they grow local rice varieties for consumption and cash crops for income. Every household raises chickens, ducks, and pigs for food, while cows and Mithun are kept as investments. In 2014, the Dzuleke community began operating homestays with the support of the Tata Trust through the Northeast Development Agency (NEDA).

Currently, only four families in Dzuleke Village operate homestays. The Dzuleke Ecotourism Board (DETB) manages the allocation of tourists, ensuring that each homestay receives guests approximately two to three times per month. According to local informants, income from the homestay business has become the main source of livelihood for these families, supplementing the seasonal income from growing vegetables. However, during winter, when snow prevents vegetable cultivation, the homestay business becomes even more critical as a steady source of income.[16]

In 2014, the Dzuleke community established the Dzuleke Ecotourism Board (DETB) to manage local tourism. The community earns income through village entrance fees, where all visitors, both foreigners and Indians, must pay 20 rupees per person. DETB also collects a 10-percent tax on income from homestays, village entrance fees, and tent rentals. This tax revenue is used for community development projects, such as building public bathrooms, installing trash cans, hiring cleaners, and purchasing new camping equipment.

Some villagers earn additional income by working as guides. However, tourism in Dzuleke faces several challenges, particularly communication issues. The community has no direct connection with the Department of Tourism, resulting in insufficient financial support. Corruption, poor infrastructure, bad road conditions, and weak telephone signals further hinder tourism, as tourists struggle to contact the

[16] Interviewed with homestay owner on 14 May 2023.

community directly. Additionally, the village faces challenges in disposing of non-biodegradable waste.

For the Dzuleke's CCA Project, the state does not provide financial support. Despite the lack of a budget, the community has initiated tree plantation efforts using local species, with villagers participating as volunteers. The community opposes government control of the forest, as state regulations often restrict tourism activities and access to resources, which the villagers rely on.[17]

Benreu Village has received state support to develop into a tourist destination. This includes the construction of a large resort, Mt. Pauna Tourist Village, which was eventually transferred to the community. Only eight households run homestays, with the World Bank providing infrastructure support like restrooms. Hiking, bird-viewing, and butterfly watching are the main activities for visitors. However, tourism development faces several challenges, particularly in infrastructure. Tourists find it challenging to get in touch with the community directly due to unfinished road work and restricted telephone signals. Furthermore, despite operating since 2019, homestay owners have failed to adequately promote their establishments, even though many villagers frequently use social media sites like Facebook.

Benreu has yet to establish a tourism management organization. Tourists do not pay entrance fees, and homestay owners and guides do not contribute taxes to the community. As a result, the community lacks its own budget for managing tourism and conservation projects, relying instead on support from government agencies and NGOs for forest conservation activities.

A popular community product in Benreu Village is rhododendron juice, which is used to treat various ailments, including inflammation, diarrhea, dysentery, headaches, diabetes, arthritis, fever, high blood pressure, and cancer. Rhododendron flowers could also be developed into a kind of tea. During the winter, women collect mustard leaves to make a paste used in chutneys (a local condiment made with chili paste) and pickles. Chamomile flowers, which grow abundantly along the roads, have the potential to be developed into tea, but this has not yet been pursued. Food processing activities could be further developed through SHGs, which provide women with investment funds.

In summary, communities in both Northern Thailand and Nagaland have established community forest areas and implemented rules and regulations to ensure the sustainable use of forest resources. With limited state support for conservation activities, such as fire management and monitoring, these communities have adopted community-based Payment for Environmental Services (PES) schemes to cover costs and fund social welfare and infrastructure development. Notably, communities in Nagaland, benefiting from higher income generated through ecotourism, have been able to establish community funds. However, these funds alone are insufficient to sustain conservation efforts, prompting communities to seek support from NGOs and their networks. Additionally, local communities in both regions have diversified their livelihoods through activities such as ecotourism, weaving, the

[17] Interviewed with the Dzuleke community's leader on 14 May 2013.

cultivation of local herbs, and livestock farming, thereby reducing their reliance on forest products.

4.5 Social Networks for Community Forest Management

Communities in northern Thailand's forest areas have formed watershed networks with other communities facing similar challenges, collaboratively driving policy initiatives. This networking has strengthened these communities, enabling them to negotiate more effectively with government agencies. They have also received support from organizations like the Northern Peasant Federation (NPF), the Northern Development Foundation (NDF), and the Sustainable Development Foundation to map out and clearly define agricultural boundaries, preventing further land clearance. These plot maps have become essential tools for negotiating land rights with the government.

Additionally, these communities have partnered with the Northern Development Foundation and both domestic and international educational institutions to research agroforestry methods, rotational cultivation, forest management, climate change, and food security for highland farmers. The research findings challenge common misconceptions by showing that rotational cultivation does not contribute to global warming.

The role of community leaders has been crucial in this process. These visionary leaders introduce new knowledge and practices, excel in building networks with government agencies, private organizations, educational institutions, NGOs, and businesses, and are proactive in organizing activities that enhance the community's visibility.

For instance, the leaders of Hin Lad Nai have cultivated strong relationships with various organizations, receiving substantial support from the Meditation Fund of Phra Methee Wachirodom and collaborating with renowned chefs who organized community events. These chefs also facilitated the sale of community products, ensuring that profits returned to the community. During the 2019 forest fires, leading chefs prepared and sold food to raise funds for firefighting equipment. Additionally, the strategic use of media to publicize these activities has enhanced the community's visibility and garnered further support from the broader society.

The Hin Lad Nai Community serves as a learning center for researchers, students from both domestic and international educational institutions, as well as private organizations and other communities. Visitors come to study the community's livelihood, culture, traditions, and indigenous knowledge related to resource management, with the goal of expanding this management model to other communities.

Ban Huay E-Khang offers ecotourism activities, including hiking, homestays, and health-focused services like herbal medicine lessons, steaming, and massages. A female community leader has built networks with both domestic and international groups, such as Taiwan study groups learning natural dye weaving and Danish students exploring the navel forest and community livelihoods. All activities are collaborative efforts, with earnings partly going to the community fund, ensuring that income generated from local resources is reinvested in their upkeep. Additionally, the community has created nature trails and lodgings to promote educational tourism, which aids tourists in comprehending the support for the community's way of life.

Communities in Nagaland have partnered with government agencies and NGOs to promote tourism, alternative livelihoods, and forest conservation. The Ministry of Tourism and Culture supports ecotourism, while the Forest Department collaborates with NEPED, the Japan International Cooperation Agency (JICA), Germany's Kreditanstalt für Wiederaufbau (KfW) Development Bank, and the Food and Agriculture Organization (FAO) to establish CCAs and support alternative livelihood projects. Aaranyak, an Assam-based NGO, plays a key role in biodiversity conservation by educating and empowering communities.

These partnerships have greatly enhanced conservation efforts in Nagaland. For example, the KfW-supported Forest & Biodiversity Management in Himalaya (FBMH) project strengthens CCA management in 35 villages, promoting long-term sustainability. The project also encourages alternative income sources, such as rubber and coffee plantations, to reduce dependence on swidden farming and hunting. To mitigate human-animal conflicts, the Forest Department funds Community Game Watchers and provides essential monitoring equipment.[18]

The JICA-supported Nagaland Forest Management Project involves 185 villages, aiming to improve forest cover, biodiversity, and community livelihoods while strengthening local institutions. Its key activities include promoting sustainable agroforestry, offering training programs, and supporting infrastructure development like water harvesting systems and farmhouses. All initiatives are implemented with the community's informed consent. Additionally, JICA empowers local communities by supporting Self-Help Groups (SHGs) in each village.[19]

In 2014, 22 CCAs in Nagaland established the Nagaland Community Conserved Area Forum (NCCAF) to collaborate on biodiversity conservation. This network provides a platform for sharing knowledge, ideas, and skills, with the goal of unifying the voice of all CCAs in Nagaland and ensuring that their initiatives are recognized at national and international levels. The NCCAF's objectives include promoting the plantation of indigenous species, conserving forests, wildlife, and biodiversity, advancing ecotourism, supporting scientific research, and raising awareness about sustainable resource use (NCCAF website, 2023). The Foundation for Ecological Security (FES), a non-profit organization, supports the NCCAF and helps communities across India restore soil, water, and forest resources.

Members of Khonoma Village have formed SHGs, each with about 10 members, to receive financial support from government organizations and NGOs. Women's weaving groups, for instance, generate income from weaving, and the earnings are

[18] Interviewed with the administrative personnel from the Nagaland Forest Department on 8 May 2023.

[19] Interviewed with the administrative personnel from the Nagaland Forest Department on 8 May 2023.

reinvested into the group, allowing new members to borrow funds for their own investments. This revolving fund aims to expand group membership. The weaving group, established 15 years ago, has received support from various institutions and primarily produces woven scarves. Originally, large scarves sold for 4000 rupees, but the group has since developed smaller scarves priced at 500 rupees each to make them more affordable. These products are typically sold in souvenir shops in Kohima and within Khonoma Village.[20]

The Dzuleke community has received support from Tata Trust through the Northeast Development Agency (NEDA) to improve their houses and gain knowledge on running a homestay business. Additionally, government agencies, including the Department of Fisheries, have contributed to tourism development by building weirs and constructing houses for tourists, providing the community with a source of income for further development. The Department of Fisheries has also launched a pilot Rainbow Trout Farming Project in the community. If successful, this project will offer the community an opportunity to generate income and increase their food sources through fish farming.[21]

The Benreu community has been developed into a tourism village with infrastructure support from government agencies and the World Bank, enabling families to earn additional income through homestay businesses. As part of the Paona Conservation Area and the NCCAF network, the Benreu community has also received support from the FES and NEPED to implement resource conservation activities and promote alternative livelihoods.

Ngwalwa Village is supported by government agencies to grow cash crops such as betel nuts (areca nuts), spices, vegetables, and fruits. Ngwalwa Village is widely known as the "Betel Nut Village" due to its reputation for producing high-quality betel nuts. In 2023, NEPED distributed 100 betel nut seedlings to each family, enabling farmers to begin harvesting within five to six years and earn around 80,000 rupees (US$952) annually. According to field data, the village generates an annual income of approximately 3,500,000 rupees (US$41,666) from betel nut cultivation. Betel nut brokers often reserve the trees in advance and handle the harvest and transportation to Assam for sale.[22]

In addition to cultivating betel nuts for over 50 years, Ngwalwa villagers grow king chilies in jhum areas, selling them for 200–300 rupees (US$2–3.6) per kilogram. Bananas are the primary fruit crop, with several large plantations on private land. Other key crops include *yongjak* (stink bean), various vegetables, and fruits. The community has also adopted beekeeping, with honey from domesticated bees selling for 1000 rupees (US$12) per liter and honey from stingless bees, known locally as *Teliane dui*, fetching 2000 rupees (US$24) per liter. Some villagers collect honey from the forest, earning about 100,000 rupees (US$1200) per family.[23]

---

[20] Interviewed with the Knonoma weaving Self-Help Group on 9 May 2023.

[21] Interviewed with the Dzuleke community's leader on 14 May 2013.

[22] Interviewed with the Ngwalwa village chairman on 6 May 2023.

[23] Interviewed with the Ngwalwa women's group on 6 May 2023.

The World Bank supports Ngwalwa Village through Self-Help Groups (SHGs), providing a revolving fund of 3000,000 rupees (US$35,714) annually. A Nagaland state agency also promotes sericulture, offering training and 20,000 rupees (US$238) per family to establish silk farms, with the state purchasing the produced silk.[24]

The 25 SHGs in Ngwalwa Village, each with ten or so members, use the money for projects like planting trees and loans to build rental homes. The community does not yet have a system in place to manage conservation areas, and tourism development has not yet been pursued. Most village programs concentrate on planting large trees, such as bamboo, palm, and rubber trees, which usually stop after harvest with little further work. The Nagaland Forest Department has not supported conservation efforts. Although the community is not entirely aware of the detrimental impact of cash crop production on biodiversity and soil quality, it is becoming more noticeable.

In summary, this comparative study underscores the critical role of social capital in facilitating the livelihood strategies of communities in both regions. All communities have established networks with NGOs, the private sector, and government agencies to support conservation activities, infrastructure development, and alternative livelihoods. However, implementing joint forest management in Nagaland presents unique challenges, as forest land is owned by the communities themselves, making cooperation essential. The case of Ngwalwa exemplifies this issue, where local priorities are focused on cash crops and tree plantations, leading to reduced emphasis on forest conservation. This reflects the power of the community to prioritize monetary gains while avoiding CCA projects to bypass strict rules and regulations governing access to forest resources.

4.6 Indigenous Knowledge as Negotiation

Local knowledge is resilient, enabling communities to cope with and adapt to social, political, and environmental changes. This study highlights how the Karen community in Thailand and communities in Nagaland have preserved their traditional knowledge while integrating it with scientific knowledge. Through this fusion, they have developed resource management strategies and created new, flexible identities that are well-suited to an evolving context.

The rituals and local wisdom of Karen communities began to fade with the advent of modern knowledge and their conversion to Christianity. However, Karen communities are actively working to revive and pass on this wisdom to younger generations, teaching them about seeds, herbs, and rituals related to their traditional agricultural systems. Women in these communities play a crucial role in preserving various seeds. The Baan Mae Saan community maintains over 30 rice varieties and more than 70 types of vegetables in their rotational fields.

[24] Interviewed with the Ngwalwa Self-Help Group on 6 May 2023.

The integration of local and scientific knowledge is clearly demonstrated by the youth group of Hin Lad Nai. These young people have had the opportunity to exchange knowledge with students and academics from both domestic and international institutions who visit for research or study. This exposure has enabled them to acquire academic knowledge, enhance their use of technology for online business, and improve their English language skills.[25]

The youth group of Huai Hin Lad Nai has built a network with other villages in Lampang and Chiang Mai provinces by organizing camps. These camps provide opportunities for youths studying in urban areas to exchange knowledge with those who remain in the community, facilitating the transfer of indigenous knowledge and culture to those living outside the community. According to the youth group leaders, it is crucial for young people to understand the history of previous generations' struggles to coexist with the forest. The community has also demarcated land-use boundaries and established regulations for resource conservation, enabling them to negotiate effectively for the stewardship and sustainable utilization of forest resources.[26]

Similarly, the women in Huay E-Khang have created video content to share their knowledge about herbs and healthcare on Facebook through the page "Ethnic Women's Community Interpreter." They also published a booklet on using herbs to treat diseases in children and adults, a resource based on community knowledge that is distributed in schools and health centers. Looking ahead, they plan to develop this indigenous knowledge into ecotourism ventures that could generate income for the community.[27]

Ban Huay E-Khang has initiated activities to revitalize indigenous Karen knowledge among the youth, organizing events multiple times each year. These initiatives aim to educate young people about traditional practices, including food, seeds, rituals, and the production processes of rotational farming. To preserve and transmit this indigenous knowledge, particularly in relation to food security, community leaders—supported by various networks—have invited elders to share Karen folktales and impart their expertise on plant varieties and medicinal herbs found in rotational farms. This effort plays a vital role in preserving and revitalizing indigenous knowledge.

During the government's Zero Burning policy, aimed at combating PM 2.5 pollution and global warming, Mae Lan Kham villagers faced challenges as they were unable to burn their rotational farms according to the growing season. To work around this, villagers used information from NGO networks to identify when satellites would monitor hotspots, allowing them to burn their fields undetected.[28]

Since 2021, through the proposed Protection and Promotion of the Livelihoods of Ethnic Groups Act, the indigenous peoples in Thailand have been advocating for

[25] Interviewed with the Hin Lad Nai youth group on 29 May 2022.

[26] Interviewed with the Hin Lad Nai youth leader group on 29 May 2022.

[27] Interviewed with the Huay E-Khang women's group on 6 March 2022.

[28] Interviewed with the Mae Lan Kham village head on 13 March 2022.

their rights to access forest resources. This legislation aims to preserve traditional cultures and livelihoods, encourage the participation of ethnic groups in national development, and ensure sustainable access to natural resources. It seeks to reduce inequalities and prevent discrimination against ethnic communities (Policy Watch, 8 February 2024). The draft has garnered support from civil society, political parties, and academic institutions. However, it has faced challenges in the House of Representatives due to concerns that the definition of "Indigenous Peoples" may conflict with national security interests (IMN Voices, 25 September 2024). Despite these obstacles, the proposed law represents a significant hope for the indigenous communities who strive to protect their livelihoods and maintain their presence in protected areas.

Like the Karen communities in northern Thailand, the Nagas preserve their indigenous knowledge by passing it down through generations, traditionally through the Morung house—a dormitory where boys learned about livelihood and socialization from elders. In the past, Morung served as an institution of culture and learning. However, with the conversion to Christianity, boys and girls began attending Christian schools. Despite this shift, Morung remains an important institution for cultural, traditional, and etiquette training, with young people often visiting the centers. The existence of Morung has helped preserve indigenous culture, even as Christianity has influenced the community since 1890.

Morung has since evolved into the Age Group House, continuing its role in cultural transmission. Peer groups are formed within each clan, divided by an age gap of five to ten years. In Khonoma Village, each of its three clans has its own peer group led by a mentor who provides guidance and discipline. These peer groups are named after their mentors, reflecting the age and clan of the members.

Peer groups in Khonoma compete to raise funds for community development projects, such as building sidewalks, organizing village feasts, and hosting group gatherings. During school breaks, groups of 30–100 youths often work in the fields, earning money that contributes to a fund for community development. In 1992, eight peer groups collaborated to build walkways and stairs around the village. The senior groups are particularly influential. For example, businesspeople from the *Thenoma* clan have built large houses for younger members and donated land in Dimapur and a rental house in Kohima to the community, generating regular income.[29]

The Morung social structure and learning system have greatly strengthened the community, with youth groups playing a crucial role in community development, infrastructure projects, and conservation efforts. These groups actively raise funds to support various projects and are key players in monitoring CCAs. The integration of traditional knowledge with the new education system in the villages allows for the use of cultural practices to promote forest conservation projects.

Despite the influence of Christianity, traditional Naga beliefs related to resource conservation continue to endure. For the Nagas, cosmology encompasses a

[29] Interviewed with Khonoma's senior member of the Age Group House on 21 May 2023.

three-dimensional relationship between nature, humans, and the supernatural, all of which are intricately interconnected. They perceive the world as comprising numerous valuable elements—such as mountains, forests, rivers, and valleys—that require careful stewardship. The earth, regarded as the home of animals, trees, and plants that provide sustenance, is treated with profound respect and reverence.

According to Naga legends, animals and humans share a common ancestor, with tigers regarded as man's older brothers. If a villager kills a tiger or leopard, it is treated with the same gravity as taking a human life, and the person must undergo a period of strict isolation, typically lasting around 30 days, and perform rituals to honor the tiger's spirit.

Naga beliefs hold that spirits inhabit every aspect of their environment, including houses, barns, rice fields, forests, and rivers, and they are represented by the sun, moon, and stars. These spirits are believed to influence health, happiness, success, and abundance, making their worship essential for ensuring survival and prosperity. For example, the Trans-Dikhu-Doyan tribes believe that after death, a person's spirit enters a hawk, which flies over the village to carry souls to *Asilim*, the Land of the Dead. As a result, consuming hawks or eagles is forbidden (Nienu, 2015: 139–146).

Following animistic traditions, hunters in Khonoma once sought blessings at a cliff known as the "Face of Woman." A smiling face was seen as a sign of good fortune in hunting, while a sad face indicated likely failure. Although most villagers have converted to Christianity, many still maintain their belief in spirits. Rituals for blessings before and after planting are still common, though they have evolved. Instead of traditional animal sacrifices, villagers now invite priests to bless their crops, asking God for a bountiful harvest. These rituals are performed individually by each household rather than collectively.[30]

In Benreu village, a respondent shared a story about "*Siperai*," a spirit believed to protect the village and its forest. According to the legend, whenever a storm, earthquake, or other natural disaster occurs, people seek refuge in Benreu. Villagers continue to narrate the story of *Siperai* to younger generations even after converting to Christianity. *Siperai* was said to be a man who married a woman from the village and asked for soil from his father-in-law's land. The next morning, a large amount of soil had disappeared, taken by *Siperai* to build the largest mountain, Paona. Villagers believe that *Siperai*'s home is atop this mountain. If someone cuts down trees or branches on the mountain, they must perform a ritual called *kena*, involving the sacrifice of a pig to feed the entire village. Additionally, those who reach the mountain's peak must leave behind a coin or a piece of clothing; otherwise, they risk becoming lost in the forest or unable to return home.

Khonoma has favorable climate and soil conditions for growing medicinal plants. In recent years, many households have begun cultivating medicinal plants that are not native to the area to boost their income. However, the indigenous knowledge of Khonoma's people regarding medicinal plants continues to be passed down orally, with no written documentation (Chase & Singh, 2013). Only a few traditional

[30] Interviewed with the Khonoma's guide on 9 May 2023.

healers possess this knowledge, and the younger generation shows little interest in preserving it, placing these traditions at risk of fading away.

To ameliorate the gradual decline of their indigenous knowledge over the generations, communities in Tuensang and Peren Districts organized village storytelling and drawing competitions across 13 villages in November 2022. This was done in collaboration with NEPED and funded by KfW Germany Bank. Adults competed in storytelling, sharing traditional livelihoods, folktales, and personal experiences with nature, while children participated in drawing competitions on topics such as traditions, legends, history, and nature. Additionally, a competition related to village infrastructure provided communities with the opportunity to draft project proposals for village development, addressing societal, economic, and environmental needs. This initiative not only revitalized their indigenous knowledge and traditions but also fostered trust between communities and state agencies, creating a sense of participation in development projects (The Naga Republic, 6 December 2022).

In addition to playing a key role in conserving for the community forest, the youths of Khonoma village are also involved in entrepreneurship, particularly in homestay businesses, tourism companies, and as community guides. NGOs have been instrumental in helping them develop their potential. For instance, Peter, a 31-year-old from the village, learned about the environment and biodiversity through a training course supported by Assam's Aaranyak NGO and Green Hub. The training equipped him with skills in videography, flora and fauna identification, and expertise in biodiversity, making him one of the few specialists in the Khonoma community. In 2019, Peter began creating videos to share the community's conservation stories, and he has produced content for National Geographic and various NGOs.[31]

In summary, the focal communities in northern Thailand and Nagaland have effectively utilized traditional beliefs and indigenous knowledge for forest conservation, despite their conversion to Christianity. Local traditions, such as Morung, play a crucial role in reinforcing indigenous identity and supporting communities in their efforts toward infrastructure development and conservation activities. Indigenous Peoples have successfully integrated traditional knowledge with contemporary practices to enhance their capacities and facilitate the transfer of knowledge to younger generations. This integration of knowledge is particularly important for the Karen communities in Thailand, as it strengthens their position in negotiating with the state regarding forest governance.

[31] Interviewed with Peter on 10 May 2023.

4.7 Discussion

This research explores sustainable community forest management in northern Thailand and Nagaland, India. Despite differences in land ownership systems between the two regions, the findings indicate that community-based forest management in both countries encounters significant challenges in environmental conservation. Effective community forest management cannot function independently but requires collaboration with state authorities and NGOs. Building partnerships through community-based institutions is essential to securing the necessary support for sustainable practices.

In Thailand, conservation forests are governed by the state, specifically the Royal Forest Department, which faces challenges due to limited budgets and ineffective policies. As a result, forest-dependent communities have developed their own resource management systems. This initiative allows them to establish regulations and define clear land-use boundaries for farming, forest utilization, and conservation. These communities showcase their sustainable livelihoods while actively conserving forests to legitimize their rights to reside within forested areas.

In Nagaland, land ownership is held by communities, clans, and families, with the state lacking authority to manage forests without local consent. While this private and common property regime allows for independent resource access, it also presents challenges. Social and economic changes have led to deforestation and hunting for trade, posing significant issues for community forest conservation.

Communities in both Thailand and Nagaland face challenges related to forest resource degradation, resource competition, and conflicts both within and between communities, as well as with the state. In response, Karen communities in Thailand and those in Nagaland have developed community-based resource management mechanisms using the principle of Payment for Environmental Services (PES). In this system, the community acts as both the user and steward of the resources, receiving compensation for their forest protection efforts. PES initiatives enhance the capabilities of local institutions and generate social, economic, and environmental benefits (Brownson et al., 2019). This dual role strengthens community forest management while delivering tangible benefits to the community.

However, community-based PES initiatives cannot depend solely on a strong conservation ethos for successful implementation. Several factors can hinder the effectiveness of PES schemes. For instance, in Benreu Village, the remote location and inadequate infrastructure make it challenging to collect entry fees from tourists, prompting the community to rely on NGO support for conservation activities. Similarly, the case of Ngwalwa highlights that communities may prioritize income generation through cash crops, often at the expense of focusing on conservation efforts.

In addition, the Karen in Thailand use community-based PES as a tool to construct their identity as innovative forest guardians, helping to secure their land rights. Similarly, engaging in conservation initiatives, the Nagas aim to transform their image from headhunters to forest protectors, leveraging this new identity to promote ecotourism.

The Community Forest Management (CFM) model serves as a platform for communities to build social networks with academics, NGOs, and government agencies. Both regions benefit from the support for alternative livelihoods and conservation activities, such as payments for guards, fire management, self-help groups (SHGs), and cattle banks. In Thailand, the Karen communities also receive support for land rights advocacy from the networks. Moreover, this model has the potential to reduce conflicts within communities, as well as between communities and government agencies.

The establishment of community conservation areas, while beneficial, can have unintended negative consequences, such as self-exclusion (Hall et al., 2011). Local groups, including women, artisans, and pastoralists who rely heavily on forest resources, may face marginalization due to the rigid community rules and governance structures. To foster broader participation and mitigate these impacts, livelihood diversification is critical. The formalization of Community Conserved Areas (CCAs) in Nagaland and community forests in Thailand aims to gain state recognition, which is vital for securing financial support from government institutions and NGOs. This external support is essential for developing sustainable alternative livelihoods, such as ecotourism, homestays, handicrafts, and livestock farming. These initiatives not only reduce dependency on forest resources but also enhance community resilience and socio-economic well-being.

It is crucial to mention that Joint Forest Management (JFM) cannot fully replace self-organized community forest management systems. The evidence from a village in Orissa, India, shows that the introduction of JFM has led to a decline in common property rights and weakened reciprocal relationships with neighboring villages. Furthermore, the administration of forest resources became a tool for political influence and monetary gain, undermining the core principles of equitable resource management (Nayak & Berkes, 2008). JFM often relies on a superficial understanding of communities, viewing them as small, homogeneous units with clearly defined boundaries (Karlsson, 2011). This "one-size-fits-all" approach fails to account for the complex social dynamics of forest-dependent communities and undermines the shared social value of forest benefits across village boundaries. By implementing Joint Forest Management (JFM) and similar Community-Based Natural Resource Management (CBNRM) models, the state can exert greater control over community initiatives, potentially leading to inequality among villages (Sarin, 1997). In contrast, the Dzuleke community in Nagaland demonstrates that communities with land ownership are less reliant on external support and enjoy greater autonomy. However, they face challenges in sustaining conservation initiatives and effectively monitoring reforestation efforts.

The Community Forest Management (CFM) is often oversimplified, assuming that all villagers are actively engaged in conservation practices. However, the indigenous peoples living in forests do not necessarily prioritize conservation (Li, 2002). For instance, many Karen communities participate in cash-cropping and forest product harvesting (Walker, 2001: 160). Similarly, in Nagaland, the indigenous peoples grow cash crops and implement tree plantation projects—such as those involving teak, firewood, and bamboo—in *jhum* fields. These projects typically last only three to five years and often lack mechanisms for monitoring their long-term success (Tiwari, 2022). The reforestation project in Ngwalwa village, Peren district, illustrates how a group sought state support for tree plantations with a primary focus on income generation rather than conservation. Once the bamboo was harvested, no further tree planting occurred. Additionally, the Ngwalwa community members have chosen not to engage in CCA formalization to avoid state influence, which could potentially restrict their utilization of forest resources.

This study highlights the resilience of indigenous knowledge, demonstrating its role not only in forest conservation but also in its integration with scientific knowledge to enhance well-being and promote alternative livelihoods. The Naga's social structure and the Morung learning system have strengthened the community, enabling youth groups to play a crucial role in community development and the success of forest conservation projects. However, indigenous knowledge in both regions is at risk of decline. While the Indigenous Peoples actively work to revitalize and transmit their knowledge to younger generations, they also leverage social media to promote and disseminate this cultural heritage. Additionally, indigenous knowledge is vital for the Karen communities in Thailand, as it has been strategically utilized to enhance their negotiating position with the state on matters of forest governance.

The governments of Thailand and Nagaland have sought to discourage indigenous practices, such as rotational farming systems, which have been practiced for generations in both regions. These practices are perceived as unsustainable by the state (Erni, 2015; Rawat et al., 2018). As a result, development policies in both regions have encouraged local communities to transition from traditional rotational farming to cash crops and tree plantations. This shift has led to increased water and land use, exacerbating deforestation and environmental degradation (Cramb et al., 2009; Delang, 2002; Yano & Lanusosang, 2013). However, the focal Karen communities have resisted these policies by continuing their rotational farming practices, emphasizing livestock farming rather than adopting cash crop cultivation. This resistance highlights their commitment to sustainable practices and forest conservation.

In contrast, the Thai state could draw valuable lessons from Nagaland, where rotational farming systems have been sustained for generations, enabling communities to conserve forest cover without direct state intervention. This highlights that rotational farming can serve as a sustainable land-use practice, balancing ecological preservation and livelihood needs.

This study highlights the crucial role that communities play in conserving forests in both northern Thailand and Nagaland, though the approaches and contexts vary between the regions. However, for these efforts to be successful, it is vital for government agencies to provide support in terms of funding and legal frameworks to facilitate the establishment of forest communities.

Thai government agencies should grant long-term land use rights to local inhabitants in protected areas and increase budget allocations to support community-based PES schemes. With secure land rights, communities could expand their conservation efforts to include REDD+ projects and directly receive funding from international institutions. This would not only strengthen community engagement in forest conservation but also reduce the government's manpower and resource burden.

It is essential for the Nagaland Forest Department to provide greater support to communities in establishing CCAs. Expanding CCA corridors to neighboring communities could help reduce conflicts and address the issue of isolated conservation areas. Additionally, the Nagaland Forest Department, NEPED, and international organizations like JICA and the German Development Bank should expand their support for alternative livelihoods that reduce dependency on forest resources. Additionally, fostering constructive dialogue between the state and communities is crucial to promoting more sustainable community forest management practices.

4.8 Conclusion

This research highlights the complexities of sustainable community forest management in northern Thailand and Nagaland, India, uncovering both shared challenges and region-specific dynamics. Both regions grapple with critical issues such as deforestation, resource degradation, and socio-economic pressures.

In Thailand, forest governance is state-centric, with conservation forests managed by government agencies like the Royal Forest Department. Communities in protected areas face challenges due to the lack of secure land rights, often relying on advocacy to negotiate access to forest resources. In contrast, Nagaland operates under a decentralized governance system, where communities, clans, and families own the land. This autonomy allows communities to independently manage their forests, though it also presents difficulties in ensuring consistency across villages. While Thailand struggles with community-state conflicts, Nagaland's governance model avoids direct state intervention but faces challenges in achieving conservation strategies.

In both regions, community-based forest management mechanisms are central to conservation efforts, supported by robust social networks. In Thailand, communities have developed PES practices to finance conservation activities, infrastructure, and social welfare, relying on collaboration with NGOs, academics, and other institutions. Similarly, in Nagaland, communities utilize community-based PES mechanisms, but their dependence on local resources and land ownership reduces reliance on external support. However, the sustainability of these initiatives in Nagaland is hindered by limited resource-sharing across villages. In both cases, social networks play a crucial role in mobilizing resources and facilitating alternative livelihoods.

Indigenous knowledge remains a cornerstone of forest management in both regions, though it faces distinct pressures. In Thailand, the indigenous peoples, such as the Karen, employ traditional practices like rotational farming and their ethnic identity to resist state policies and assert their rights to forest lands. This knowledge is further incorporated into community-based PES initiatives, enhancing sustainable practices and reinforcing their identity as forest protectors.

In Nagaland, indigenous knowledge is deeply intertwined with social structures, such as the Morung learning system, which strengthens community cohesion and promotes conservation practices. Participation in community forest management has transformed their identity from headhunters to forest protectors, fostering eco-tourism development. However, economic transitions, including the shift toward cash crops and short-term income generation, threaten the preservation of traditional knowledge in both regions.

This research demonstrates that while Thailand and Nagaland employ different governance systems and management practices, both regions underscore the importance of community-based approaches to forest conservation. Thailand's reliance on community-based PES highlights the potential for locally-driven solutions to fund and sustain conservation efforts, even within a state-centric framework. Nagaland's governance empowers communities with autonomy, though it necessitates greater focus on long-term sustainability and resource-sharing mechanisms. Preserving indigenous knowledge, strengthening social networks, and fostering inclusive community-led initiatives are critical to achieving sustainable forest management in both contexts.

References

Brownson, K., Guinessey, E., Carranza, M., Esquivel, M., Hesselbach, H., Ramirez, L. M., & Villa, L. (2019). Community-based payments for ecosystem services (CB-PES): Implications of community involvement for program outcomes. *Ecosystem Services, 39*, 100974.

Cairns, M. (2007). *The Aldar managers: The cultural ecology of a village in Nagaland, N.E. India.* Ph.D. Dissertation. Australian National University.

Chase, P., & Singh, O. P. (2012). People's initiative for conservation of forests and natural resources: A success story of Khonoma village forest, Nagaland. *Nebio, 3*(3), 61–67.

Chase, P., & Singh, O. P. (2013). Ethnomedicinal plants used by Angami tribe of Nagaland, India. *Indian Journal of Tropical Biodiversity, 21*(1), 29–42.

Community Development Knowledge. (2021). *Community forest situation in Thailand.* Retrieved June 10, 2024, from https://dcommunity.co/knowledge/สถานการณ์-ป่าชุมชน-ไทย/

Cramb, R. A., Colfer, C. J. P., Dressler, W., Laungaramsri, P., Le, Q. T., Mulyoutami, E., Peluso, N. L., & Wadley, R. L. (2009). Swidden transformations and rural livelihoods in Southeast Asia. *Human Ecology, 37*, 323–346.

Das, D. (2018). *The politics of Swidden farming: Environment and development in eastern India.* Anthem Press.

Delang, C. O. (2002). Deforestation in northern Thailand: The result of Hmong farming practices or Thai development strategies? *Society & Natural Resources, 15*(6), 483–501.

Department of Economics and Statistics, Government of Nagaland. (2024). *Nagaland economic survey 2023–2024.* Kohima. Retrieved July 22, 2024, from https://statistics.nagaland.gov.in/storage/statistical_data/2024/4181710318206.pdf

Department of Environment, Forests & Climate Change, Government of Nagaland. (2021, January 31). *Status of forests.* https://webtest.nagaland.gov.in/forest/status-of-forests/#:~:text=The%20recorded%20forest%20area%20of,and%20Unclassed%20Forest%20constitute%2093.56%25

Engel, S., Pagiola, S., & Wunder, S. (2008). Designing payments for environmental services in theory and practice: An overview of the issues. *Ecological Economics, 65*(4), 663–674.

Erni, C. (2015). *Shifting cultivation, livelihood and food security. New and old challenges for indigenous peoples in Asia.*

Forest Land Management Office. (2023). *Final report for preparation of information on the forest areas in 2023*. Retrieved April 6, 2024, from https://www.forest.go.th/land/รายงานโครงการจัดทำข้อ-12/

Forsyth, T., & Walker, A. (2014). Hidden alliances: Rethinking environmentality and the politics of knowledge in Thailand's campaign for community forestry. *Conservation and Society, 12*(4), 408–417.

George, J., & Yhome, K. (2008). Community forest management: A case study of Nagaland, India. In *Governing shared resources: Connecting local experience to global challenges, 12th Biennial Conference of the International Association for the Study of Commons, Cheltenham* (pp. 14–18).

Green News. (2021, 1 July). Seven Years of NLPB — What Has Progressed, What Hasn't, and the Latest Challenges. Retrieved from https://greennews.agency/?p=24238 Accessed on 30 November 2024

Green News. (2023, August 31). *Separation of the community area before announcing the area of Tham Pha Thai National Park, the Northern Peasant Federation and P-move called for it.* Retrieved November 30, 2024, from https://greennews.agency/?p=35384

Green News. (2024, November 12). *Cabinet approves 2 draft forestry laws amid opposition - No more legal action against villagers*. Retrieved November 30, 2024, from https://greennews.agency/?p=39290

Hall, D., Hirsch, P., & Li, T. M. (2011). *Introduction to powers of exclusion: Land dilemmas in Southeast Asia*. National University of Singapore Press and University of Hawaii Press.

Hares, M. (2009). Forest conflict in Thailand: Northern minorities in focus. *Environmental Management, 43*, 381–395.

Hirsch, P. (1990). Forests, forest reserve, and forest land in Thailand. *Geographical Journal, 156*, 166–174.

Hirsch, P. (1997). Community forestry revisited: Messages from the periphery. In M. Victor, C. Lang, & J. Bornemeier (Eds.), *Community forestry at a crossroads: Reflections and future directions in the development of community forestry. Proceedings of an International Seminar, Bangkok, Thailand, 17–19 July 1997*. RECOFTC Report No. 16.

Hirsch, P., & Lohmann, L. (1989). Contemporary politics of environment in Thailand. *Asian Survey, 29*(4), 439–451.

IMN Voices. (2024, September 25). *The House of Representative withdrew the draft—Rejected the definition of "indigenous peoples"*. Retrieved November 30, 2024, from https://imnvoices.com/?p=5160

Kanazawa, H., Hoshikawa, K., & Nawata, E. (2006). The Changes in Land Use and Agricultural Systems of Karen People in North Thailand—A Case Study of Sam Sop Bon Village—. *Japanese Journal of Tropical Agriculture, 50*(3), 133–141.

Karlsson, B. G. (2011). *Unruly hills: A political ecology of India's northeast*. Berghahn Books.

Kikhi, K., & Ghosh, J. (2016). Intricate Interface of land, identity and development in Nagaland state. In L. Rymbai et al. (Eds.), *Politics of location and identity in Northeast India*. Heritage Publishing House.

Laungaramsri, P. (2005). Swidden agriculture in Thailand: Myths, realities and challenges. *Indigenous Affairs, 2*, 6–13.

Leblond, J. P., & Pham, T. H. (2014). Recent forest expansion in Thailand: A methodological artifact? *Journal of Land Use Science, 9*(2), 211–241.

Li, T. M. (2002). Engaging simplifications: Community-based resource management, market processes and state agendas in upland Southeast Asia. *World Development, 30*(2), 265–283.

Longkumer, A. (2015). 'As our ancestors once lived': Representation, performance and constructing a National Culture amongst the Nagas of India. *Himalaya, 35*(1), 51–64.

Martemjen. (2017). *Biodiversity conservation, indigenous knowledge and practices: A Naga perspective*. Notion Press.

Nayak, P. K., & Berkes, F. (2008). Politics of co-optation: community forest management versus joint forest management in Orissa, India. *Environmental management, 41*(5), 707–718.

Nienu, V. (2015). *Naga cultural milieu: An adaptation to mountain ecosystem*. Dorylus Publishing Group.

Papu, S., & Nathani, N. (2020). India's first green village–Khonoma. *International Journal of Environmental Science and Development, 11*(1), 21.

Policy Watch. (2024, February 8). *Draft of the protection of Ethnic groups' livelihoods Act provides the land rights for farming and access to natural resources*. Retrieved November 30, 2024, from https://policywatch.thaipbs.or.th/article/life-14

Rawat, M. S., Zhimo, V. V., & Imchen, N. (2018). Deforestation in Nagaland, North-East India: Causes, effects and subsequent environmental degradation-A preliminary review. *Global Journal of Current Research, 6*(3).

RECOFTC. (2021, September 17). *Special report from the online seminar "7 years of NCPO policy, progress and challenges"*. Retrieved July 22, 2024, from https://www.recoftc.org/thailand/stories/รายงานพิเศษจากเสวนาออนไลน์-"7-ปี-นโยบาย-คทช-ความก้าวหน้าและข้อท้าทาย"

Roth, R. J. (2008). "Fixing" the forest: The spatiality of conservation conflict in Thailand. *Annals of the Association of American Geographers, 98*(2), 373–391.

Royal Forest Department (RFD), Community Forest Management Office. (2022, November 21). *Community Forest Act 2019*. Retrieved March 23, 2024, from https://www.forest.go.th/communityexpert/2022/11/21/พรบ/

Salam, M. A., Noguchi, T., & Pothitan, R. (2006). Community forest management in Thailand: Current situation and dynamics in the context of sustainable development. *New Forests, 31*, 273–291.

Sarin, M. (1997). Grassroots initiatives versus official responses: The dilemmas facing community Forest Management in India. In M. Victor, C. Lang, & J. Bornemeier (Eds.), *Community forestry at a crossroads: Reflections and future directions in the development of community forestry. Proceedings of an international seminar, Bangkok, Thailand, 17–19 July 1997.* RECOFTC report no.16.

Sarma, A. (2022). Can shifting cultivation ensure local food sovereignty? *A Tradition in Transition, 56.*

Sarma, J., Imti, T. T., & Yaden, S. (2019). Domestic firewood consumption pattern and degradation of Forest cover: A case study of Mokokchung town, Nagaland. *International Journal of Science and Research (IJSR), 8*(9), 7.

Seub Nakhasathien Foundation. (2020, August 27). *Report on the Thai forest situation for the year 2019–2020*. Retrieved August 10, 2024, from https://www.seub.or.th/document/สถานการณ์ป่าไม้ไทย/รายงานสถานการณ์ป่าไม้ไ-6/#:~:text=มีการสำรวจเสร็จใน,เนื้อที่ประมาณ%204%2C295%2C501.24%20ไร่%20โดย

Sumi, K.H. (2025). Naga Community Conservation of Biodiversity for A Sustainable Future. *International Journal of Innovative Research in Technology, 12*(6):3157–3165.

Thai PBS. (2019, November 27). *Retracing the path of 28 years: Thailand's first community forest law*. Retrieved June 10, 2024, from https://www.thaipbs.or.th/news/content/286506

The Morung Express. (2022, January 13). *Nagaland reports 235 sq km loss in forest cover: ISFR 2021*. Retrieved July 22, 2024, from https://morungexpress.com/nagaland-reports-235-sq-km-loss-in-forest-cover-isfr-2021#:~:text=As%20per%20the%20report%2C%20Nagaland,among%20the%20highest%20in%20India

The Naga Republic. (2022, December 6). *NEPED introduction FPA rural development methodology in Nagaland*. Retrieved November 30, 2024, from http://www.thenagarepublic.com/news/nagaland-news/neped-introduces-fpa-rural-development-methodology-in-nagaland/

Theunuo, T. (2016). Poetics and politics of ethnic identity in the colonial and post-colonial Northeast India. In L. Rymbai et al. (Eds.), *Politics of location and identity in Northeast India*. Heritage Publishing House.

Thong, J. S. (2023). *Naga customary law*. Tseminyu District.

Tiwari, B. K. (2022). Shifting cultivation. In J. J. Wouters & T. B. Subba (Eds.), *The Routledge companion to Northeast India*. Routledge.

Trakansuphakorn, P. (2008). Space of resistance and place of local knowledge in Karen ecological movement of northern Thailand: The case of Pgaz K'Nyau villages in Mae Lan Kham River basin. *Japanese Journal of Southeast Asian Studies, 45*(4), 586–614.

Vandergeest, P. (1996a). Property rights in protected areas: Obstacles to community involvement as a solution in Thailand. *Environmental Conservation, 23*(3), 259–268.

Vandergeest, P. (1996b). Mapping nature: Territorialization of forest rights in Thailand. *Society & Natural Resources, 9*, 159.

Vandergeest, P. (2003). Racialization and citizenship in Thai forest politics. *Society & Natural Resources, 16*(1), 19–37.

Walker, A. (2001). The 'Karen consensus', ethnic politics and resource-use legitimacy in northern Thailand. *Asian Ethnicity, 2*(2), 145–162.

Walker, A. (2004). Seeing farmers for the trees: Community forestry and the arborealisation of agriculture in northern Thailand1. *Asia Pacific Viewpoint, 45*(3), 311–324.

Wittayapak, C., & Baird, I. G. (2018). Communal land titling dilemmas in northern Thailand: From community forestry to beneficial yet risky and uncertain options. *Land Use Policy, 71*, 320–328.

Yano, K., & Lanusosang, T. (2013). Globalisation and its impact on agriculture: An overview of Kohima District, Nagaland, India. *International Journal of Bio-resource and Stress Management, 4*(4), 651–654.

Chapter 5
Bio-Cultural Harmony: Jhum, Livelihood and Forest Conservation in Nagaland

Sentitula

Abstract Forest conservation in Northeast India is inextricably linked to the social and cultural fabric of indigenous communities. For centuries, customary laws and regulations have played a pivotal role in preserving the region's green cover, safeguarding its biodiversity, and supporting livelihoods. Effective conservation strategies must recognize and integrate these traditional systems of governance to ensure long-term success. In Nagaland, *jhum*, a traditional agricultural practice, deeply rooted in the region's cultural heritage, has faced criticism for its perceived negative environmental impacts. However, recent studies highlight its potential for sustainable land use management when practiced with appropriate techniques. The current study explores the historical context of *jhum* cultivation in Nagaland, its cultural significance, and its role in shaping the region's forests. It also examines the challenges posed by climate change, land degradation, and limited market access for *jhum* products. *Jhum* improvement strategies applied in test villages have demonstrated success in fostering community-based forest management and diversifying livelihoods. The ongoing efforts and comparative analyses of the implemented forestry model indicate that slash and burn is beneficial to soil health and agroforestry. Nagaland shows promise in achieving a harmonious balance between *jhum* cultivation, livelihood and forest conservation, ensuring a sustainable future for the region in the wake of climate-change.

Keywords *Jhum* · Livelihood · Forest conservation · Nagaland

---

Sentitula (✉)
Mokokchung, Nagaland, India

Department of Environment Forest and Climate Change, Government of Nagaland, Kohima, Nagaland, India

Y. Santasombat (ed.), *Bio-Cultural Diversity, Ethnicity, and Local Knowledge*,
https://doi.org/10.1007/978-981-95-4242-0_5

5.1 Introduction

Northeast India is a hotspot of diverse ecology and culture. Biodiversity, climate regulation, and livelihoods (Chakrabarti, 2024; Akhilraj & Sendhil, 2024) rely much on indigenous communities. For centuries, their unique systems of forest governance based on customary laws and traditions have helped protect and conserve forests. In recent decades, however, the region has witnessed significant changes in forest management practices, driven by population growth, economic development, and government policies.

Although, India's total forest and tree cover increased by 1540 square kilometers (sq. km) between 2019 and 2021 which is (24.62% of the geographical area), the five northeastern states (Arunachal Pradesh, Manipur, Meghalaya, Mizoram, and Nagaland) experienced a loss (India State of Forest Report, 2021) (Fig. 5.1). The major threats to forest cover include illegal logging particularly in Assam and Nagaland, conversion of forest land for agriculture and settlements, extreme weather events etc. The introduction of modern forestry concepts and the establishment of state-managed forest areas have led to both positive and negative impacts on the region's forests. Successful conservation initiatives, such as community-based forest management and protected areas, have contributed to the increase in forest cover in some states. In this regard, the evolution of *jhum* cultivation or shifting agriculture has been pivotal in the forest management strategies of the region.

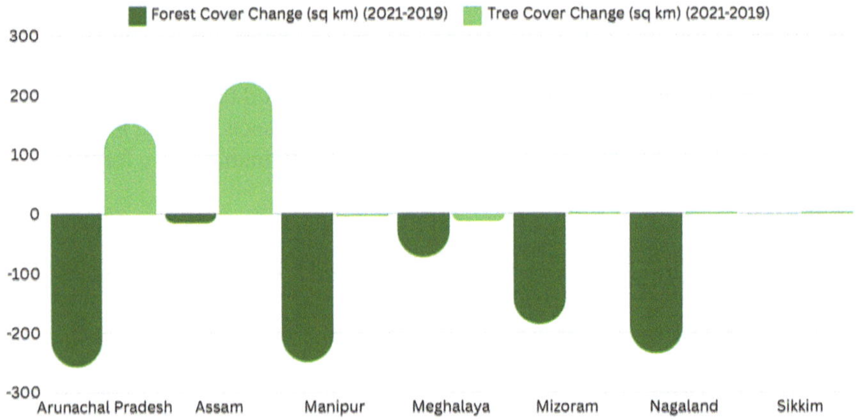

Fig. 5.1 Change in forest cover and tree cover in Northeast India (ISFR, 2021)

5.2 Historical Intricacies of *Jhum* or Shifting Cultivation

Jhum cultivation, a traditional agricultural practice in Northeast India, has a long history dating back thousands of years. As noted by Heinimann et al. (2017), it may have originated in Southeast Asia and spread to other regions. In Northeast India, *jhum* has been deeply intertwined with social cohesion and community building (Edake, 2013). While *jhum* cultivation has faced criticism for its perceived negative impacts on the environment, recent studies have highlighted its potential for sustainable land use management. When practiced with appropriate fallow periods and land management techniques, *jhum* can contribute to biodiversity conservation, soil health, and carbon sequestration (Aier & Changkija, 2003; Fox, 2000; Nielsen et al., 2006).

However, *jhum* cultivation is not without its challenges. Climate change, including rising temperatures, erratic rainfall, and glacier retreat, has significantly impacted the region's agriculture, affecting water availability, crop yields, and soil erosion (IPCC, 2021; Nepal & Bhatta, 2019; Sharma et al., 2022). These challenges have led to crop failures, reduced agricultural productivity, and food insecurity for tribal farmers in Northeast India.

To address these challenges and promote sustainable land use, innovative approaches and policy interventions are necessary. Diversifying into cash crops and exploring alternative livelihood opportunities can help reduce pressure on *jhum* lands and ensure sustainable land and livelihood management. By understanding the complexities of *jhum* cultivation and implementing appropriate strategies, Northeast India can continue to practice this traditional agricultural system while promoting sustainable development and conserving its valuable forests. This paper explores the complex interplay between *jhum* cultivation, traditional livelihoods, and forest conservation in the northeastern state of Nagaland.

Nagaland, is a distinctive Indian state established through the Sixteen-Point Agreement signed in July 1960 between the Prime Minister of India and representatives of the Naga People's Convention. This agreement paved the way for Nagaland's creation as the 16th state of the Indian Union on December 1, 1963. Nagaland's cultural and governance systems are shaped by its historical agreement with the Indian Union. Unlike other Indian states, Nagaland's customary laws and procedures hold supremacy over national statutes, thanks to Article 371(A) of the Constitution. This cultural distinctiveness is further highlighted by the diverse tribal governance systems, ranging from near-dictatorship to extreme democracy.

For example, the Konyak tribe follows an autocratic system, with 'commoners' and powerful Chiefs (Anghs). The Sema tribe has a hereditary chieftainship system, where the Chief's words are considered law. In contrast, the Ao tribe has a 'republican' system, with citizens actively involved in governance through the Village Council. The Angami tribe practices an 'extreme democratic' system, prioritizing consensus.

Each Naga Village was like an 'independent sovereign republic,' owning and governing its resources, planning development activities, maintaining law and order, delivering justice, and securing defense. Villages also forged diplomatic relationships and made treaties to share or jointly manage resources in contiguous areas.

5.3 Forest Management in Nagaland

In 2019, Nagaland reported a total forest cover of 12,486.40 sq km constituting 75.31 % of its geographical area and tree cover of 362 square kilometers. The total growing stock in Nagaland's forests is 57.71 million m^3. Twenty percent of the growing stock is classified as Tree Out of Forest (TOF), while 64.25 percent of forests is the growing stock in recorded forest area. Almost 10 percent of the forests in Nagaland are classified as very dense and have a canopy cover of more than 70 percent, while moderately dense forest with a canopy cover of 40 percent to 70 percent is estimated at 36 percent of the total forests. By 2021 the forest cover declined by 235.4 square kilometers (forest cover constituting 73.90% of its geographical area) and while the tree cover increased by three square kilometers (ISFR, 2019, 2021). Over 80% of forests are community-owned and managed by traditional village councils with only 11.7% owned by the state (Table 5.1). The conservation of common pool resources by the local communities are initiated by various local institutions, youth organizations/women groups, religious/cultural groups and village councils.

The decline in forest and tree covers underlines the need to integrate modern forest management practices with community efforts to address the challenges. Nagaland's forests also offer a rich variety of Non-timber forest products (NTFPs), including wild fruits, vegetables, mushrooms, and medicinal plants. Favorable

Table 5.1 Key statistical details on forests in Nagaland

	Statistic	Value	Source
1	Land ownership (government)	11.70%	Nagaland Forest Department
2	Nagaland land ownership (community)	88.30%	Nagaland Forest Department
3	Number of JFMCs	941	Nagaland Forest Department
4	Working schemes prepared	114	Nagaland Forest Department
5	Forest cover	73.90%	SFR 2021
6	Growing stock in recorded forest area	30.28%	SFR 2021
7	Growing stock in trees outside forest area	15.60%	SFR 2021
8	Community reserves	189 in 1089.93 sq. km	Nagaland Forest Department
9	Community conservation areas	407 nos	Nagaland Forest Department

agro-climatic conditions make medicinal herbs and plants abundant in the hilly regions. Several commercially important medicinal plants grow in the state, offering opportunities for income generation. Nearly 60% of the population living near forests relies on NTFPs for sustenance.

A three-tier system consisting of Working Groups: State Forest Development Agency (SFDA), Forest Development Agencies (FDA), and Joint Forest Management Committees (JFMCs) oversees forest management. The Nagaland Forest Act, 1968, and other regulations govern forest management and tree felling. The state government has implemented initiatives to promote sustainable forest management, including the establishment of Joint Forest Management Committees (JFMCs) and Working Schemes for scientific forest management. Despite these efforts, the state faces challenges from human-induced forest fires and the impact of jhum cultivation.

Nagaland has 407 documented Community Conservation areas. The Forest department having realized the importance of community involvement in management, protection and conservation of Forest and Wildlife outside Protected area, has initiated the declaration of Community Reserves under the Wildlife Protection Act 1972 and many villages has taken up this initiative of declaring their area as community Reserves and as of now 189 Community Reserves has been notified by the Government of Nagaland.

5.4 *Jhum* Cultivation: A Way of Life in Nagaland

In Nagaland, *Jhum* cultivation is practiced by over 60% of the population, covering more than 50% of the state's land area. It provides approximately two-thirds of the state's food requirements. *Jhum* land is typically owned by the entire village community, with individual rights for temporary cultivation. Over the years it has evolved as a relevant and adaptive food production practice due to the region's challenging climatic and geographical conditions. Farmers have developed various soil and water conservation measures to sustain *jhum* cultivation. The average fallow period for *jhum* in Nagaland is eight to nine years, which is higher compared to other northeastern states. Naga tribes possess indigenous knowledge of managing fallow periods and regenerating tree species.

5.4.1 Jhum Improvement Approaches in Nagaland: A Case Study

It has been noticed that *Jhum* areas have been abandoned due to variety of reasons and this provides emerging space for forestry in village landscape particularly around *jhum* land—with *jhum*, in fallow, and in discontinued *jhum* blocks and this space can be utilized to expand area under tree cover, improve tree density, and diversify sources of income at household and community level. This will also aid livelihood and food security and mitigate and adapt to climate change vulnerabilities.

And therefore, under the Nagaland Forest Management Project (NFMP), *jhum*-improvement approaches have been adopted instead of *jhum*-alternative, in areas where *jhum* cultivation is practiced, to boost ecological, cultural and economic return from *jhum*. This project is a comprehensive initiative aimed at improving forest ecosystem, supporting income generation, and promoting sustainable development in Nagaland. The project is supported by the Japan International Cooperation Agency (JICA) and implemented by the Department of Environment, Forest, and Climate Change, Government of Nagaland.

The NFMP designs **five forestry models/plantation models** to address different land use scenarios and achieve its objectives:

1. ***Jhum* Agroforestry (JAF):** This model involves integrating tree species into existing *jhum* cultivation systems to enhance biodiversity and provide additional income sources for farmers.
2. ***Jhum* Fallow Forestry (JFF):** This model focuses on planting trees in fallow *jhum* lands to promote forest regeneration and prevent soil erosion.
3. ***Jhum* Conversion to Forestry (JCF):** This model involves converting abandoned *jhum* lands into forest plantations to increase tree cover and improve the ecological condition of the area.
4. ***Jhum* Conversion to Conservation Area (JCC):** This model aims to establish community-managed conservation areas in former *jhum* lands to protect biodiversity and promote sustainable land use practices.
5. **Protection or Expansion of Existing Community Conservation Area (PEC):** This model focuses on protecting and expanding existing community-managed conservation areas.

Table 5.2 Longmisa JAF Model

Species planted	Average Ht. (ft) as on Oct, 2023	Survival rate
Duabanga grandiflora (Khokon)	15	95%
Choerospondias axillaris (Ait)	16	
Melia azedarach (Asa)	21	
Chukrasia tabularis (Aichangtong)	16	
Phoebe sp. *(Melen/Moo)*	17	

Table 5.3 Sungratsu JAF Model

Species planted	Average Ht. (ft) as on Oct 2023	Survival rate
Duabanga grandiflora (Khokon)	11	90%
Choerospondias axillaris (Ait)	21	
Toona ciliata (Sazü)	15	
Hibiscus macrophyllus (Akha)	10	
Baccaurea ramiflora (Tangshi)	17	

5.4.2 Comparative Study of Plantation Models

Longmisa JAF Model—Plantation year for Longmisa village was 2021 (Table 5.2) covering an area of 56.98 Ha. From the JFMC nursery, saplings were distributed to 32 beneficiaries doing *jhum* cultivation in the model area. Besides that, saplings were also distributed to villagers undertaking plantation outside the model area. Artificial plantation (AR) along with naturally regenerated saplings (ANR) were maintained. Gap plantations were done in subsequent months and year.

Sungratsu JAF Model—Plantation year for Sungratsu village was 2021 (Table 5.3) covering an area of 55.8 Ha. From the JFMC nursery, saplings were distributed to 86 beneficiaries cultivating the model area. Artificial plantation along with naturally regenerated saplings were maintained. Gap plantations were done in subsequent months and year.

5.4.3 Comparative Analysis of Jhum *Improvement Models*

Based on observations in Nagaland, the *jhum* improvement models show varying effectiveness (Fig. 5.2):

1. ***Jhum* Agroforestry Model (JAF):** This model appears to be the most successful in current *jhum* areas. The burning of biomass during *jhum* practices contributes to soil fertility and reduces weed competition, leading to comparable growth rates for both artificially planted (AR) and naturally regenerated (ANR) saplings.
2. ***Jhum* Fallow Forestry Model (JFF):** Due to several years of fallow, JFF areas can be challenging to manage. High rainfall and favorable conditions promote weed growth, making maintenance difficult. Here, ANR saplings tend to outperform AR saplings.

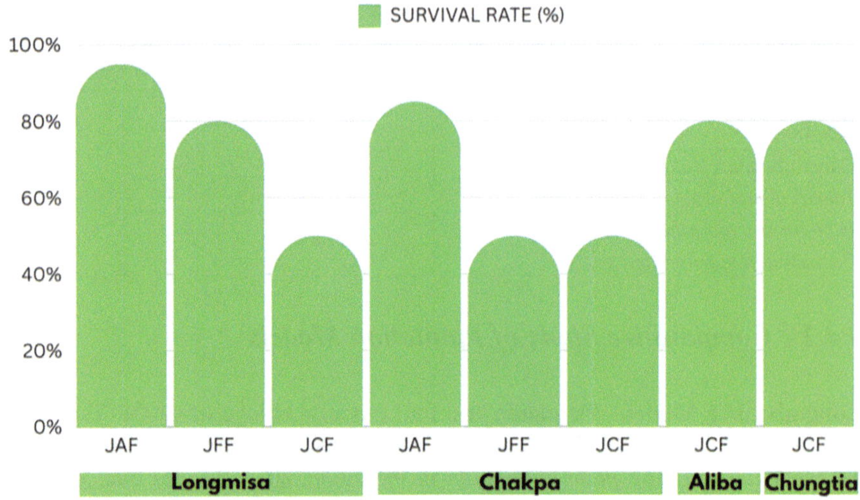

Fig. 5.2 Comparative study of the three models. *JAF*, *Jhum* Agroforestry Model; *JFF*, *Jhum* Fallow Forestry Model; *JCF*, *Jhum* Conversion to Forestry Model

3. ***Jhum* Conversion to Forestry Model (JCF):** In JCF areas with longer fallow periods and abundant natural regeneration, artificial regeneration struggles with high mortality rates. However, controlled burning and clearing brushwood before planting, as observed in Aliba and Chungtia villages, can significantly improve survival rates (reaching 80%).

These findings suggest that the JAF model performs better than JFF and JCF models in most cases. *Jhum's* traditional slash-and-burn practice, though not ideal for long-term sustainability, provides initial benefits for plantations by enhancing soil fertility and suppressing weeds, leading to better overall plantation cover.

5.5 Livelihood Enhancement among *Jhum* Farmers

According to Sharma et al. (2022), Colocasia cultivation has the potential to complement *jhum* cultivation in Nagaland by providing a more stable and reliable source of income for farmers. Colocasia is a high-yielding crop that can be grown on a variety of soils and is tolerant to drought and flooding. It can also be harvested multiple times per year, providing a continuous source of income for farmers. Also known as taro, it is a tuberous root vegetable that is native to Southeast Asia. It is a hardy crop that is well-suited to tropical and subtropical climates, making it a potential alternative to *jhum* cultivation in Nagaland. It produces abundant crops even in marginal soils, is rich in vitamins, minerals, and carbohydrates. Cassava, another tuberous root vegetable, is drought-tolerant, produces abundant crops, can be used for food, animal feed, and industrial products.

5.5.1 Colocasia Cultivation and Anishi Production in Sungratsu Village

The Sungratsu Self-Help Group (SHG) in Mokokchung district, Nagaland successfully integrated colocasia cultivation with traditional *jhum* farming practices. This case study explores the process of cultivating colocasia and producing Anishi, a traditional fermented taro leaf product, highlighting the economic benefits and potential for sustainable livelihoods.

The SHG prepared the land for colocasia cultivation from December to March. Colocasia saplings were sown in the prepared fields during March and April. Regular weeding was carried out in May to maintain crop health. Colocasia leaves were harvested from June to November and processed into Anishi, a fermented product.

Anishi is a unique Naga dish made from fermented taro leaves. The process involves several stages:

- **Harvesting:** Fresh taro leaves are collected from the fields.
- **Fermentation:** The leaves are fermented in a traditional process.
- **Pounding:** The fermented leaves are pounded into a paste.
- **Drying:** The paste is dried to form Anishi.

The Sungratsu SHG generated a total output of Rs. 3,30,000 from Anishi production (in one cropping) resulting in a profit of Rs. 1,80,590. This demonstrates the potential for sustainable and profitable livelihoods in Nagaland through value-added agricultural products like Anishi.

This case study highlights the successful integration of colocasia cultivation with *jhum* farming practices in Nagaland. The production of Anishi, a traditional fermented product, offers a valuable source of income for local communities. By diversifying agricultural activities and exploring value-added products, farmers can promote sustainable livelihoods and economic development.

5.5.2 Cassava Plantation (Yaongyimsen SHG)

The Yaongyimsen SHG in Mokokchung district, Nagaland successfully implemented a cassava farming project alongside their traditional *jhum* cultivation practices. The SHG planted 3000 cassava saplings on 1 hectare of land. Members worked together on weeding and cleaning the cassava fields. Cassava was harvested regularly and processed into chips. Each SHG harvested approximately 14,000 kgs of cassava tubers (in one cropping). The cassava tubers were sold at Rs. 10/- per kg, generating a profit of Rs. 1,40,000 per group.

This case study demonstrates the potential for profitable cassava farming. The Yaongyimsen SHG's success highlights the benefits of diversification and community-based agricultural initiatives.

5.6 Conclusion

The success of forest conservation depends on a coordinated approach that involves government agencies, NGOs, local communities, and individuals working together towards a common goal. In Northeast India, illegal logging, land encroachment, and shifting cultivation practices contribute to the decline of forest cover. Climate change, with its impacts on rainfall patterns and temperatures, also poses a major threat to the region's forests. A range of strategies are being implemented in the region including Community-Based Forest Management (CBFM) for empowering local communities to manage and protect forests sustainably, establishing national parks, wildlife sanctuaries, and community reserves to safeguard biodiversity, promoting sustainable agriculture practices that reduce deforestation and land degradation, raising awareness among local communities about the importance of forest conservation and sustainable livelihoods and implementing effective policies and regulations to address illegal logging, land encroachment, and other threats to forests.

In the process of transitioning *jhum* lands, the key benefits must be considered over misconceived notions. *Jhum* or shifting cultivation offers the following advantages:

1. **Nutrient replenishment:** The burning of vegetation during the slash-and-burn phase releases ash, which contains essential nutrients like potassium and phosphorus. These nutrients can improve soil fertility, promoting crop growth and productivity (Das et al., 2021).
2. **Weed control:** The burning process also destroys weeds, reducing competition for resources and improving crop yields (Oliver Springate-Baginski, 2013).
3. **Pest control:** Some pests may be killed or reduced by the fire, helping to control populations and prevent crop damage (Morton et al., 2020).
4. **Soil erosion prevention:** When practiced with appropriate fallow periods, *jhum* cultivation can help prevent soil erosion by allowing the land to recover and regenerate natural vegetation (Nielsen et al., 2006).
5. **Biodiversity conservation:** If managed sustainably, *jhum* cultivation can create a mosaic of habitats, promoting biodiversity and supporting ecosystem services (Aier & Changkija, 2003).
6. **Improved soil structure:** The burning process can break down organic matter, improving soil structure and water infiltration (Fox, 2000).

For maximizing the benefits of *jhum* cultivation while minimizing its negative impacts, key approaches supported by relevant research may be summarized as below:

1. **Longer fallow periods:** Increasing the fallow period between *jhum* cycles allows for natural forest regeneration and soil recovery. Studies have shown that longer fallow periods can improve soil fertility and biodiversity (Fox, 2000; Nielsen et al., 2006).

2. **Agroforestry:** Incorporating trees into *jhum* fields can provide shade, reduce soil erosion, and enhance biodiversity. Agroforestry systems can also generate additional income for farmers through the sale of forest products (Giri & Rawat, 2016).
3. **Soil Conservation Practices:** Implementing soil conservation measures, such as contour farming, terracing, and cover cropping, can help prevent soil erosion and maintain soil fertility (Sharma et al., 2022).
4. **Community-Based Forest Management (CBFM):** Empowering local communities to manage their forests can promote sustainable land use practices and reduce deforestation (Rivera, 2025).
5. **Diversification of Livelihoods:** Encouraging farmers to diversify their livelihoods beyond *jhum* cultivation can reduce pressure on forest lands and improve food security (Shaw et al., 2022).
6. **Sustainable *Jhum* Practices:** Promoting sustainable *jhum* practices, such as avoiding excessive burning and using appropriate tools, can minimize environmental damage (Oliver Springate-Baginski, 2013).
7. **Policy Support:** Implementing supportive policies that promote sustainable *jhum* cultivation and address the challenges faced by *jhum* cultivators is essential. This includes providing technical assistance, financial incentives, and access to markets (Schröder et al., 2024).

The case studies demonstrated in Nagaland show *jhum* cultivation has remained an integral part of its biocultural heritage. As a vital component of the region's economy, conservation and livelihood cannot be possible without involving *jhum*. The roadmap adopted aims to achieve a holistic balance between traditional practices, community involvement, and sustainable land use strategies. As climate change continues to impact the region, Nagaland strives to innovate solutions keeping indigenous knowledge as the foundation of the conservation strategies.

The government initiatives of *Jhum* Agroforestry Model and community-based forest management programs have shown promise. The vision is to shift the reins of conservation back to the local communities, to restore that sense of stewardship. To achieve this vision, longer fallow periods, agroforestry, and soil conservation measures (Saikia et al., 2024) would strengthen community-based forest management programs and empower local communities to participate in decision-making, improve market access for *jhum* products and support value addition to enhance income generation. Supportive policies that promote sustainable *jhum* cultivation and address the challenges faced by *jhum* cultivators will facilitate the process. Farmers require provisions for technical assistance, financial incentives, and training (Schröder et al., 2024) to diversify livelihoods through agroforestry, livestock rearing, and handicrafts. Community-based initiatives such as JFMCs may be strengthened and investments must increase on research to understand climate change impacts and develop effective adaptation strategies (Borah et al., 2024).

Traditional Practices and Community Involvement Are Essential for Effective Conservation and Livelihood Promotion
Sustainable management of *jhum* and forests cannot be achieved in isolation. It is pertinent to recognize the value of traditional practices and fostering community-led initiatives for sustainable future where *jhum* cultivation and forest conservation go hand in hand, mutually supporting each other.

Slash-and-Burn: A Model of Promise in Nagaland
The slash-and-burn approach, often associated with negative environmental impacts, can be a sustainable land management practice when implemented responsibly. In Nagaland, this method involves clearing a small area of forest, burning the debris, and cultivating crops for a few years before allowing the land to fallow. When practiced with appropriate fallow periods and careful land management, slash-and-burn can contribute to forest regeneration, soil health and biodiversity conservation. By balancing agricultural needs with the restoration of forest ecosystems, this approach can offer a viable option for sustainable livelihoods in the region.

Balancing Livelihood and Conservation
Livelihood can be supported in a coordinated manner as demonstrated by the case studies of the Sungratsu SHG, Yaongyimsen SHG for Colocasia and Cassava cultivation and the plantation models. The Anishi project, which focuses on the sustainable production and marketing of fermented taro leaves, is an excellent example of how livelihoods can be supported in a coordinated manner. By providing training, resources, and market access to local communities, the project has helped to revitalize traditional food production and create new economic opportunities. The Anishi project demonstrates the potential for sustainable development that integrates cultural heritage, local livelihoods, and environmental conservation.

By addressing these challenges Nagaland can ensure a sustainable future for *jhum* cultivation, promote biodiversity conservation, and improve the livelihoods of its people serving as a model for the rest of the northeastern states.

References

Aier, M. E., & Changkija, T. (2003). Jhum cultivation and its impact on the environment in Nagaland, India. *Journal of Human Ecology, 43*(3), 167–174.

Akhilraj, M., & Sendhil, R. (2024). Forest Management Practices in India in the Context of Sustainable Development Goals–Trends and Prospects. Sustainable Development Goals: Challenges and Initiatives, Ed. Panch. Ramalingam, Shaheed Khan, andPradeep Kumar, 374–387.

Borah, D., Rout, J., & Nooruddin, T. (2024). Traditional knowledge-based sustainable agriculture in the Eastern Himalayas in India. In A. Borthakur & P. Singh (Eds.), *Addressing the climate crisis in the Indian Himalayas* (pp. 57–78). Springer. https://doi.org/10.1007/978-3-031-50097-8_4

Chakrabarti, A. (2024). *Indigeneity, development and sustainability: Perspectives from Northeast India*. Springer Nature.

Das, A. K., Mudi, R., Behera, P. K., & Sarma, K. K. (2021). Mapping shifting cultivation in the Indian Himalayas using remote sensing and GIS. *Journal of Mountain Science, 18*(1), 1–17.

Edake, S. N. (2013). *A tradition in transition: Understanding the role of shifting cultivation for sustainable development of Northeast India*. Unpublished doctoral dissertation, Tezpur University.

Fox, J., Truong, D. M., Rambo, A. T., Tuyen, N. P., Cuc, L. T., & Leisz, S. (2000). Shifting cultivation: a new old paradigm for managing tropical forests. *BioScience, 50*(6), 521–528.

Giri, G. S., & Rawat, P. S. (2016). Traditional agricultural practices in the Indian Himalayas: A review. *International Journal of Current Microbiology and Applied Sciences, 5*(1), 1–10.

Heinimann, M., Mertz, O., Frolking, S., et al. (2017). The future of shifting cultivation in Southeast Asia: A review. *Global Environmental Change, 45*, 1–12.

India State of Forest Report 2019. (2019). Forest Survey of India.

India State of Forest Report 2021. (2021). Forest Survey of India.

IPCC. (2021). *Climate change 2021: The physical science basis. Contribution of Working Group I to the Sixth Assessment Report of the Intergovernmental Panel on Climate Change*. Cambridge University Press.

Morton, J. F., Borah, D., Edwards, M., et al. (2020). Shifting cultivation in Northeast India: A review. *Journal of Human Ecology, 67*(1), 1–10.

Nepal, S., & Bhatta, R. (2019). Climate change and rural out-migration in the Himalayas.

Nielsen, U., Mertz, O., & Noweg, G. T. (2006). The rationality of shifting cultivation systems: labor productivity revisited. *Human Ecology, 34*(2), 201–218.

Rivera, R. (2025). Integrating Agroforestry and Community-based Forest Management (CBFM) for Sustainable Land Use and Social Equity: A Mini Review. *Asian Journal of Research in Agriculture and Forestry, 11*(4), 68–77.

Saikia, B., Nookathoti, T., & Hiremath, C. (2024). Alder-based shifting cultivation in Nagaland – A theoretical perspective. In K. Kanga, S. K. Singh, K. Shevkani, V. Pathak, & B. Sajan (Eds.), *Transforming agricultural management for a sustainable future* (pp. 157–180). Springer. https://doi.org/10.1007/978-3-031-63430-7_7

Schröder, L. S., Bhalerao, A. K., Kabir, K. H., Scheffran, J., & Schneider, U. A. (2024). Managing uphill cultivation under climate change–An assessment of adaptation decisions among tribal farmers in Nagaland state of India. *Journal of Environmental Management, 349*, 119473.

Sharma, A., Shashni, S., & Rathore, S. (2022). Traditional knowledge system for sustainable agriculture practices of rural communities of North-Western Himalaya, India. In S. C. Rai & P. K. Mishra (Eds.), *Traditional ecological knowledge of resource management in Asia* (pp. 229–254). Springer. https://doi.org/10.1007/978-3-031-16840-6_12

Shaw, A. K., Natori, Y., & Edake, S. (2022). *A tradition in transition: Understanding the role of shifting cultivation for sustainable development of North East India*. TERI.

Springate-Baginski, O. (2013). Shifting cultivation in Northeast India: A traditional practice in transition. In K. Maikhuri, R. S. Rao, & P. Rao (Eds.), Shifting Cultivation, Livelihood and Food Security: New and Old Challenges (pp. 21–46). New Delhi: TERI Press.

State Forest Report, Forest Survey of India. (2021). India State of Forest Report 2021. Ministry of Environment, Forest and Climate Change, Government of India.

Chapter 6
Rice-Cum-Fish Agricultural Practices, Cultural Landscape, and Agroecology in Nagaland: A Reflection for Northern Thailand

Warong Wonglangka, Repakaba Tzudir, Pandin Ounchanum, Aumpika Amloy, Supakul Ruangwitthayanusorn, Phailin Thongthammachat, and Aranya Siriphon

Abstract This article explores traditional agricultural practices, focusing on the integration of wet rice cultivation and fish farming in Nagaland. Through a case study of Kigwema Village in Kohima city, the research examines the evolution and persistence of these practices to better understand Nagaland's cultural landscape and integrated agroecology. Our fieldwork employs an agroecological perspective, using a multidisciplinary approach that combines geographical landscaping, vernacular architecture, and anthropological studies to capture both environmental and cultural dimensions. This comprehensive approach provides deeper insights into the interaction between agricultural practices and cultural landscapes, highlighting how integrated agricultural ecosystems with diverse land use, plant species, and biodiversity have been sustained. The findings suggest that, despite challenges from modern technology, Nagaland's wet rice cultivation methods, particularly the resilient *Zabo* farming system—featuring small fishponds in wet rice fields—demonstrate the enduring value of merging traditional and modern knowledge in Naga agriculture. These Naga-based customs, which are still practiced today, could inform agricultural methods in Northern Thailand, where rice-cum-fish systems are just as significant. The study addresses the adaptability of integrating local knowledge with the natural environment and offers recommendations that could guide similar practices in Northern Thailand.

W. Wonglangka (✉) · P. Ounchanum · A. Amloy · S. Ruangwitthayanusorn · P. Thongthammachat
Faculty of Architecture, Chiang Mai University, Chiang Mai, Thailand
e-mail: warong.w@cmu.ac.th

R. Tzudir
Department of Sociology, Nagaland University, Meriema, Nagaland, India

A. Siriphon
Faculty of Social Sciences, Chiang Mai University, Chiang Mai, Thailand

Y. Santasombat (ed.), *Bio-Cultural Diversity, Ethnicity, and Local Knowledge*,
https://doi.org/10.1007/978-981-95-4242-0_6

Keywords Wet rice cultivation · Fish farming · Local knowledge · Agroecology

6.1 Introduction

Nagaland, one of the eight northeastern states of India, is bordered by Arunachal Pradesh and Assam to the north, Manipur to the south, the Kachin State of Myanmar to the east, and Assam to the west. It is primarily inhabited by the Naga tribes, with a population of 2.2 million spread across 16 districts. The Naga people, historically mountain dwellers, are often referred to as "A Cultural Heritage of the Nagas" (Neli, 2022). Their expertise in rice cultivation, particularly through the rice-cum-fish agricultural practices and terrace rice cultivation, has long been a defining feature of their cultural landscape. This indigenous technology allows for the transformation of steep mountains, otherwise unsuitable for farming, into terraces for rice cultivation.

The study explores the cultural landscape of the Naga people in Nagaland, a place where wet rice cultivation has been practiced for over 600 years. We examine the evolution and persistence of these traditional agricultural practices to gain a better understanding of the region's cultural characteristics and agroecology. Adopting an agroecological perspective (Leippert et al., 2020; Barrios et al., 2020; Rosset & Altieri, 2017), we focus on how the Zabo farming system—characterized by small fishponds in wet rice fields—integrates traditional and modern knowledge in Naga agriculture to achieve sustainable food production through ecological farming at the farm and household levels. This approach helps to merge traditional practices and ways of living that shape relationships with the natural world, addressing human values and environmental concerns.

Our academic team, comprising Thai and Naga scholars, conducted fieldwork in July 2023 in Nagaland, specifically in Kohima and Prek. We employed a multidisciplinary approach, incorporating geographical landscaping, vernacular architecture, and anthropological studies to integrate physical, environmental, and cultural dimensions. This paper focuses specifically on *Kigwema* Village in Kohima, where the resilient "Zabo farming system" exemplifies the enduring value of blending traditional and modern agricultural knowledge. It demonstrates how this integration could lead to sustainable, resilient, and culturally relevant farming practices.

By leveraging its multipurpose aspects, the Zabo farming system efficiently utilizes natural resources, ensuring long-term productivity while preserving the environment. Apart from incorporating diverse crops and fish, emphasizing crop rotation, and employing organic fertilizers to enhance soil health, this system integrates modern techniques for soil conservation. Additionally, the Zabo system effectively manages water resources through the strategic placement of fishponds and rice fields. By empowering local farmers to adapt to changing environmental conditions and market demands, it helps preserve cultural identity while boosting agricultural productivity. This integration ensures that cultural heritage is maintained alongside advancements in farming, allowing for practices that reflect both traditional values and contemporary challenges.

This perspective on the Zabo farming system provides a deeper understanding of how agricultural practices and cultural landscapes interact, emphasizing the ecological biodiversity sustained by integrated agricultural ecosystems with diverse land use, plants, and species. The findings indicate that, despite challenges posed by modern technology and chemical fertilizer, Nagaland's wet rice cultivation methods—particularly the resilient Zabo farming system—illustrate the lasting significance of combining traditional and contemporary knowledge in agriculture. These practices, deeply rooted in Naga traditions and still in use today, could inform agricultural methods in Northern Thailand, where rice-cum-fish systems are just as significant. The study highlights the flexibility of integrating local knowledge with the natural environment and offers recommendations that could guide similar practices in Northern Thailand.

6.2 The Nagas of Nagaland: A Comprehensive Overview

Nagaland, one of India's northeastern states, spans 16,579 square kilometers with elevations from 610 to 3826 meters above sea level. It shares borders with Arunachal Pradesh, Assam, Manipur, and Myanmar. The state's terrain is predominantly mountainous, characterized by three parallel ranges—the Patkai, Barail, and Naga Hills—culminating in Mount Saramati, with the highest peak at 3826 meters. Nagaland experiences a monsoon climate, with annual rainfall between 1800 and 2200 millimeters. Summers see temperatures from 16 °C to 31 °C, while winters can drop to 4 °C. The Naga tribes, numbering 17,[1] inhabit the state's 16 districts,[2] each with its own language, governance, and cultural identity.

The Naga tribes speak languages belonging to the Tibeto-Burman branch of the Sino-Tibetan language family. Historically, their oral traditions were central to their cultural transmission. The Naga people are believed to have migrated from northwestern China to northeastern India and Southeast Asia 10,000 to 12,000 years ago, sharing cultural practices with Austroasiatic groups like the Kachin, such as

[1] 17 tribes include Angami, Ao, Chang, Chakhesang, Dimasa, Kachari, Khiamniungan, Konyak, Kuki, Lotha, Phom, Pochury, Rengma, Sangtam, Sumi, Yimchunger, and Zeliang, which account for 97.47 percent of the total population in the state. Additionally, there are other sub-tribes, including Kuki, which make up 1.10 percent, and Garo, Kachari, and Mikir, which account for 0.91 percent.

[2] The 16 districts include Kohima (established in 1957), Tuensang (established in 1957), Mokokchung (established in 1957), Mon (established in 1973), Wokha (established in 1973), Phek (established in 1973), Zunheboto (established in 1973), Dimapur (established in 1997), Peren (established in 2004), Kiphire (established in 2004), Longleng (established in 2004), Noklak (established in 2020), Tseminyu (established in 2021), Niuland (established in 2021), Chumoukedima (established in 2021), and Shamator (established in 2022).

megalithic structures, ancestor worship, and headhunting (Jacobs, 2012; Kire, 2019; Oppitz, 2008).

The Naga are resilient mountain dwellers who adapted their social systems to their environment. Historically, their isolated village communities, each with a distinct language, were naturally fortified by the terrain. Inheritance followed a patrilineal system, with property passing to the youngest son. Their cosmology intertwines nature, humanity, and the supernatural, with a deep respect for nature and belief in the influence of supernatural powers (Hutton, 1921; Nienu, 2015; Nshoga, 2008; Das, 2023). Seeing all things as interconnected, Naga cosmology shaped their cultural practices. This includes rituals and headhunting (King, 1879), symbols of bravery, and social status. While cultural objects signified one's position, the quantity of human heads captured during raids represented a person's social standing. British colonization in the nineteenth century disrupted the Naga way of life, particularly headhunting, which was seen as barbaric. Christianity was introduced by American missionaries in 1872, gradually replacing traditional beliefs and practices with the Christian Church and establishing schools with Westernized education (Oppitz, 2008).

In Nagaland's agricultural landscape, traditional Naga farming primarily revolves around two systems. The first is "jhum cultivation" (slash-and-burn or shifting cultivation), which is prevalent in the northern and central regions of the state. This method, also known as alder-based farming, is favored due to the steep slopes and limited flat land that are unsuitable for paddy fields (Cairns et al., 2010; Paul et al., 2022). In *jhum* cultivation, fields incorporate alder trees (specifically, Alnus species) as part of a sustainable farming system. These fields are cultivated for a few years before being left to regenerate naturally, which promotes biodiversity and sustainability. The second system is terraced wet rice cultivation. Although it is more common in the southern parts of Nagaland, it has extended into northern Manipur. This technique utilizes gentler slopes, transforming them into terraced rice fields.

Both *jhum* cultivation and terraced wet rice cultivation involve intercropping, which is the practice of planting various crops together to increase production and land utilization. The Zabo system, which features small pits within paddy fields for fish rearing, is particularly integral to terraced wet rice cultivation. This approach demonstrates efficient resource use by adding a protein source to rice production areas.

These systems, regularly practiced today, create a mutually supportive ecosystem by integrating crops and fish. They play a crucial role in promoting food diversification and security while showcasing a sustainable approach to soil and water management (Murry & Lalruatsangi, 2019; Murry & Das, 2021; Solo & Kikhi, 2021). This significantly enhances food grain production in the state, exemplifying the combination of various agricultural activities to achieve balance across economic, social, and environmental dimensions. The Naga people's harmonious relationship with their environment is reflected in their sustainable agricultural practices, along with the spiritual significance they attach to their natural surroundings (Kithan, 2014).

6.3 Case Study: Kigwema Village, Kohima

Kigwema village, located in the Jakhama region of Kohima, is about 13 kilometers south of the city.[3] Nestled at the foothills of Japfü, one of the region's highest mountain ranges, this Angami tribe village lies at 25.6063°N latitude and 94.1271°E longitude, with an elevation ranging from 1450 to 1670 meters above sea level (see Fig. 6.1). The Mezierü River, which originates from Japfü Mountain, flows through Kigwema from southwest to northeast, merging with the Dzü-ü River in a dendritic pattern.

At an average elevation of 1585 meters, Kigwema is home to approximately 4724 residents across 769 households. The village is equipped with roads, footpaths, electricity, and a mountain water system with tanks for domestic use. It features two Christian churches and three schools, and it is divided into five Khels, each with its own communal space. The village is organized into five key areas. The

Fig. 6.1 Mapping Kigwema Village, its five Khels and households, and five key areas. Photo by Thunyaporn Thoopthimthean, 2024

[3] Kohima, the capital city of Nagaland, is the second-largest city in the state after Dimapur. The city's geographic coordinates range between 25°6'N and 27°4'N latitude and 93°20′E and 95°15′E longitude. Kohima exemplifies the typical Naga settlement pattern, sitting at an average elevation of 1444 meters above sea level and covering an area of 1463 square kilometers. Kohima experiences a temperate subtropical climate with substantial rainfall, particularly between August and September, averaging 2000 millimeters annually. The Dhansiri, Dikhu, Doyang, and Zungki rivers are the main waterways flowing through the city.

Fig. 6.2 An overview of the survey in Kigwema Village, Kohima, Nagaland. Photo by Warong Wonglangka, 2023

first area is Kigwema Gate, the village entrance, adorned with designs that reflect Naga craftsmanship and culture. Next is Morung, the traditional center for learning and cultural exchange. Elevated and surrounded by stone walls, the village courtyard serves as a communal space, which is often located near important buildings and monuments honoring village leaders. The village council hall is where community meetings and decision-making take place, playing an essential role in governance and unity. Finally, the village church acts as the spiritual center of the community (see Fig. 6.2).

The surrounding forests, home to ancient trees, provide habitats for wildlife and shade for plants. Kigwema's water supply comes from a stream in a distant forested military zone, as well as from local springs and rivers. These forests play a crucial role in managing water flow to lower-lying areas. Conservation efforts include community forests, where resources are harvested under shared agreements. Firewood collection and foraging are allowed in community forests, while conservation forests are reserved for specific times of the year. Additionally, these forests serve as sacred spaces, reflecting the Naga people's spiritual and cultural values.

6.4 Agricultural Practices in Kigwema Village

In Kigwema Village, like other villages in Nagaland, agriculture relies primarily on two systems: shifting cultivation, known as "jhum, alder-based farming," and terraced rice fields. These activities typically occur on the lower slopes of the hills, which are farther from the main settlement. Due to the distance of alder-based jhum areas from the village, many households focus on terraced rice fields and gardens closer to home. As for the ***jhum* cultivation**, which is situated at a distance from the village, the Kigwema villagers reserve these fields for economic crops, primarily cash crops like potatoes and chili peppers that have high market value and desirable traits. These agricultural products are known for their size and flavor, serving as a significant income source for the Angami people, with sales extending throughout India. **Terraced Rice Fields** refers to the terraced rice paddy system that has been maintained for over 600 years, employing traditional irrigation techniques and incorporating fish farming and the Zabo method. The terrain in this area is not characterized by large slopes; rather, the residential area is steeper than the rice fields.

6.5 Land Use and Resource Management in Kigwema Village

As for the jhum fields, alder-based farming, which is particularly important for sustaining natural resources, has been practiced by the Kigwema villagers. While more land is dedicated to cash crops, integrated shifting cultivation remains prevalent, blending traditional methods with modern practices. This approach includes the use of organic matter, crop rotation, and the integration of various crops to maintain soil fertility and productivity. Alder trees are pruned into low bushes to prevent shading and enhance nitrogen fixation (Cairns et al., 2010; Paul et al., 2022). Crops such as maize, chilies, potatoes, beans, and leafy vegetables rely on rainfall and cloud moisture. Cabbage and potatoes are especially prominent due to their adaptability to local soil and climate conditions, as well as market demand. Households manage resources by rotating crops, using organic inputs, and optimizing both local and distant cultivation areas. For instance, chili peppers are intercropped with basil, pumpkins are planted alongside maize, and mustard greens are grown under trellised beans. This practice maximizes space and increases yields without the use of chemical fertilizers or pesticides, aiming to balance agricultural productivity with sustainable resource use. Despite the distance, integrated shifting cultivation supports sustainable agriculture by combining traditional knowledge with modern techniques, helping to maintain soil health and manage resources efficiently (Kehie et al., 2017; Saikia et al., 2024).

Terraced Rice Fields For over 600 years, the terraced rice paddy system of Kigwima has been employing traditional irrigation techniques and incorporating fish farming and the Zabo method. The terrain on the lower slopes of the hills nearby is suitable in terms of a fine, clayey soil that is relatively uniform and free of stones. The rice fields are organized in a terraced pattern, reflecting the local topography. Bunds constructed from clay are raised to separate different sections of the rice fields. The steepness and design of the bunds, along with the designated pathways within the fields, are implemented for ease of movement. The sections of the fields with other crops, such as potatoes, chives, cabbage, and garlic, are planted in an intercropping system. The lush green of the rice fields and the presence of other crops are planted in different stages of growth. While small machinery like walking tractors is used, chemical fertilizers and pesticides are avoided. Soil fertility is enhanced by plowing in animal manure and crop residues after harvest. Only a few rice varieties are cultivated, selected for their suitability to the local environment. After rice is planted and harvested annually, fields are prepared for intercropping with potatoes, Chinese chives, cabbages, and garlic. Weeds removed during the rice-growing period are left to decompose in the water, enriching the soil.

In Kigwema, effective water management is crucial for rice cultivation, utilizing water sourced from streams that flow from the nearby mountains and surrounding forested areas. This sophisticated strategy involves careful diversion of water from these streams to ensure an adequate supply for multiple rice fields (see Fig. 6.3). To capture the water efficiently, farmers excavate shallow trenches that channel the flow into a primary waterway. This main waterway serves as the central supply line for various rice plots, each accessing water through small channels created within their bunds. The main waterway ultimately converges into a lower stream, where excess water from the fields is drained, maintaining a balanced water system.

The clayey soil in the region facilitates the excavation of traditional irrigation systems. Farmers construct small channels and utilize metal pipes to redirect water from elevated fields to lower ones. This method also allows for the management of excess water, which is directed to lower streams, ultimately gathering in a main channel that irrigates the rice fields. Bunds, built from clay, are raised

Fig. 6.3 The rice cultivation on the sloping terrain of Kigwima Village is linked with the surrounding forested areas with a sophisticated water management strategy. Photo by Warong Wonglangka, 2023

approximately 70 centimeters above the field level, following the contour of the land. While these bunds are effective for water retention, their height, coupled with their narrow and steep nature, can make walking along them challenging in certain areas. Designated walkways are established within the fields to aid movement. Water is introduced into the fields through the openings in the bunds, which can be sealed with a mixture of clay and plant debris when necessary. Additionally, metal pipes penetrate the bunds to facilitate the diversion of water from higher fields to adjacent lower ones. When sufficient water flow is achieved, these pipes are sealed to prevent overflow. The careful management of excess water ensures that it flows into the lower stream, promoting effective water management across the fields.

Integrated fish farming is achieved when fish farming is supported by the traditional irrigation system within the rice fields. During the dry and winter months, fish are housed in deep ponds and barriers made of nets embedded in the soil. Fish can swim into the fenced plots during the monsoon season due to increased water levels.

Forest Management In Kigwema, forest management is crucial. The Angami Nagas manage their forests through community forests, forest gardens, and sacred forests, all of which are subject to rules set by the community. Located closest to the village, the community forest provides firewood, particularly from the cherry blossom trees. The forest gardens and sacred forests are located approximately ten kilometers away.

Kitchen Gardens Kitchen gardens are vital for household vegetable consumption. Plants are grown in every available space around homes, often on the ground or between stones in the walls. Climbing vegetables, like chayote, are supported on structures above pathways. Soil management is less intensive than in rice fields, with weeds commonly used to retain moisture. Livestock, such as ducks, pigs, and chickens, provides manure to enrich the soil. Fruit trees, including peaches, pears, and pomelos, are planted around homes. Water for these gardens comes from rainfall and leftover household water, with plots typically located near washing areas.

6.6 The Zabo System, Rice Field, and Integrated Agroecology

The Zabo farming system primarily involves terraced wet rice cultivation. It utilizes resources efficiently and cyclically in an integrated manner, drawing from the air, soil surface, and subsoil, which leads to diverse and sustainable outputs deeply rooted in Naga traditions (Murry & Das, 2021). Originating in *Kikruma* village in the Phek district of Nagaland, India, the Zabo or *Zabü* farming system is characterized by a unique, indigenous water harvesting method that enriches the soil and enhances agricultural productivity, particularly for irrigating rice fields (Amenla & Shuya, 2021). The term "Zabo" in the Chakhesang dialect has often been inaccurately translated as "impounding runoff water," for it refers to a small pit within a paddy field that is primarily used for fish rearing. Conversely, "Ruza" denotes a larger pond, typically covering around 0.2 hectares, intended for storing runoff water.

According to Zanehu Tunyi (cited in Das, 11 August 2023), a septuagenarian former school principal and farmer from Kikruma village, "Ruza" means impounding water, runoff water pond, or tank for irrigation in the Chokri dialect. However, researchers and reports have popularized the term as the Zabo system. Over the years, the term "Zabo" has increasingly been used in reference to "Ruza," which has led to its widespread adoption in scholarly writings. Consequently, much of the literature produced by academics refers to this agricultural practice as the "Zabo system of farming."

Both interpretations of the two main fish farming systems in the rice field—the *Zabo* system and separate fishponds, which integrate multipurpose usage—have been represented in the case study of the *Zabo* farming system in Kigwema village.

Firstly, the Zabo farming system involves digging a deep pit at the lowest point of the rice field where water is collected (see Fig. 6.4). Fish fries are released into this pit, allowing them to feed and grow alongside the rice. As the rice matures, the fish remain in the pit, particularly when the fields are harvested and the water level decreases. This pond, used from the time the fish are juveniles, helps them become familiar with their environment, serving as both habitat and refuge. Additionally, the pond acts as a water reservoir, maintaining a stable water level during dry seasons to ensure the fish's survival. Dried branches are placed in the pond to provide hiding spots for the fish. This careful management reflects a deep understanding of ecological balance and natural resource conservation.

Secondly, separate fishponds are utilized, involving large ponds that may be located away from the rice cultivation areas. Situated alongside or within other fields used for growing rice, some rice fields are specifically adapted for fish farming (see Fig. 6.5). These ponds are dedicated solely to fish farming. The concept of separate fishponds integrates fish farming within the rice cultivation area without requiring deep excavation. Instead, water is retained at a depth similar to that of rice fields. Specific fish species that thrive in shallow water are selected and farmed in these ponds, with nets used to prevent escape and protect them from predators, like snakes.

These separate fishponds can be subdivided into multiple sections, allowing for the cultivation of different species, such as herbivorous and carnivorous varieties. Alongside the fishponds, shelters are typically constructed to provide rest areas for workers during the day. The surrounding area is often used for cultivating vegetables, which can be consumed during the workday or taken home. These two methods reflect the integration of fish farming within rice cultivation systems, utilizing both in-field and dedicated pond setups to optimize resource use and support sustainable agricultural practices.

In our study, Zabo fish farming in rice fields involves three key factors: the land, the rice variety, and the fish species. Firstly, it is crucial to identify the lowest point of the land in the rice field to dig a Zabo pond, which serves as a nursery for fish fry

Fig. 6.4 The Zabo method involves digging a clay pit at the lowest point of the rice field. This pond is used as a habitat and refuge since the fish are juveniles, allowing them to become familiar with their environment. The pond also serves as a water reservoir, maintaining a stable water level during dry seasons to ensure the survival of the fish. Photo by Warong Wonglangka, 2023

at the beginning and as a habitat for the fish when water levels drop during the harvest season. The field's bund and its ridge must be strong with the capability to withstand water pressure and keep the fish from escaping. Surveys have shown that in areas without fish farming, the ridges are usually small, shallow, and narrow. In contrast, ridges in areas with fish farming are thicker, higher, and wider to accommodate the fish.

Secondly, the rice variety is essential, as it must be suitable for this integrated system to withstand prolonged flooding while providing the fish with adequate living space for growth. Additionally, it should be resistant to diseases and pests, as the use of pesticides—whether natural or chemical—can harm the fish in the field.

Lastly, the fish species chosen for this farming system must be resilient to the local environmental conditions, including cold, murky, and shallow water. These characteristics ensure that the fish can thrive alongside the rice plants, contributing to a successful and sustainable farming system.

Farmers practice Zabo fish farming by digging circular or rounded rectangular ponds with a diameter of about 1.5 meters and a depth of 70–120 centimeters at the lowest point of the field. Fish fry is then introduced into these ponds, with bamboo sticks placed inside to provide hiding spots. The fish species primarily belong to the Cyprinidae family. In addition to the native species from the northeastern river basins of India, others known for their resilience, voracious appetite, omnivorous diet, and rapid growth are introduced.

Fig. 6.5 An integrated farming system in the rice fields of Kigwema village, which includes a "separate fishpond" and the Zabo system within the rice cultivation area. Alongside the fishponds, shelters are typically constructed to provide rest areas for farmers during the day. Photo by Warong Wonglangka, 2023

During the rice-growing season, when the water level in the fields averages around ten centimeters, the pond edges are opened, allowing the fish to swim freely throughout the field to feed. Although fish do eat plants, the small mouths of the fish fries prevent them from damaging the rice plants. As the fish grow, so do the rice plants, making them too large for the fish to consume. By the time of the harvest, the water in the fields reduces, and the fish return to the ponds where they can survive with sufficient water and food, making them easy to catch for consumption. In addition to farmed fish, the rice fields naturally support small animals such as snails, crabs, and frogs, which flourish in the fields or enter with the overflow of water. In this system, flooded rice paddies are planted with spaces between the rice plants, allowing the fish to thrive naturally. The rice plants' canopy provides shade, preventing excessive weed growth, which enables the fish to swim freely throughout the fields while using the rice bunches as shelter.

The fish movement through the water and mud helps to stir up soil sediments, bringing nutrients from the bottom to the top and creating air pockets in the soil. This oxygenates the soil, promoting healthier root growth among the rice plants. Apart from consuming insect pests, the fish also help by eating weeds that would otherwise compete with the rice for nutrients. The fish waste provides essential nutrients that strengthen the rice plants, making them more resilient to disease without the need for chemical fertilizers. This results in more robust rice plants with higher yields.

Moreover, the rice plants attract natural insects, and the fields become a habitat for algae and plankton, which serve as food for the fish. The fish also consume mosquito larvae and other disease-carrying organisms, curbing the spread of waterborne diseases to humans. This integrated fish-rice farming system ultimately produces higher overall yields compared to growing rice alone, contributing to both food security and income for the farmers while reducing the environmental impact of chemical use. It was found that the common carp (*Cyprinus carpio*) is also raised in these areas. This species is either sourced from local rivers or purchased from fish markets before being released into the rice fields. Despite an attempt to raise the silver carp (*Hypophthalmichthys molitrix*) alongside the common carp, it was observed that mixed-species farming resulted in poorer weight gain for the fish.

The Zabo farming system in Kigwema village exemplifies a meticulous and sustainable approach to environmental management, integrating crops and fish within a mutually supportive ecosystem. This method illustrates "integrated agroecology" (Leippert et al., 2020; Barrios et al., 2020; Rosset & Altieri, 2017), combining traditional practices and diverse agricultural activities to balance economic, social, and environmental dimensions. Such integration fosters relationships with the natural world and addresses both human values and environmental concerns.

In contrast to other districts in Nagaland, where water scarcity limits agricultural options, Phek and Kohima benefit from more accessible water sources. This unique availability has enabled Kikruma villagers, situated in Phek's rain-shadow region, to develop the Rüza farming system, a communal sustainable water management method. Established nearly a century ago, this indigenous system utilizes elevated ponds to impound runoff, providing a gravity-fed irrigation source for rice terraces below. Communally maintained and shared among households, these ponds allow villagers to cultivate successfully despite minimal rainfall, ensuring a steady water supply even during droughts.

Beyond supporting rice cultivation, the Rüza system integrates forestry, horticulture, animal husbandry, and fishery, reinforcing food security and ecological resilience. This adaptive system demonstrates how water accessibility has shaped agricultural practices in Phek and Kohima, setting these districts apart from Nagaland's more water-limited areas. The Rüza farming in Kikruma showcases its potential as a sustainable model for other water-scarce, hilly regions, balancing environmental sustainability with community needs. This fusion of geographical adaptation and communal resource management reflects the resilience of the Chakhesang people and offers valuable insights for agricultural development in similar terrains.

6.7 Integrating Aquaculture with Rice Farming: Lessons from the Nagas for Northern Thailand

Although the rice cultivation areas in Northern Thailand and the paddy fields in Nagaland, India, are both situated in similarly sloped terrains, their land management practices for rice fields and fish farming differ significantly. In Northern Thailand, the fish found in the rice-farming system are not the result of intentional fish farming; rather, they are a byproduct of water diverted into the paddy fields for rice cultivation. Water is drawn from natural streams through the traditional irrigation system known as the "Muang Fai" system. It serves as a habitat for fish and small aquatic animals, such as shrimp and snails, which are carried into the rice fields by the diverted water.

Fish Harvesting Practices Fish harvesting in the Northern Thai rice fields occurs during two main periods. The first period coincides with water diversion into the fields, where nets are used to catch small fish and fry carried by the water flow. The second period takes place near the rice harvest season when water levels in the fields begin to decrease. Fish tend to gather in the lowest parts of the fields, making them easier to catch for consumption. The fish caught during this period are generally larger, as they have had approximately four months to grow in the rice fields.

Adaptation of Fish Farming Techniques We have observed that Northern Thai farmers in certain areas have begun to adapt the concept of raising fish in rice fields, like the Zabo system and separate fishponds of the Nagas, to enhance land use efficiency and increase food production. This involves digging deep trenches along the rice field bunds to maintain appropriate water levels and provide shelter for the fish. However, this method has not gained widespread popularity. This is because the deep trenches present obstacles during the rice harvesting process, as they disrupt the transportation of harvested rice out of the fields. This inadvertently limits the adoption of fish farming in Northern Thailand.

Considerations for Implementing the Zabo System Integrating the Zabo system and segregated fishpond techniques from the Nagas into Northern Thai-style paddy fields could offer significant benefits. However, for this integration to be successful, four key considerations would be addressed. First, selecting suitable rice varieties is essential. These varieties should be compatible with the water levels that support fish habitation, ensuring that both rice and fish can thrive together. Second, it is crucial to choose fish species that can thrive in shallow, turbid waters. Understanding their behavior and physical characteristics will enhance the likelihood of successful fish farming alongside rice cultivation. Third, effective soil management for shelter creation is important. By digging ponds at the lowest points of the fields, farmers can provide shelters for the fish. This method helps mitigate issues associated with traditional trenching along the bunds, which often obstructs rice harvesting and complicates transportation. Finally, timing the release of fish into the fields is vital.

Identifying the optimal timing requires considering both the age of the fish and the growth stage of the rice. This careful timing is essential for successfully integrating fish into the rice fields, ultimately leading to a more productive and sustainable farming system.

6.8 Discussion and Analysis: Navigating Agricultural Change in Nagaland

The cultural landscape and agricultural ecology of the Naga people in Nagaland, India, represent a diverse and sustainable agricultural system that is rooted in traditional knowledge and is dedicated to food security for local communities. Practices like the Zabo system, alder-based farming, and organic agriculture, particularly evident in Kigwima village, embody a worldview with a deep connection to the natural environment. This rich heritage has fostered a repository of local wisdom that continues to guide current agricultural approaches, albeit with ensuing challenges.

6.9 Integrating Traditional Agricultural Practices with Global Sustainability Goals

The study on rice-cum-fish agricultural practices, cultural landscape, and agroecology in Nagaland aligns with the principles of the United Nations Conference on Sustainable Development held in Johannesburg (*United Nations*, 2002), focusing on sustainable resource management, poverty reduction, and cultural preservation. The Zabo system, a model of rice-fish integration, promotes biodiversity, enhances soil fertility, and reduces dependency on synthetic inputs, supporting sustainable land use and ecosystem health. This system not only sustains food production but also diversifies income sources, thus contributing to poverty alleviation and food security.

Traditional knowledge embedded in Nagaland's agroecological practices underpins sustainable development, strengthening cultural identity and community resilience, in alignment with the summit's emphasis on indigenous knowledge. A core aspect of this approach is water management, central to the Zabo system, which uses rainwater effectively and maintains water quality, exemplifying integrated resource management. The recommendation to adapt these practices in regions like Northern Thailand underscores the summit's call for knowledge sharing and the global application of sustainable solutions tailored to local contexts.

This study further connects with the Millennium Development Goals (MDGs), especially MDG 7 on environmental sustainability, by promoting biodiversity, sustainable resource use, and reduced reliance on synthetic inputs. The Zabo system exemplifies environmental stewardship through rainwater harvesting and effective water management, supporting ecosystem health.

In relation to the Sustainable Development Goals (SDGs), Nagaland's rice-cum-fish system addresses multiple objectives. It enhances food security and economic resilience, supporting poverty reduction and hunger alleviation. By minimizing chemical inputs and optimizing resource use, the system fosters healthier environments and improves water quality, aligning with goals for clean water, health, and responsible consumption. Additionally, its low environmental impact supports climate action by reducing carbon emissions and enhancing climate resilience.

Adapting these practices in other regions, such as Northern Thailand, emphasizes the importance of knowledge-sharing and international cooperation as promoted by the SDGs. Overall, Nagaland's rice-cum-fish system illustrates how traditional agricultural practices can support sustainable development, particularly in addressing food security, environmental health, and climate resilience.

6.10 Traditional Farming Systems in Nagaland: Confronting Modern Challenges

These challenges stem from both internal and external factors. The migration of younger generations to urban areas has led to a decline in traditional farming practices. Additionally, the commercialization of agricultural products and increased state involvement in enhancing agricultural output have intensified these pressures. A survey conducted in Kohima revealed that 85 percent of vegetable vendors prefer their children to pursue stable government jobs rather than farming, while only 15 percent were open to their children entering agriculture, citing concerns about challenges in the agri-allied sector (Solo, 2022).

The push for commercialization and state initiatives to boost agricultural output have further strained traditional practices. The Government of Nagaland reported a significant increase in the cultivated area, production, and yield of cereals, pulses, oilseeds, and commercial crops for the fiscal year 2022–23 compared to the previous year, thanks to state and centrally funded programs. This trend is expected to continue into 2023–24 (The Morung Express, 2024). In light of this, government officials continue to promote the advantages of commercial farming and modern technological methods, emphasizing the need to transition away from traditional practices.

Moreover, this emphasis on high-output commercial farming has led to increased reliance on synthetic pesticides (Nath et al., 2024) and a rise in the use of fertilizers over the years. Climate change also poses significant threats to crop yields, exacerbating issues like increased weed growth, crop diseases, and pest infestations. In 2021, Nagaland faced a severe infestation of Fall Armyworm (FAW), affecting over 3000 hectares of maize across 334 villages. This outbreak, worsened by warm temperatures and dry spells, devastated maize crops and disrupted other agricultural products (Solo, 2022). The average annual rainfall does not sufficiently meet the village's water requirements, prompting villagers to implement the Ruza system for irrigation to address persistent water scarcity (Das, 2023). Additionally, resource overexploitation—such as deforestation and soil degradation—compromises the effectiveness of these systems.

6.11 Adaptability and Resilience in Traditional Agriculture: Preserving Cultural Heritage

Despite the influx of monoculture farming systems and modern methodologies into Naga communities through state development projects, the Naga people have learned to adapt these new methods selectively to complement their traditional agroforestry systems, rather than completely replacing them. This approach is driven by their ongoing focus on food security.

Traditional farming practices persist in rural areas of Nagaland, where cultivators strive to uphold these customs despite pressures to adopt modern methods. Agriculture has historically been central to the identity of the Naga people, serving not only as a means of livelihood but also as a cultural cornerstone. While modern advancements have influenced farming techniques, the community has not fully abandoned its traditional systems. This adherence reflects a deliberate effort to preserve cultural heritage and maintain a connection to ancestral ways of life, even as the community navigates contemporary agricultural demands.

6.12 A Reflection for Northern Thailand

Reflecting on the agricultural systems in the Lanna region, it is evident that ethnic groups and Lanna communities once possessed a wealth of knowledge aligned with their worldview and relationship with the natural environment, similar to the Naga people. However, recent trends have seen Lanna agricultural communities significantly reduce their traditional knowledge and shift toward monoculture farming for commercial purposes. This shift has led to challenges in managing resources, soil, water, and toxic chemicals that harm ecosystems.

While some groups, such as certain Tai Yuan communities, still maintain traditional water management systems like the "Muang Fai" for shared irrigation, they face widespread issues due to state-run irrigation systems and changes in natural water sources. These challenges highlight significant questions regarding agricultural systems connected to the Lanna people's natural landscape and culture.

How can Northern Thai farmers learn from the Naga people? Should agroforestry systems and integrated farming practices, like those of the Nagas, be applied in their context? The Naga people have selectively adapted new techniques to enhance their traditional agroforestry systems, prioritizing food security while preserving their cultural heritage. This commitment reflects their historical reliance on agriculture as a vital aspect of their identity and livelihood, allowing them to maintain a connection to ancestral practices amidst contemporary challenges.

The integrated agroecology aspect is not merely an agricultural tool, but it seeks to address structural problems associated with monoculture farming, the roles of the state and private sector in managing food systems, and the cultivation of genetically modified crops. These factors have diminished the importance of food security for Lanna communities. Integrated agroecology emphasizes the role of farming communities as knowledgeable resource managers. These communities have accumulated understanding from nature and possess the potential to seek new knowledge, rather than relying solely on state assistance. Shifting this perspective can lead to a deeper appreciation of the original role of Lanna farming communities, which have long held local knowledge that enables them to integrate new practices harmoniously with their natural landscape and culture. These perspectives and strategies would help distribute wealth generated from production while preserving ecosystems, in line with agroecological principles. The primary goal is to emphasize food security, the role of farming communities, and the health of ecosystems in sustainable agricultural practices.

6.13 Conclusion

This study investigates the agricultural practices and cultural landscape of the Naga people in Nagaland, India, highlighting their historical expertise in rice cultivation through methods like the Zabo system, which combines wet rice farming with fish rearing. Using the case study of Kigwima village, we see that the Naga community has practiced these sustainable agricultural methods for over 600 years. It exemplifies resilience and adaptability in the face of modern pressures, which include urban migration, the commercialization of agriculture, and the changing environment.

The study reveals complex and diversified farming systems characterized by efficient land use and adaptability to the natural landscape. These systems rely on social relationships and labor allocation at both individual and community levels to support established livelihoods. The Zabo farming system exemplifies an integrated approach that utilizes small fishponds within rice fields, promoting ecological balance and enhancing food security. This method not only preserves traditional practices but also incorporates modern techniques, illustrating a harmonious relationship with the environment. It emphasizes the importance of local knowledge in resource management, allowing the Naga people to maintain their cultural identity while navigating contemporary agricultural challenges.

Furthermore, the research draws parallels between Naga practices and those in Northern Thailand, where similar agricultural systems exist but face challenges due to shifts toward monoculture and commercial farming, as well as climate changes. By adopting agroecological principles and integrated farming practices akin to those of the Naga, Northern Thai farmers could enhance food security and sustainability. The findings suggest a need for a shift in perspective toward valuing traditional knowledge and sustainable practices, advocating for a model that balances economic, social, and environmental dimensions. This integrated approach could serve as a valuable framework for addressing the challenges faced by both Naga and Northern Thai agricultural communities, fostering resilience and preserving cultural heritage in the process.

Acknowledgments This article is dedicated to Rawiwan Oranratmanee, the head of the research team, who has passed away. We wish her eternal rest and peace, knowing she will always remain in our hearts. Gratitude goes to Prof. Rawiwan Oranratmanee.

References

Amenla, I., & Shuya, K. (2021). Zabo (Zabü) farming of Kikruma village, Nagaland, India. In Michigan State University Extension East Lansing, Michigan, USA & National Institute of Agricultural Extension Management (MANAGE) Hyderabad, India (Eds.), *Innovations in agricultural extension* (pp. 1–11). Michigan State University Extension.

Barrios, E., Gemmill-Herren, B., Bicksler, A., Siliprandi, E., Brathwaite, R., Moller, S., Batello, C., and Tittonell, P. (2020). The 10 Elements of Agroecology: enabling transitions towards sustainable agriculture and food systems through visual narratives. *Ecosystems and People, 16*(1), 230–247.

Cairns, M., Keitzar, S., & Yaden, T. A. (2010). Shifting forests in Northeast India: Management of *Alnus nepalensis* as an improved fallow in Nagaland. In *Voices from the forest* (pp. 341–378). Routledge.

Das, B. (2023). *Ruza, a traditional water harvesting system for the water-scarce mountains*. Mongabay. Retrieved August 23, 2024, from https://india.mongabay.com/2023/08/ruza-a-traditional-water-harvesting-system-for-the-water-scarce-mountains/

Hutton, J. H. (1921). *The Angami Nagas*. Macmillan.

Jacobs, J. (2012). *The Nagas hill peoples of Northeast India*. River Books.

Kehie, M., Khamu, S., & Kehie, P. (2017). Indigenous alder-based farming practices in Nagaland, India: A sustainable agricultural model. *Journal of Traditional and Folk Practices (JTFP), 5*(2), 82–152.

King, C. D. (1879, June). Among the Naga Hills. *The Baptist Missionary Magazine, LIX*, 144–146.

Kire, E. (2019). *Walking the roadless road: Exploring the tribes of Nagaland*. Rupa Publications.

Kithan, L. N. (2014). Indigenous system of paddy cultivation in terrace and jhum fields among the Nagas of Nagaland. *International Journal of Scientific and Research Publications, 4*(3), 1–4.

Leippert, F., Darmaun, M., Bernoux, M., Mpheshea, M., Müller, A., Geck, M., Herren, M., Irungu, W., Nyasimi, M., Sene, J. M., Sow, M., Sylla, I., & Termote, C. (2020). *The potential of agroecology to build climate-resilient livelihoods and food systems.*

Murry, N., & Das, S. (2021). Zabo farming system—A sustainable farming based on traditional knowledge for natural resource management practiced by tribal in Nagaland, India. *International Journal of Agriculture, Environment and Biotechnology, 14*(2), 203–205. https://doi.org/10.30954/0974-1712.02.2021.13

Murry, N., & Lalruatsangi, K. (2019). A review study on sustainable indigenous farming system practice in Nagaland. *International Journal of Human Resource Management and Research (IJHRMR), 9*(6), 11–16.

Nath, U., Puzari, A., & Jamir, T. (2024). Toxicological assessment of synthetic pesticides on physiology of *Phaseolus vulgaris L.* and *Pisum sativum L.* along with their correlation to health hazards: A case study in south-west Nagaland, India. *Journal of the Saudi Society of Agricultural Sciences, 23*(4), 300–311. https://doi.org/10.1016/j.jssas.2024.01.001

Neli, V. (2022). Terrace rice fields: A cultural heritage of the Angami Nagas. In G. K. Chophy & S. K. Chaudhuri (Eds.), *The cultural heritage of Nagaland* (pp. 315–330). Routledge.

Nienu, V. (2015). *Naga Cultural Milieu*. Thomson Press.

Nshoga, A. (2008). *Traditional Naga village system and its transformation*. Anshah House.

Oppitz, M. (2008). The Nagas: An introduction. In M. Wettstein (Ed.), *Naga identities: Changing local cultures in the northeast of India* (pp. 11–28). Snoeck Publishers.

Paul, D., Alfred, J. R. B., Darlong, V. T., Vatsauliya, P. K., & Nongmaithem, A. (2022). Diversity of soil mesofauna under traditional jhum and sedentary agroecosystems in Meghalaya. *A Tradition in Transition, 198*, 198–218.

Rosset, P. M., & Altieri, M. A. (2017). *Agroecology: science and politics* (pp. 160-pp).

Saikia, B., Nookathoti, T., & Hiremath, C. (2024). Alder-based shifting cultivation in Nagaland–A theoretical perspective. *Journal of Climate Change, 10*(2), 19–26.

Solo, V. (2022). *Agriculture: A withering profession*. Nagaland Post. Retrieved August 23, 2024, from https://nagalandpost.com/index.php/2022/11/02/agriculture-a-withering-profession/

Solo, V., & Kikhi, K. (2021). An overview of the farming systems in Nagaland. *Journal of Pharmacognosy and Phytochemistry, 10*(1S), 238–243. https://doi.org/10.22271/phyto.2021.v10.i1Sd.13492

The Morung Express. (2024, March 6). *Substantial progress in production of agriculture crops in Nagaland*. Retrieved August 22, 2024, from https://morungexpress.com/substantial-progress-in-production-of-agriculture-crops-in-nagaland

United Nations. (2002). *Report of the World Summit on Sustainable Development, Johannesburg, South Africa, 26 August–4 September 2002*. United Nations. https://digitallibrary.un.org/record/478154?ln=en

Chapter 7
A Historical, Gender-Role, and Feasibility-Based Investigation from the Meitei Perspective in Manipur

Pakdeekul Ratana, Bedavati Laishram, and Moirangthem Boby Singh

Abstract Historical studies tend to focus on male roles during political and economic events. However, there is a dearth of research on women and their gender roles in historical and sociocultural contexts. This chapter focuses on the stories of women who play key roles in shaping the society, culture, and economy of a unique region: Manipur in northeastern India, the sole residence of the Meitei. This study will examine the history and context of Manipur to highlight the role of gender across historical dimensions. It seeks to analyze the relationship between the genders and explore their respective gender roles in the distinctive backdrop of the Manipur Kingdom. Additionally, the chapter addresses how important it is to have a space for examining gender roles, which can vary depending on the context, ethnicity, and the local women's culture in different regions. Furthermore, it proposes prospects for social and cultural connections and economic development through tourism, which could foster relationships between northern Thailand and northeastern India. Due to the close cultural ties between the two areas, this partnership may create a new market.

Keywords History · Gender roles · Manipur · Meitei · Tourism

7.1 Introduction

Located in northeastern India, Manipur has several physical features. The Meitei population primarily resides in the central Imphal Valley, which features a levelled and fruitful landscape. This region is also home to the largest freshwater lake in northeast India, Loktak Lake. The Manipur hills surround the valley, enhancing the

P. Ratana (✉)
Dr., Faculty of Humanities at Chiang Mai University, Chiang Mai, Thailand
e-mail: pakdeekul.r@cmu.ac.th

B. Laishram · M. B. Singh
Dr., Department of History at Manipur University, Imphal, Manipur, India

Y. Santasombat (ed.), *Bio-Cultural Diversity, Ethnicity, and Local Knowledge*,
https://doi.org/10.1007/978-981-95-4242-0_7

state's diverse topography with their harsh landscape and abundant forests. The region is bounded by Nagaland to the north, Assam state to the west, Mizoram to the southwest, and Myanmar (Burma) to the east and south (Singh, 2014, p. 1). Different ethnic communities have known the land by several names. The Burmese refer it as Kathe, the Shan as Ka-se, the Cachar as Moglai, and the Assamese as Mekheli (Devi, 2011, p. 2). Contemporary Manipur spans 22,327 square kilometers. It is known as "Little Paradise on the Earth," "Switzerland of India," "A Flower on the Lofty Heights," or "Takane No Hana" by the Japanese, and "Jewel of India" by Pandit Jawaharlal Nehru, the first Prime Minister of India. Manipur is renowned for its picturesque landscapes, rich traditions, vibrant culture, and fascinating history.

Due to its geographical location, Manipur is considered the easternmost frontier of Indian culture and civilization. Manipur has a distinctive historical background spanning over two millennia as a sovereign and autonomous state governed by the Ningthouja dynasty. The city of Imphal, with Kangla[1] as its center, served as the seat of authority. Hence, Manipur is credited with one of the longest-ruling dynasties in South and Southeast Asia.

Manipur's terrain consists of two separate regions: mountains and valleys. The hill region comprises over 90% of the state's land area. Manipur can be legitimately known as a state characterized by its hilly terrain. Hills encompass the Imphal Valley in all directions. The valley's altitude above sea level exerts a substantial influence on the climate and consequently shapes the culture of its inhabitants. Since the early days, there have been plenty of lakes in the valley, the biggest being Loktak Lake in the Bishnupur district. Keibul Lamjao National Park, an integral part of Loktak Lake, is the home of the endangered Sangai (Cervus Eldi Eldi) and the only floating national park in the world.

The different hill villages have been settled by various tribes, such as Tangkhul, Kabui, Mao, Maram, Khongjai, Anal, and Maring, since the early days. The Government of India reports that Manipur has a multiethnic population comprising 34 scheduled tribes who are predominantly residing in the hill districts, excluding the Meitei community (National Commission for Scheduled Tribes, 2023). The region is divided into two geographical categories. First, in the Imphal Valley, the different ethnic groups include the following: the Meitei, the state's majority, who practice Hinduism and Sanamahism; the Meitei-Pangal, who are Muslims; and the Meitei Bamon, who are Brahmins. Second, in the hill areas, the Chin–Kuki, Mizo, and Naga communities, mainly Christians, are the main settlers (Government of India, 2012). However, Meitei is the dominant community among all other communities settled in Manipur.

The varied landscape influences Manipur's climate, ecology, and socio-economic activities. It has served as a gathering spot for numerous ethnic groups and cultures. Other appellations, such as Tillikoktong Ahanba, Kanglei Pungmayol, Kangleipak,

[1] Kangla, also known as Kangla Fort or Kangla Palace, is located near the Imphal River and prominently represents Manipur's grandeur. The term "Kangla" in the Manipuri or Meitei language refers to dry land. Kangla served as the historical capital of Manipur, where several Meitei monarchs ruled the destination from this fort.

and Meitrabak, were previously rendered to the land. The name "Manipur" was given to this land in the eighteenth century after King Garibniwaz (1690–1748) adopted Hinduism as the state religion.[2]

Manipur enjoys a delightful climate characterized by mild summers that create a pleasant and comfortable atmosphere. The areas around the lakes experience high humidity, adding to the region's unique charm. The monsoon season brings heavy rainfall, transforming the landscape into lush greenery (Emilia, 2023). Autumn in Manipur has pleasant weather, which makes it a perfect season for discovering the wonders of nature. Winter delivers a temperate environment that is not extremely severe, providing an ideal equilibrium. The excellent weather conditions greatly support and promote successful agricultural activity.

The Meitei people, native to the Manipur valley, have lived in the region for a long time. Essentially, they have preserved their unique identity despite foreign influences, as they have lived there since prehistoric times. The kingdom of Manipur, ruled by the Ningthouja dynasty, served as a cultural and political center, and interactions with neighboring cultures have enriched the unique heritage of the Meitei community. Throughout history, the Meitei community has demonstrated a remarkable capacity to safeguard their language, traditions, and cultural practices. Rural communities with lower population densities have emerged as individuals sought permanent settlements in advantageous places that support commerce and sustenance.

The Meitei society possesses a wealth of folklore and enthralling mythology, fostering a profound unity among its members. The people of Manipur have distinct customs, culture, beliefs, language, religion, and a vibrant art scene. The religions present in Manipur during the 2011 census were Hinduism, Islam, Christianity,

[2]The eighteenth century was a significant development with Meidingu Pamheiba's adopting Vaishnavism, a sect dedicated to Vishnu, instead of the original Sanamahi religion of the Meitei people. Shantidas Goswami, a Bengali Brahmin from Sylhet, was invited to Manipur. The adoption of Hinduism and the arrival of Bengali Brahmins resulted in modifications to the governance structure and cultural society of the kingdom, including the establishment of the Sanskrit designation "Kingdom of Manipur" (Sudhir, 2012) and the designation of Manipur as a state religion. For the first Manipuri king, adopting the title "Maharaja" before his name, rather than the traditional "Ningthou" or "Meidingu," was a significant departure from ancient Meitei customs. He altered his original name from "Pamheiba" to "Gharibniwaz." Besides, he changed numerous traditional customs and rituals, including the ancestral practices from burial to cremation by the river in the Hindu tradition, which subsequently evolved into the cremation tradition of the Meitei people. He destroyed all sacred objects and scriptures related to the Sanamahi deities, and due to their opposition, he exiled the scholars and priests of the Sanamahi religion to a remote village. King Gharibniwaz's religious reformation in the Manipur kingdom aimed to establish his authority and power as a formidable warrior-king. His conflicts with neighboring kingdoms and significant territorial expansion are evident. Chandidas Gosvami, a Brahmin, persuaded him that Vaishnavism was more appropriate for his role as a warrior king, akin to Lord Rama, a distinguished monarch, than the Sanamahi belief. Consequently, he established a novel governmental structure by integrating Hinduism with political authority. It obliterated the original Samanahi religion (Parratt, 1989; Bharucha, 1991; Sebastian, 2021). During the late Middle Ages, the Manipur kingdom experienced turmoil stemming from a succession struggle among the sons of King Gharibniwas. This conflict ultimately facilitated the British entry into Manipur, establishing it as a British colony.

Sikhism, Jainism, and Buddhism. Christianity and Hinduism are the two main religions in the state. Nevertheless, the Meitei people of Manipur persist in maintaining their ancestral faith, Sanamahism. Sanamahism continues to be practiced in every Meitei home, with a designated sacred space dedicated to the worship of the deity Sanamahi. The Lai Haraoba festival is a culturally significant event that combines mythology, ceremonial acts, and community involvement. It is dedicated to the worship of traditional deities and spirits, promoting a sense of unity with a shared cultural identity. Despite the cultural differences between the residents of the hill and valley areas, they actively participate in and value each other's cultural festivals. Besides, Mera Houchongba[3] exemplifies solidarity since it brings together hill dwellers and valley residents in celebration.

The Meitei people in Manipur possess a multitude of significant historical landmarks that are intricately connected to their past. With great historical and cultural importance, the Kangla Fort is one such example. As the capital of the former Manipur Kingdom under the Meitei monarchs, it functioned as the primary hub for political and cultural activities. Throughout the history of Manipur, this place has held significance in Meitei culture, cosmology, and religious practices. The Ima Keithel (Mother's Market) is a unique all-women market in Imphal that is not only a bustling economic center but also holds cultural importance. It reflects the economic activities and entrepreneurship of Meitei women. The market is well known worldwide for being the only market for womenfolk. Govindajee Temple, a golden-domed temple, is a historic center for Vaishnavites in Manipur, attracting many devotees. Holding great significance in Manipur history, Moirang is renowned for its Keibul Lamjao National Park (the world's only floating national park), Loktak Lake, and Ibudhou-Thangjing religious sites. INA Museum at Moirang is the first place where the Indian National Army's tricolor flag was hoisted on April 14, 1944. Maibam Lotpa Ching at Nambol is a historic site in Nambol where the fierce battle between British and Indo-Japanese forces happened in WWII. The Imphal War Cemetery has 1,600 commonwealth burials from the Second World War Battle of Imphal and Kohima. The Khongjom War Memorial at Khongjom, Thoubal district, is the location where the last battle of the Anglo-Manipur War of 1891 was fought, after which the British annexed Manipur, which later became a princely state in 1947.

This chapter presents the findings of an empirical study conducted in Manipur, which is in northeast India. The study is designed to investigate Manipur's historical, sociocultural, and gender-based perspectives by multiple data collection methods, including a literature review conducted in collaboration with the Department of History, Manipur University, in Manipur. The evidence pertains solely to the historical study of the kingdom and the Meitei people, who form the predominant population in Manipur. In contrast, differing from the evidence associated with the Meitei people, the Chin–Kuki, Mizo, and Naga ethnic groups possess distinct folktales and

[3] Mera Houchongba is Manipur's cultural festival of solidarity. All indigenous ethnic groups, including the Meitei and Naga tribes, celebrate it. It occurs on the fifteenth lunar day of the Mera month in the Meitei calendar.

legends that narrate the stories of their respective ethnicities. Additionally, this research employs a comprehensive interview method to unravel the feasibility of tourism connectivity between Manipur and northern Thailand from both sides' perspective. The interview process involved a sampling of 12 individuals, including local scholars in arts and culture, entrepreneurs, academics, local politicians, and community leaders.

7.2 History of Manipur

During China's Warring States period (403–221 BCE), it was believed that many people from China, Tibet, and Southeast Asian regions like Burma, Laos, and Vietnam migrated to Manipur for suitable land for settlement. Historical evidence indicates the existence of various ethnic groups and lineages in the valley of Manipur: the Angom, Luwang, Mangang, Khaba, Nganba, Sarangthem, Leishangthem, Chenglei, Haorok Konthou, Manting-Mara, Lela Khongnang, Lokha-Haokha, Chakpas, Khem, Heirem Khunjan, Moirang, Thanga-Kambong, Urok-Usai, and Lokha-Lokhu. According to the ancient Manipuri text Poireiton Khunthok,[4] Poireiton migrated from an advanced South Asian society to Manipur, probably Nepal. Nongda Lairen Pakhangba was believed to be the founder of the Manipur Kingdom. In the struggle for the throne, Nongda Lairen Pakhangba defeated Khaba Nongjengba of the Khaba Clan with the support of the Angoms and Luwangs. He engaged in a struggle with Poireiton. He ultimately achieved victory over the rival who recognized his superiority. Manipur was under the lineage domain of seven Ningthous (or chiefs' principalities), and the credit for forming seven clans in Meitei society was given to Luwang, Khuman, Mangang (later Ningthouja), Angom, Moirang, Khaba-Nganba, and Sarang-Leishangthem (Chenglei).

Establishing the Meitei kingdom with Kangla as the seat of power was a significant part of Manipur's history. The Royal Chronicle of Manipur, named "Cheitharol Kumbaba," has given a detailed account of the 76 kings of the Ningthouja dynasty, reigning from 33 CE to the middle of the twentieth century. Though there were many opinions among historians regarding the chronology of the first king of Manipur, the common understanding was that his ascension and reign were from 33 CE to 154 CE. On the day of the coronation—known as Phambal Tongba or Men Tongba at Kangla—the chief of the four leading clans (Angom, Luwang, Moirang, and Khuman) draws water from Nungjeng and pours it on the king and queen. Occupying a higher ritual position, the Angom Ningthou (chief) Pureilomba presented the coronation robe.[5] In the struggle for the throne, the Angoms sponsored

[4] Poireiton Khunthok is an ancient Manipuri text that deals with Poireiron's adventures in his colonizing mission.

[5] This is because the Angom clan significantly contributed to establishing the Meitei state, with Puleiromba assisting Pakhangba in ascending to the throne of Kangla. Notwithstanding their historical alliance with Ningthouja, the Angoms preserved their autonomy and significantly contrib-

Nongda Lairen Pakhangba to be the ruler of the Kangla Kingdom. The King also organized a game of Sagol Kangjei, alternatively modern-day polo, on the day of Loukhatpa (which means marriage) of Laishna to please the queen. He divided the land into four panas for efficient administration: Ahallup Pana, Naharup Pana, Khabam Pana, and Laipham Pana.[6] Nongda Lairen Pakhangba introduced the institution of "Lallup," a kind of service to the state according to which every able male had to attend the state every 10 days out of 40 (Singh & Singh, 2023, p. 758).

7.2.1 Administration of Early Manipur

Nongda Lairen Pakhangba,[7] the inaugural Meitei sovereign of the Ningthouja dynasty, ascended the throne of the Kangla of Kangleipak[8] in 33 CE. He had his council of ministers, traditionally known as Ningthou Pongba Tara, and 64 Phamdous (nobles). To establish systematic administration in the country, Nongda Lairen Pakhangba introduced five administrative departments: Laipham Shanglen, Kuchu Shanglen, Khabam Shanglen, Naharup Shanglen, and Ahallup Shanglen.[9] These departments catered to the needs of the people or groups from various areas and facilitated appropriate resettlements in acceptable locations, significantly enhancing social cohesion.

The Pacha Loishang,[10] another important department concerned with women's welfare was established by the first historical king of Manipur, Nongda Lairen Pakhangba. It was presided over by Queen Laisna. The strength of this court was 11, and the ten wives of the council of ministers were its members. This court determined the punishment inflicted on female criminals (Singh, 1976, p. 25). The highest punishment was traditionally known as Khongoinaba. However, unlike the male,

---

uted to administrative matters. The Angom alliance was founded on marital connections, with no restrictions on marriage. The Angoms were beneficent patrons of the Meitei monarchs, whereas the Shelloi Langmais, an independent tribe from the Nongmaiching Hills, sustained a connection with the Ningthouja. The Angoms rendered exemplary service to the Meitei state and were accorded a prestigious position in the Meitei court.

[6] The four panas included Ahallup pana, Naharup pana, Khabam pana, and Laipham pana

[7] Nongda Lairen Pakhangba stands for Nongda or God-sent; Lairen or mythical snake-empowered; and Pakhangba or "Pa" means father combined with "Khangba," meaning "one who knows the real father."

[8] The Meitei language, or Manipuri, is associated with the Manipur region.

[9] Five administrative departments: Laipham Shanglen was to administer the court of justice; Kuchu Shangen was another court of justice; Khabam Shanglen dealt with the welfare of Khabam, who came together with Poinaota; Naharup Shanglen meant the department of Naharup, people who migrated together with Poireiton; and Ahallup Shanglen means the department of Ahallup, who are people coming with Thongaren, elder brother of Poireiton and the chief of Khamnung at the tri-junction of Tibet, Burma, and India.

[10] Pacha Loishang or a women's court. The court dealt with rape, sexual assault, domestic violence, wife beating, polygamy, divorce, and other crimes against women (Arambam, 2021, p. 11).

women criminals were spared from imprisonment and death. Queen Laisna thus took a significant role in the state administration. The introduction of this separate department of womenfolk was a substantial contribution by Nongda Lairen Pakhangba. It shows the degree of respect and honor for the status of women in early Meitei society.

Khuyoi Tompok, son of Nongda Lairen Pakhangba and Laisna, ascended the throne of Kangla in 154 CE as the second king of Manipur. He was credited for introducing the traditional Pung, a drum used during wartime, dance, music, and other socio-religious events. The invention of the drum by Khuiyok Tompok symbolized the technological development and cultural attainment phase of the country. Senbung (gong) made during his reign was a significant cultural link between Manipur, upper Burma, and Thailand. The Sharungba and the Ngarungba were the two nobles among the 64 Phamdou (nobles) responsible for procuring and maintaining animals and fish. Besides, another development of his reign was the institutionalization of Ningolakpa to oversee and educate the Ningol (maidens) in social customs and habits.

Naothingkhong was the seventh-century king who ruled the land from 663 to 763 Naothingkhong married several wives from the neighboring clan's principality, and his chief queen was a Khuman princess. This demonstrated that polygamy was practiced by the Meitei society during those days. Both forms of marriage were prevalent: elopement (Naothingkhong and Pitanga) and engagement (Naothingkhong and the Khuman princess). The custom of matrilocal residence (staying in the wife's parental home) was predominant. According to the Cheitharol Kumbaba, Shamlung, who was Pong King Shanpha's younger brother, stayed at the Apong Ingkhol (homestead of Pong) for 10 years. He was saved by his devoted wife, Langmai Pitanga, daughter of Wamei Lanhaba of the Achong Yumlembam clan and Shelloi Langmai resident. She left the world when Naothingkhong became the King of Meitei, who subsequently received Khuman Princess Keirunghanbi as the chief queen. After the death of his first wife, Pitanga, Meidingu Naothingkhong repented. He then ordered the worship of Pitanga as Yumjao Lairembi, an ancestral deity of the royal household (Singh, 1975, p. i). In commemorating his beloved consort, the king performed an annual religious ceremony at Chingoi (a streamlet rising in the Nongmaiching hill facing north and falling into the Iril River) at Nongmaiching hill. Even today, thousands of people take a holy dip in the Chingoi River every year on the 28th day of the Manipuri month, Lamda, also known as the Baruni Khurumba. The life and society of the Meiteis during Naothingkhong's reign were discovered from the text Naothingkhong Phambal Kaba (Singh, 1995, p. i).

Meidingu Yanglou Keiphaba (r. 969–984) was remembered in the history of Manipur mainly for introducing Khoi Mayek in embroidery phanek, the sarong of Meitei women. Meidingu Irengba (r. 984–1074) was a famous king of Manipur who introduced the Heikru Hidongba Festival, a boat race festival performed usually on the Bijoygovinda Thangapat (moat). He also established "Haomacha Loishang," a department of tribal affairs.

Meidingu Loiyumba's (r. 1074–1122) reign was an essential period in Manipur's history in which the kingdom's consolidation was witnessed. Giving "sagei" (surname according to their occupation) was started by Meidingu Loiyumba. Known as Loyumba Shilyen (Loyumba's Decree), it was an important historical document for reconstructing Manipur's social and economic history during this period. All laws of the land were codified during the reign of Meidingu Loyumba. The decree was issued in the year 1110. According to this decree, Loiyamba divided the population of the land according to their occupation, whereby one occupation was assigned to one yumnak (surname).[11] This decree was accepted as the first written constitution of the Meiteis. Maibas (shamans) of 101 families were assigned the duties of priests at the abodes of traditional deities called "Umanglais," executing traditional healing activities, performing rites, offering sacrifices, and preparing birth and death ceremonies.

Meidingu Ningthoukhomba was a powerful king who ascended the throne of Kangla in 1432 after the vigorous reign of his father, Meidingu Punshiba, for 28 years. Ningthoukhomba was described as a worthy son of a worthy father. The chronicles recorded the military exploits of Ningthoukhomba. During his reign, the state formation of Manipur was completed, and Moirang was subjugated to the Meitei king. Linthoingambi, the queen of Meidingu Ningthoukhomba, had an illustrious career as a queen. In 1443, the king went on an expedition to attack Kabo Akla. During his absence from invading Kabo, the Tangkhuls of Tuisem village attacked the Kangla palace to capture the paddy stored at the state granary. The brave and intelligent Queen Linthoingambi dressed like her husband, the king, and displayed courage and skill against the raiding tribespeople. She was well-trained in Meitei warfare and an expert in cavalry, with a skillful use of different weapons. The queen and the Meitei army defeated the Tangkhul raiders, who came fully armed with spears and their captive, Hen Puringba, who was the chief of the Tushem village of Tangkhul. Thus, the palace was saved by the brave Queen Linthoingambai. She reportedly raided the Tangkhul villages of Sunaching and Mongba Hiyanglam and invaded the Kabui Yangla in the east, Khongyang Kumphei, and Leishokpung in the south (Kabui, 2003: 141). She also conquered Kharam in the south and defeated the Luhuppas. The chronicle Ningthourol Lambuba recorded all these military exploits of the queen.

King Ningthoukhomba of the Meitei dynasty gained authority over the influential Moirang principality. The kingdom engaged in commerce with foreign nations and maintained diplomatic ties with the neighboring regions of Cachar and Burma. His illustrious rule enticed many immigrants from both the Eastern and Western regions. The Kabaw Shan established themselves in the valley by intermarrying with the Meitei women and assimilating into the Meitei society. Having succeeded his father (Ningthoukhomba), Meidingu Kyamba began his reign and marked the beginning of the medieval history of Manipur. Once again, the rule of Kyamba (r. 1467–1508) brought about social transformation due to the movement of many social factions. Under his reign, King Kyamba created two administrative divisions, known as Ahallup pana and Naharup pana, by selecting two lakpas to supervise

[11] Yumnak means surname, which is given based on profession, post, and progeny.

these divisions. The introduction of a lot of new systems in Manipur also marked his reign. By the end of the year, he introduced the announcement of Cheithaba, who was authorized and selected by the king to bear the misfortunes of the kingdom. With this new system, the Cheiraoba, the advent of the New Year ceremony, was also initiated. He started keeping the royal chronicle of Manipur, Cheitharol Kumbaba, which recorded the significant events in the country from the year 1485. The reign of Meidingu Kyamba brought a substantial change in the destiny of Manipur, becoming a power in Southeast Asia and defeating the King of Kyang Khambat.

Khagemba's reign (1597–1652) was known for the presence of several eminent scholars like Apoimacha, Yumnam Tomba, Salam Tomba, Khaidem Temba, and Konok Thengra. Several manuscripts and books were written during this period. Khagemba established several markets in different parts of Manipur. He patronized several games and promoted Sagol Kangjei, or polo, as the national game of Manipur. The king practiced many marriage alliances with Cachar and Kabo to bring peace to them. The king also brought many social reforms. He shifted the burial ground from the Ingkhol (the homestead) to the village boundary. The king was referred to as "Laiyingthou" or "Lainingthou," which signifies a divine king.

Meigingu Charairongba (r. 1698–1709) was the first Manipuri king who was converted to Vaishnavism. On April 9, 1704, Banamati Krishna Acharya introduced the king and his followers to the Nimandi religion. In 1707, Charairongba constructed the Sri Krishna Temple at Guru Aribam Bamon Leikai, and it still exists today. He also promoted the settlement of migrated Brahmins from other parts of India and constructed several traditional Meitei temples.

Meidingu Garibniwaz (r. 1709–1748) ascended the throne of Kangla after his father, Charairongba, died. He was the most remarkable king of Manipur. Manipur attained the zenith of her glory over the military, religious, cultural, and literary domains. He had a good administration with 13 different departments. He also maintained a strong army that was well-trained in cavalry with Arambai weapons. Moreover, he did many works to promote indigenous religion, including the construction of Sanamanhi Temples. Santidas Gosain persuaded the king to convert from Nimandi to Ramandi. The king, along with hundreds of subjects, initiated the conversion, which was known as "Nongkhrang Iruppa." Many traditional sacred texts on Sanamahism (puya) were burnt, and this was historically known as "Puya Meithaba" (the burning of puya).

The First Anglo-Manipuri Treaty was signed on September 14, 1762, at Chittagong between Haridas Gossain, an agent of Maharaja Jai Singh, and Mr. Henry Verelst, the Chittagong Factory's chief, on the East India Company's behalf. Basically, the treaty was to contend with the rising power of the Konboung dynasty in Burma with the help of a British company. Maharaja Bheigyachandra was one of the most fabulous kings of Manipur, who introduced the Indian classical dance, Ras Leela. The king's installation of the Shri Shri Govindaji idol was a landmark event in the history of Vaishnavism in Manipur. In 1776, Bheigyachandra installed the image of Govindaji (Sri Krishna) by cutting a jackfruit tree from Kaina, as depicted in his dream. The remaining part of the fruit constructed the following important

Hindu idols: Bijoy Govinda; Arambam Nitananda, R.M.C. Road; Gopinath, Ningthoukhong; Abdaita, Lamangdong, Bishnupur; Bangshibudon Khangabok; and Sri Radhakrishna, Kongba.

The royal chronicle of Manipur, the Cheitharol Kumbaba, records the construction of the mythical animal Kanglasha (a Meitei mythical dragon lion) in front of Uttra[12] at Kangla during the reign of Maharaja Chourajit (r. 1803–1813). Marjit, the younger brother of Chourajit, wanted to be the King of Manipur and dethroned the latter with the help of Ava warriors in 1813. He did significant developmental work at Kangla, such as the boundary walls of the citadel and the mud-walled Kangla. Manipur becomes a tributary state of Burma. At the coronation of Prince Bagyidaw, Bodopaya's son, all the tributary kings, including King Marjit, were invited to pay homage to the new King, but Marjit did not attend due to the insecure throne. Finally, Bagyidaw attacked Manipur in 1819 under the commander Mahabandula. Cheitharol Kumbaba records the Burmese king's capture of three lakhs of Manipuris as war captives. This dark period in Manipur history was known as "Chahi Taret Khuntakpa" or the "Seven Years' Devastation."

In this Chahi Taret Khuntakpa, Prince Herachandra, known as "Taku Ningthou" (Dacoit King), carried out a lot of guerilla warfare with his army, who were skilled in the use of horses, arambai,[13] and guns. The Ava armies were instilled with fear of the frequent arambai attacks of Herachandra's armies, who were hiding in the trees. A term called "Awa Mityeng" (gaze of Ava) was popular when the Ava armies used to look up the trees. The Chahi Taret Khuntakpa ended with the liberation of the Burmese from the valley of Manipur by the Manipur levy forces of Gambhir Singh (r. 1821 and r. 1825–1834) with the help of English forces. It was followed by the Treaty of Yandaboo in 1826, signed between Burma and the English, which recognized Gambhir Singh as the raja of Manipur. Prince Nara Singh and Lt. P.B. Pemberton took great responsibility for the liberation (Cady, 2009, p. 323).

The British trained Manipur Levy armies. With the help of these trained armies, he conquered Samsok in 1825. Gambhir Singh attacked Thibomei (Kohima), Khipu-Mei village, in 1832. In 1826, the ambassador of Khambat (Kabo Valley) arrived in Manipur, as recorded in Cheitharol Kumbaba. Gambhir Singh then shifted his capital from Bishenpur to Canchipur (Langthabal). The Burmese government paid the compensation until India's prime minister left Kabaw Valley permanently for Myanmar in 1953. In 1835, the office of political agency was established in Manipur, and Lt. Gordon became the first political agent.

[12] The Kangla Uttra Sang, also known as "Uttra Sanglen," is situated in Kangla, Imphal, Manipur. The sacred governance structure of Sanamahism, a traditional Meitei religion, is being mentioned. The structure is constructed using conventional Meitei architectural motifs, featuring the prominent and sacred "chirong" (which means "horns" in the Meitei language), a V-shaped ornamentation that embellishes the roof of the building.

[13] Arambai is a dart-like weapon with a pointed tip dipped in poison and a feathered tail. The Meitei warriors possessed this dreaded and covert weapon. The weapon was extensively used by Manipuri king, Garibniwaz in the expedition against the neighboring kingdoms, especially the Burmese.

Due to the frequent aggression of Burma, the people of Manipur migrated to Assam, Tripura, and Bangladesh. The deported Manipuris to Ava were cultivated as cavalry and cultivators and sold as enslaved people.

After Gambhir Singh's death, the administration was under the regent Nara Singh. Maharani Kumudini, the dowager queen of Gambhir Singh, who wanted to handle the administration of Manipur, envied the popularity of Regent Nara Singh. Kumudini instigated Prince Nabin and his party to assassinate Nara Singh, but Nara Singh narrowly escaped. She, along with her son Chandrakirti, fled to Cachar.

Maharaja Chandrakirti (r. 1831–1886) was one of the famous kings of Manipur. He becomes the King of Manipur for the second time on July 13, 1850, defeating Maharaja Debendra. He sent an emissary to Cachar to bring back his mother, Kumudini, who was welcomed by the people of Manipur with great pomp and show. The king encouraged trade and commerce with the neighboring states and kingdoms. Numerous items were imported from Cachar and Burma.

A vital assembly known as Jilla Darbar took place on August 10, 1874, where the King and Lord Northbrook, the ruler of India, convened. Many distinguished individuals accompanied Maharaja Chandrakirti. Gifts were given, and Chandrakirti complied with the British request by transferring control of Kohima to the British authority. The king was bestowed as "Knight Commander of the Order of the Star of India" (KCSI).[14] The monarch advocated for the advancement of Western education. In 1872, he founded a school in Imphal that used English as the primary language of instruction. Manipur established another notable educational institution, Johnstone Middle School, which has since been renamed Johnstone Higher Secondary School.

7.2.2 *Manipur and its Relations with the Neighbors in History*

In the past, Manipur was one of the trade routes between the East and the West. For more than 2,000 years, Manipur has remained a critical point of South and Southeast Asian economic and cultural exchanges. As a cradle of human civilization, Manipur has been a host to diverse ethnic groups who have come and settled in different periods. Manipur thus showcased a pluralistic culture amidst heterogeneous groups. Multiracial, multireligious, and multilingual are the basis of the culture and civilization of Manipur. Historian Gangmumei Kabui observed that Manipur, a gateway of India to Southeast and East Asia, was influenced by the geographic situation and ecology, which shaped Manipur's history and cultural development (Kabui, 2003, p. 70).

Many examples demonstrate the fostered ties between Manipur and its neighboring areas. For instance, numerous routes in Manipur linked the Surma Valley of

[14] Knight Commander of the Order of the Star of India (KCSI) was one of the three classes of the Most Exalted Order of the Star of India, an order of chivalry founded by Queen Victoria in 1861.

Assam with Myanmar (Burma). People from various regions relocated and established themselves in Manipur at different points in Manipur's historical timeline. The individuals from the Western region were referred to as "Nongchup Haram," while those hailing from the Eastern region were designated as "Nongpok Haram." They entered the Meitei society by establishing new yumnaks (surnames) or being accepted into existing yumnaks. Their contributions exerted a profound influence on Manipur society. They carried cultures from the west and east.

According to history, numerous Manipuri kings have attempted to improve their diplomatic ties with their neighbors. Historical ties between Manipur and the surrounding neighborhoods (constituting the early states) exposed how Manipur's government was affected by social, political, and economic factors. The following are some examples of the connections between Manipur and its neighboring countries.

1. *Amidst the Complex Dynamics of the Manipur–Burma Relationship.*

In 1563, the kingdoms of Manipur and Burma established an official partnership. Documented in the Cheitharol Kumbaba, the Meitei king, Mungyamba (r. 1562–1597), has invaded the twin cities of Mungkhong and Mungyang across the Ningthee River, expanding the territorial boundaries of the state beyond the Ningthee River. Hitherto, the river Ningthee, known as the Chindwin River in Burma, marked the easternmost boundary of Manipur. This victory is the first-ever recorded contact with Burma in the Cheitharol Kumbaba. The relationship that developed henceforth was virtually marked by hostility and aggression.

Diplomatic marriage partnerships were established between Burma and Manipur in their historical interactions. The historical archives chronicled the marriage alliance between the royal lineages of Manipur and Burma, with the first instance dating back to the eleventh century. The records revealed the marriage union between King Kyansittha (r. 1084–1112) and a princess from Manipur. According to Cheitharol Kumbaba, Queen Tourangbi, who was wedded to the King of Ava, departed for her place of birth in the year 1637. In 1704, the Ava king dispatched envoys to propose a matrimonial alliance with Princess Chakpa Makhao Ngambi of Manipur. The marriage partnership benefitted both kings in maintaining a mutually beneficial connection. An example of this was when King Sane (r. 1698–1714) of Burma subjected his Manipuri wife to torture and demoted her from the position of "Chief Queen" (Singha, 2016, pp. 875–876). The frequent military campaigns by King Garibniwaz against Burma ensued. Manipuri King Garibniwaz was directed to invade Burma as a mark of avenging the ill-treatment of his sister. This was an important reason for the overshadowing of the matrimonial alliance.[15]

[15] The conflict between the Kingdom of Manipur and the Kingdom of Burma, referred to as Ava, originated from a sequence of events linked to familial relationships and political disputes. After the demotion of his sister, Awa-Leima, who was married to the Burmese King, Manipur King Charairongba experienced significant resentment. Before his death, he assigned his son, Pamheiba (later Garibniwaz), the responsibility of avenging this affront to their family. The tensions intensified following the arrival of a delegation from Burma, accompanied by Samsok emissaries, in Manipur to seek the hand of a Meitei princess in marriage. Although King Garibaniwaz received

The diplomacy of marriage was unsuccessful in preventing terrible hostilities between Manipur and Burma during the eighteenth and early nineteenth centuries. This was the case even though matrimonial alliances were a well-established practice. One of the prime factors that caused the inevitable conflict between Burma and Manipur was the Shan state in the Kabaw Valley. The royal chronicle is replete with accounts of military expeditions in the Shan state of the Kabaw Valley. The Shan had remained a formidable power, but after the decisive defeat of the Shan in 1470 by the Manipuri King Kiyamba (r. 1467–1508), the Shan power began to decline. This resulted in the Shan state being attacked by the Burmese from the south, by the Manipuris from the west, and by the Chinese from the north. The resulting conflicts naturally embroiled the Burmese and the Manipuris, too.

Notwithstanding the conflicts, other contact forms also developed between Burma and Manipur. Records of trade and other social exchanges soon became a regular feature. The reign of King Khagemba (r. 1597–1652) witnessed not only military conflict with Burma but also increasing contact with the Burmese, the predominant people of Burma. Some of the Burmese, who had been war captives, were even employed as musketeers by Khagemba.

A turning point in the history of Manipur–Burma relations began with the reign of Charairongba (r. 1697–1709). Until the reign of Paikhomba (r. 1666–1697), Manipur did not experience an alarming setback from the Burmese depredations. The situation, however, began to reverse subsequently, with the Burmese becoming a power to reckon with under the Toungoo dynasty (1510–1752). Charairongba maintained a diplomatic alliance with the Burmese and attempted to establish friendship through matrimonial diplomacy. The king gave his daughter, Chakpa Makhao Ngambi, to the King of Ava (Burma), Taninganwe, in marriage. This policy stabilized economic communication and social and cultural exchanges for a while. However, the political ties between the two kingdoms did not remain as Charairongba had presumed.

News of the Toungoo ruler's cruelty and the ill-treatment of his daughter offended the king, who entrusted his son and successor, Garibniwaz, to avenge the misconduct and insult by the Burmese ruler. Garibniwaz succeeded Charairongba in 1709 and ruled until 1748 CE. He was the most illustrious ruler of Manipur, and the kingdom witnessed tremendous transformation under his victorious reforms and military prowess.

During the reign of Garibniwaz, also known as Pamheiba, the Toungoo dynasty of Burma was slowly disintegrating, and the Manipur king had successfully subdued several Burmese raids at the Kabaw Valley. According to the Cheitharol Kumbaba, as many as seven military expeditions were sent by King Pamheiba to drive the Burmese out of the Kabaw areas, which they had continued encroaching on since the previous regime. Many Burmese were reportedly captured, along with many guns and horses. Pamheiba had been awaiting an opportune time to fulfill his

---

the Samsok people courteously, their presence intensified his anger, escalating tensions between the two kingdoms.

father's wish, to avenge the insult of the Manipuri princess, which implied an insult to the state, and thus, he prepared for it.

The Toungoo ruler, King Taninganwe (r. 1714–1733), had sent his envoys to the court of Pamheiba with the proposal to marry the Manipuri princess, Tampha Wangamlon. Pamheiba conceded to the proposal and sent the message through the Burmese emissaries to receive the bride at an appointed time the following year at the Chindwin River, which marked the boundary between the two kingdoms. The following year, the Manipuri king stationed his forces at the appointed place, masquerading as a marriage troupe. The sequel was the successful offensive against the Burmese armies. Many men and women were brought to Manipur as captives. The retaliating Burmese forces could not forge a victorious campaign against the mighty Manipuri king, and in the subsequent years, several expeditions were sent to overpower the Burmese.

The next phase of importance in the history of Manipur-Burma relations is in the nineteenth century. Burma under the Konbaung dynasty (1752–1886), established by King Alaungpaya (r. 1752–1760), witnessed a radical political transformation. King Alaungpaya was hailed as the new national hero as he consolidated most of the scattered states, especially North Burma, and reinstated the strength of Burma under his control. The military expeditions that were now sent against Manipur comprised superior weaponry, and thus, for the first time, Manipur was faced with renewed Burmese vigor and advanced armament. The result was palpable. Manipur came under the aggressive and rugged hands of the Burmese. Many men and women were taken as prisoners, including the members of the royal house.

2. *Marriage Alliances Between the Royal House of Manipur and the Ahom Kingdoms.*

Like other world monarchies, Manipur has followed the tradition of matrimonial alliances with neighboring kingdoms since the early days. The kings of Manipur adopted the marital alliance policy as a peaceful diplomatic step with their neighboring kingdoms, such as Ahom, Takhels (Tripura), Cacharis, and Burmese. Such marriages helped negotiate various conflict situations. In such marriages, royal women were used as hostages in state politics. Such marriages were a common phenomenon in the pre-colonial periods as an external policy to develop friendly relations with the neighboring kingdoms. All these marriages were recorded in royal chronicles like Cheitharol Kumbaba, Ahom Burunji, Tukhungiya Burunji, and Rajmala. Other literary sources supplemented all these records.

The royal houses of the Manipur and Ahom kingdoms had a long history of matrimonial alliances. Chronicles like the Burunji and the Cheitharol Kumbaba provide a detailed record of these ties since their early days, bringing a cordial relationship between them. Tukungiya Burunji provides a clear account of Bheigyachandra's diplomatic moves and ultimate success in the Ahom, which helped liberate Manipur from the Burmese. King Hsinbyushin (r. 1763–1776), who had the ambition to conquer Manipur and Thailand, invaded Manipur in 1764 at the head of a well-equipped army. Failing to save the palace, King Bheigyachandra (r. 1759–1760 and r. 1767–1798), also known as Jai Singh, fled to Cachar got the support of Kritichandra

Barbarua, a powerful minister and kingmaker of the Ahom court. Jai Singh sought the help of the Ahom king. The Ahom king, Rajeshwar Singh (r. 1751–1769), sent a giant army to rescue his palace. Manipur had always been in cordial relations with Assam.

King Jai Singh proposed a matrimonial alliance with the king of Ahom and gave his daughter, princess Kuranga Nayini to King Rajeshwar Singh in marriage in 1768. It was the most remarkable marriage alliance between the Ahoms and Manipuris. Manipuris were settled in villages of Assam after marriage. The Ahom king gave six elephants to King Jai Singh. With skills befitting a king of a cultured country, Jai Singh effected a friendship with the Ahom ruler, who agreed to send an expedition to Manipur to drive away the Burmese. Cheitharol Kumbaba mentions that in 1767, Jai Singh invaded Manipur with a force of eighty thousand Assamese. Finally, the Burmese were driven out of Manipur with the help of Ahom forces.

According to the royal chronicle of Ahom Burunji, Kuranga Nayani was made the chief queen of Ahom. After the death of Rajeshwar Singh, the Moamarias, a Vaishnavite sect of upper Assam, revolted against Lakshmi Singh, brother and successor of Rajeshwar, and occupied the capital of Ahom. Kuranganayini took a commendable role in regaining the throne of Lakshmi by killing Ragha Moran, one of the principal leaders of the rebellion. She was publicly acknowledged for her service in the Ahom kingdom. Manipur king, Jai Singh again sent a Manipuri force of 700 armies to help the Ahom king, Gaurinath Singh (r. 1780–1794) against the fresh Moamaria Revolt of 1783.

In 1786, a more serious rebellion broke out as a civil war. Ahom king, Gaurinath Singh sent messages to the kings of neighboring kingdoms, including Manipur, for help in suppressing it. There were references to the matrimonial alliance between Manipur and the Cachari/Kachari kingdom of Assam. Kacharis were called Mayangs by the Manipuris in those days. Those Manipuri princesses married to the Kachari kingdom were known as Mayang Leima (Kachari Queen). Cheitharol Kumbaba mentioned the marriage of Mayang Leima Monshambi in the year 1557.

The cultural amalgamation resulting from these alliances contributed to the shared heritage of the two regions. It also fosters mutual understanding and cooperation. These alliances facilitated the cultural exchange between the two states.

In summary, these marriage alliances were about political strategy and played a crucial role in shaping cultural bonds and historical narratives between the two kingdoms.

3. *Matrimonial Alliance Between Manipur and Tripura Kingdoms.*

Tripura, a northeastern state of India, was an essential kingdom at the time. Tripura had a cordial relationship with the Manipur kingdom, though they occasionally engaged in conflicts. There were regular matrimonial alliances between the two kingdoms.

From the early period, Manipur was called by the Tripuris as Mokhali or Mekhali, whereas the Tripuris were known as Takhels by the Meiteis. In the pre-colonial period, the two kingdoms encountered each other in many aspects, such as trade, matrimonial alliances, diplomatic ties, and dwelling. The political ties were filled with cooperation, occasional conflicts, and warfare. The royal chronicles and Takhel Ngamba, a Manipuri historical account of the war against the Tripura, give details about the two kingdoms. The alliance between Burma and Tripura invaded Manipur in 1723. Commander Pancham Rai and many other Tripuris were taken as war captives in this war during King Garibniwaz. The Manipuri king took many war booties, including muskets, shields, bows, arrows, gongs, and clothes, and Tripura was hugely defeated.

The first matrimonial alliance between the royal houses of Tripura and Manipur came when Tripura king, Raja Taidakshin, grandson of King Trilochana, married a princess of Meitei. Akoicham Yangnu was also married to Tripura king, Raja Yosodhar Manikya (r. 1600–1623). The relationship between Tripura and Manipur improved during the reign of King Jai Singh. It was due to the change in the Tripura policy of Manipur under the king. In 1761, Tripura sent an elephant to the Manipur king as a gift. The Meitei king, Jai Singh, gave one of his daughters, Princess Hariseswari, in marriage to the Tripura king Rajdhar Manikya II (r. 1785–1804) during his pilgrimage visit to Tripura. This marriage brought a friendly relationship between the two kingdoms. After the marriage, the Tripura king presented 15 boats and seven hundred silver coins as a gift to the Manipur king for the journey.

The new queen, Hariseswari, installed the idol of Sri Radha Madhav,[16] brought from Manipur, in the royal compound. The people of Tripura treated the temple as an important religious site. The priests, musicians, and attendants of the queen from Manipur were allowed to settle in a village called Mekhlipara (village of Manipuris) near the royal palace.

During the "Chahi Taret Khuntakpa" (a dark historical event of Manipur in which Burmese forces devastated the Manipur Valley for 7 years), several Meitei people migrated and settled permanently in Tripura. There were other instances of royal marriage in Tripura, who were from a typical family.

The kings encouraged the Manipuri princesses to construct Meitei temples, organize festivals, and promote their culture. Many Meitei princesses were given the title of Chief Queen (Maharani) in the royal palace. The kings used to visit each other at their coronation ceremonies. However, no historical evidence supports the claim that the Meitei kings married Tripuri princesses. Nevertheless, the marriage alliances allowed the two kingdoms to maintain amicable ties.

The alliances brought great political and cultural significance. The Meitei culture was thus promoted and flourished in Tripura through such relations. The Meiteis in the royal house and dwellings in many places of Tripura were the descendants of Manipuri queens who were married to the kings and nobles and migrated at different periods of history. The Meiteis have been an essential community in Tripura. It is believed that the establishment of Agartala, the present capital of Tripura, was closely associated with the Meitei princess, Purnakala, also known as "Bamon Leima" or Brahmin Queen.

---

[16] The Hindu deity Radha is a well-known goddess in Hindu mythology, and Madhava is another name for Krishna.

7.3 Gender in Manipur

7.3.1 *Historical Background and the Socio-Economic Context of Manipur*

The Meitei society was carefully organized around the pana system, consisting of seven salais. Everyone in the Meitei society was assigned to one of these divisions, called pana, which was a crucial element of their social identity because the pana played an essential role in shaping the social, political, and economic life of the Meitei. The pana system underwent reorganization during the reign of King Loyumba (1074–1122), expanding from four to six divisions. Initially, there were four panas: ahallup, naharup, khabam, and laipham. King Loyumba organized his kingdom into six lup divisions: Luplel-Lup, Thouja Lup, Khongcha Lup, Lupkhuba Lup, Chingja Lup, and Khaija Lup (Singh, 2015, p. 192). According to Hareshwar, the term "lup" was the former name of "pana," a term introduced by Shans (Goshwami, 2019, p. 176). These six lups were perhaps the basis for the development of the pana system at a later historical period. Subsequently, the number grew to six with the addition of two more panas: hidakphanba and potsangba. He codified the division of labor based on Yumnak. This system was recorded in the Loyumba Shilyen, a royal edict for the socio-economic organization of this kingdom. Membership in a specific pana was crucial for social recognition, as those without a clear affiliation were excluded from participating in community events and public competitions.

Initially, the pana system was structured according to geographical and military requirements. Panas were initially meant for military associations, so they became constituent parts of the state's militia. In peacetime, social events and traditional games were conducted based on pana. Women were debarred from the panas. However, it transformed and became focused on sageis or yumnaks, mostly linked to trades or professions.[17] As a result, the organization became more efficient and focused on its functions rather than its geographical boundaries. In a later period, panas played a significant role in the country's economic life.

Each pana regularly provided a contingent of men to serve under the lallup, a compulsory state service. Traditional games and festivals such as "Sagol Kangjei (polo), Khong Kangjei (wrestling along with hockey), Mukna (wrestling), Yubi Lakpi (snatching of Yubi like modern rugby), and Hiyang Tanaba (boat race) were organized according to these panas, underscoring the pana organization's central role in the Meitei social structure. Women also played traditional games like Hiyang Tannaba (boat race), Thang-Ta (sword-spear), and Kang Sanaba (the biggest indoor

[17] The pana system was a significant unit within Meitei society, dividing the population into pana divisions. Meiteis were not considered if they were not affiliated with any pana. The pana's role was unique, as individuals with uncertain panas were prohibited from participating in social games and competitions. Pana was initially territorial for administrative and military purposes but later shifted to sageis or yumnaks, each with specific trades or professions, leading to a functional basis.

game of Manipur, in which a creeper seed, Kang, must hit the Kangkhil on the floor). As most of the traditional games had a military attitude, women did not play most of these traditional games."

Lallup was a feudal service rendered to the state by the able males, which meant war organization (Hodson, 1908, p. 59).[18] The system works as a military service during the war and the state's economic and other services during peace, every 10 days out of 40 days. It works through the administrative and social divisions. For this, males were given military training in warfare skills in areas where the lallup system was in operation. Kings honored those who could show merit in war and hunting by offering high positions in the administration and land, and even by providing princesses. The lallup-excluded villages paid tributes in salt, fish, pottery, coins, iron and weaponry, and timber. The system had its direct objective of mobilizing the services of adult males between the ages of 17 and 60 years for polity formation and consequent consolidation of the Manipur monarchy. In contrast, the women were forced to take the householder's role in the latter's continued absence from home in war campaigns.

Thus, it was concluded that there had been a division of labor since the early days of Manipur society. Males performed physically hard work, such as waging war, hunting, and tilling the grounds, whereas women performed domestic household chores, weaving, spinning, husking, and cooking. Again, the lallup system compelled women to perform the exclusive marketing duty of looking after and educating their children. Loiyamba systematized the administrative division of the land by creating six lups (divisions): Khongcha lup, Nongmai lup, Angouba lup, Leichol Lakpa Tolong Khomba lup, Khurai lup, Lipphambam lup, and Hangjeng lup. The introduction of these six administrative units was a significant contribution to the strengthening of the lallup service.

7.3.2 Meitei Women in the Public Spheres: Political and Social Movement

In the history of Manipur, women have not been silent spectators. Besides engaging in domestic work, the role played by Manipuri women in the socio-economic and political spheres since the early days is worth mentioning. They have strengthened their collective spirit of womanhood, and through solidarity, they have fought fearlessly against injustices. The anti-imperialist Women Uprising of 1904 and 1939, locally known as "Nupi Lan," has been a glaring example. After Manipur was defeated at the hands of the British, the Manipur seat of power, Kangla, was occupied by the British. Churachand, a 6-year-old child, was appointed by the British government as the King of Manipur. However, it was against the wishes and expectations of the local people. According to the traditional convention, people believed

[18] The word "lal" means war, and "lup" implies organization.

that ex-Maharaja Surachandra (r. 1886–1890) or his eldest son would be selected as the new raja. However, the actual power lies in the hands of the newly appointed political agent, Major Maxwell. He took harsh measures under the new British government, including collecting paddies and animals from the surrounding villages to be kept at the British residency. The disarmament policy in which arms and ammunition were collected, and the banning of practicing the Meitei martial art Thang-Ta[19] under the Arms Act of 1878, further kindled the masses. The Kanglasha in front of the Uttra was demolished. The British used the standing army of Manipur in several works, such as the collection of rations from far-off places and road construction works, for free.

Under the new British administration, the lallup system was abolished. Above this, a house tax was imposed. Villagers were expelled beyond the reserve areas. British authorities tried to acquire a sum of two and a half lakhs of rupees from the state of Manipur as a token of punishment for the disturbances of 1891. The British export and import policy further increased the burden on the people of Manipur. Under the new policy, thousands of mounds of rice were exported for free to Kohima. As a result of this policy, the people of Manipur began to face rice scarcity in Manipur. People were very much annoyed at the autocratic policies of the British authorities. All these measures were treated as a great insult to the Manipuris. These became an unbearable burden to the masses. Anti-imperialist movements broke out. On July 6, 1904, Khwairamband Market was burnt. The bungalows of British captains Nattal and Dunlop were set on fire on the 15th of the same month. As a punishment for such crimes, political agent Major Maxwell ordered the rebuilding of the burnt bungalows by collecting teaks from Kabaw Valley. People were shocked at the merciless attitude of the British authorities and decided to disobey the British order. Six Raj Kumars were arrested, as they were accused of being the main instigators behind the revolt against the British. Maxwell ordered the expulsion of these six principal instigators after a court constituted by British authority gave a similar statement. The agitated womenfolk of Manipur then fought against the British. Many women received bodily injuries and were hurt. Khwairamband Bazar was closed, and thousands of women launched big demonstrations at the political agent's office. Even the police could not control the protesting women. Law and order in Imphal were halted entirely, and the British authorities finally decided to order the reconstruction of the bungalows. This women's movement was a significant milestone in the history of Manipur. Again, the event would remain as the highest watermark of Manipur womanhood when those unlettered females, by dint of their single-minded determination, undeterred tact, and sheer grit. It brings about a change in administration. The revolution exhibited a culture of self-confidence, determination, willpower, and strength among the collective women. It led women to come forward and revolt against the unjust order, resist injustice, and show bravery when the situation called for it.

[19] Thang-Ta is an ancient Meitei martial art. It is an armed technique that mainly uses swords, called "thang," and spears, known as "ta," and is dedicated to fighting skills and worship.

The 1939 Women's Uprising (Anisuba Nupi Lan) was mainly due to the widespread famine in Manipur. It was a fact that the famine was due to the mismanagement and maladministration of the government. As the Maharaja permitted rice export outside Manipur, the Marwari capitalists exported much rice, and the price of a sangbai[20] was one rupee. Marwari installed 14 rice mills where bulks of rice were collected from different places in Manipur. The people of Manipur were inflamed over the extreme hardships caused by the government. Several taxes were imposed, such as Chandon Senkhai and Potthang Begari. The population remained silent in opposition to the king's administration. The people have long been frustrated by the oppression imposed by British power. The agony of hunger had also sparked the first drive for the challenge. Women of Manipur were offended by the oppressive governance of the British government. They started their agitation on December 12, 1939, by prohibiting the exportation of rice. Ima Keithel was deserted for 10 days, which halted all the affairs of trade and commerce. Hijam Irabot and other members of the Mahasabha were arrested for instigating the agitation. It was an uprising against the hardship of British imperialism and the aristocratic rule of the British and capitalists. The king of Manipur brought some considerable changes in the administration of the state to stop the intensity of the movement, with instruction from the governor of Assam. The women's agitation of 1939 profoundly impacted every aspect of the lives of the people of Manipur (Fig. 7.1).

The twentieth-century women's uprisings called Nupi Lan are seen to have revived in the form of a women's movement by holding flambeaus in their hands. The movement caught the limelight when the womenfolk voiced opposition to the government taking up a policy to permit the sale of liquor in the state. In 1975, every locality in Manipur formed the Nisha Bandh to control the society's sales,

Fig. 7.1 Nupi Lan monument. *Note*. From Nupi Lan movement: fighting spirit of Manipuri Women, by Northeast Bullet, 2023a, 2023b, https://www.northeastbullet.com/nupi-lan/

[20] A large traditional bamboo basket of Meitei is used for measuring the quantity of grain.

Fig. 7.2 Meira Paibi (women torchbearers). *Note.* From Meira Paibi: spirit of Manipuri Women, by Northeast Bullet, 2023a, 2023b, https://www.northeastbullet.com/meira-paibi/

consumption, and evil effects. Meira Paibi[21] emerged as an active response to such problems (Lukram, 2019, p. 176). The Meira Paibi used the torch to fight against the social injustices in various forms of Manipuri society. The Nisha Bandh movement was integrated with the Meira Paibi movement. They played a pivotal role in fighting against the social injustices. They stood against the Armed Forces Special Power Act (AFSPA), 1958, which was enforced to maintain law and order (Sudhir, 2013, p. 74). With this heavy militarization, people's fundamental rights were grossly violated and killed. Meira Paibi has been revolting to remove such draconian laws. In the villages and localities, Meira Paibi's main activities are vigilant night duties, as human rights and social crimes used to be committed in the dark. Irrespective of their social background, women participated as Meira Paibi in mass rallies, sit-in protests, hunger strikes, road blockades, and night torch rallies (Fig. 7.2).

[21] Meira Paibi groups are structured at three interrelated levels: state, district, and locality. No precise figures exist regarding the number of Meira Paibi groups. However, it is estimated to be in the hundreds or thousands (Salam, 2023, p. 4). They are connected through district-level groups, which link to state-level organizations. Current socio-political issues in Manipur are primarily categorized among state-level Meira Paibi groups. For instance, Apunba Manipur Kanba Ima Lup (AMKIL), established in 2000, led the mobilization and protests of the ceasefire agreement between the National Socialist Council of Nagaland (NSCN-IM) and the Government of India (GOI). All Manipur Women's Social Reformation and Development Samaj, also known as "Nupi Samaj," leads the protest movement against drugs and alcohol. The categorization of issues serves the function of designating a Meira Paibi group responsible for organizing around a specific socio-political matter.

7.3.3 Meitei Women in the Belief and Religious Spheres

The Meitei women play significant political roles and substantially impact society's socio-economic and religious dimensions, distinguishing them.

After King Garibniwaz (r. 1690–1748.) converted the state religion of Manipur to Vaishnavism in the eighteenth century, there was a significant change in the culture and religious practices. The conversion of the state religion has far-reaching effects on every aspect of society. The Meitei community began to adopt Hinduism along with the indigenous Sanamahism. The approach entails the worship of deities from both Hinduism and Sanamahism, creating a spiritual landscape that reflects the intricate interweaving of these two traditions. The local gods were identified with Hindu gods and goddesses. Despite cultural assimilation, there has yet to be a complete transformation in the traditional Sanamahi religion. This fascinating amalgamation is a testament to the cultural richness and history within one's community.

Meitei, since time immemorial, worshipped the spirit of deities related to a female medium known as "maibi," performing dance rituals in a trance-like state believed to be performed by ancestral and local deities (Lais). This practice serves as a form of worship and cultural expression, emphasizing the role of women as conduits between the spiritual and human realms in Meitei traditions. Most of the spiritual processions are led by females. Females led in the birth, wedding, and worshipping events. Furthermore, through their responsibilities in family upbringing, women contribute to transmitting spiritual knowledge, thought processes, beliefs, culture, and wisdom across generations. Consequently, women, in their roles as mothers, serve as agents of cultural and ideological reproduction by instilling societal expectations in their children, regardless of gender. Every village has a different abode of gods and goddesses, where several dignified female deities, such as Thongak Lairembi, Tangkhul Nurabi, Haoreima Sampubi, and Phouoibi, were worshipped. Maibi performs offerings and prayers at different times of the year for the welfare of the land and the people. Some rituals are Sajibu Lakyen Taiba, Kongba Leithong Phatpa, Sangsatpa, Thoudou Langdon Touba, and Sanamahi Cheng Hongba.

During the spring, people in villages and urban areas performed different rituals with great joy and happiness. For instance, people wore a ceremonial dress on the first day of the Lai Haraoba ceremony. They worshipped the spirits by inviting them to water places, such as ponds, canals, or rivers, as the Meitei people believed that water bodies were the places where spirits resided. Special items were offered to the water bodies. While the maibi sings by uttering the invocation and dance, the "pena" (a traditional stringed musical instrument) player plays the instrument. The drummers play the drums of the temple. The maibi says the oracle amid the sound of the bell ringing in her hand. This ritual is known as Lai Ikouba. Simultaneously, ancestral worship is also practiced by honoring offerings, establishing a connection between the living and the departed. This blend of Hindu and indigenous practice reflects the community's harmonious co-existence of religious traditions.

Lai Haraoba, the merrymaking of the sylvan deities, is a traditional religious ritualistic festival of the Meitei, usually celebrated from April to May in different places in the Manipur Valley. The celebration continues for about two weeks with rites and rituals accompanied by songs and ritual dances to please the guardian gods and goddesses of the village. The Meitei community has rituals related to female spiritual possession, particularly during the Lai Haraoba festival. During the Lai Haraoba festival, many rituals are performed. The crucial rituals include Lai Ekouba, Lai Lam Thokpa, Kangjei Sanaba, Laibou Shaba, Sharoi Khangba, and Tangkhul Nurabi Saba. If the rituals were performed without disturbances, such as natural disasters, the people believed the village would be prosperous and protected from evil spirits and unwanted happenings.

Maiba (priests) and maibi (priestesses) play distinctive roles in performing the rites, rituals, and prayers during the Lai Haraoba celebration. It is believed that Lord Sanahami and Lord Pakhangba taught the amaiba[22] and amaibi[23] the particulars of the religious rituals to be offered to the deities when ordinary mortal humans faced sorrow and suffering. If proper prayers and offerings are made to the gods and goddesses, the devotees will be relieved from their sufferings. At the same time, peace, progress, and prosperity will prevail in their lives.

The contribution of Meitei women to Manipur society is diverse. They occupy a unique place in social life. The historical roles of women in the Meitei community have undergone a significant transformation. Historically, the queen headed the medicine department called Ametpa Loishanga, established by King Mungyamba (r. 1562–1597). Traditionally, women were primarily engaged in domestic activities, contributing to the household and family responsibilities. However, during war or economic necessity, women assumed economic roles, notably in the marketplace, showcasing their adaptability and resourcefulness. Women's role in the socio-economic sector was immense. In contemporary times, women actively participate in all spheres of society, breaking traditional norms and barriers. They have entered diverse fields, including education, politics, entrepreneurship, and other professions.

7.3.4 Meitei Women as Entrepreneurs

Many individuals prioritize government employment. Their primary professions may center around public service administration, education, healthcare, and industries associated with government employment. Nevertheless, much of the population is involved in agricultural pursuits. Weaving is a highly significant profession.

---

[22] Amaiba (plural Amaibas), often known as maiba, are males well-trained to perform certain rituals in traditional devotional activities.

[23] Amaibi (plural Amaibis), also known as a maibi, is an indigenous shamanic priestess of the Sanamahi tradition in Manipur.

Fishing is a significant occupation. A significant proportion of individuals are engaged in entrepreneurship.

Historically, the community commercial location lay at the center of towns, where people from all over the region would gather and exchange or buy goods with neighboring communities and regions. Trade networks were likely established through overland routes and river connections. Most of the villages have small markets. Major agricultural and commercial products are transported to major markets, like the Ima market. Present-day good transport facilitates marketing.

Although the traditional Manipuri society tended to function on a patriarchal model in statecraft, certain aspects of the society had space for women to function autonomously. One such example was the local market or the bazaars in different areas of the state. The state is famous for the Ima Market. The Ima Keithel (known as the Mother's Market) is a unique all-women market in Imphal that is not only a bustling economic center but also holds cultural importance. It reflects the economic activities and entrepreneurship of Meitei women. The market is well-known worldwide for being the only market for womenfolk.

The nature of society and the lallup system yield a societal structure in which women play a pivotal role in the state's economic landscape. This role is exhibited clearly in the local community market or bazaars. There was an intense yearning for each family to be self-sufficient. This yearning made women take up weaving in their free time. Through time, weaving became a profession closely connected with women.

Women's active participation in the market's economic spheres is not ipso facto for an active and dynamic atmosphere of women's entrepreneurship. However, in recent times, women have started entering various entrepreneurship ventures. Leading ventures are in the handloom and handicraft, where weaving and the food processing and beverage industries are a mainstay. They have shown their managerial skills, showcasing their resourcefulness and adaptability. Education and technology development have dramatically impacted female entrepreneurs' growth. Further development in digital and financial literacy, with particular attention to budding female entrepreneurs, can profoundly impact increasing women's participation in entrepreneurship.

Some of the leading women entrepreneurs of the state are as follows: (1) Hanjabam Shubhra Devi is the proprietor of Meira foods. The business is a food processing industry that deals in pickles and fermented items. It aims to promote the state's indigenous foods, especially pickles and fermented items. Shubhra Devi's achievements include being honored as the Outstanding Entrepreneur at FICCI (Federation of Indian Chambers of Commerce and Industries) 2018 and receiving special recognition from the National Small Industries Corporation. (2) Tongbram Bijiyashanti—Prime Minister Narendra Modi acknowledged her twice on his Maan Ki Baat program. Bijiyashanti successfully extracted eco-friendly fabric from lotus stems and is the proprietor of Sanajing Sana Thambal. (3) Moirangthem Muktamani Devi received the coveted Padma Shree Award (the fourth-highest civilian award in India) in 2022. She is well known for her hand-knitted woolen shoes, which are manufactured in the Mukta Shoe Industry and were founded and managed by

Muktamani herself. (4) Arambam Renu Devi is the proprietor of Renu handloom and handicraft. The company employs around 4500 artisans and has showrooms in Bangladesh and Guwahati. The rise of digital platforms and online connectivity has the potential to profoundly change Manipur's economy. This technological shift can create fresh business opportunities, foster entrepreneurship, and streamline global trade. The increased connectivity can amplify trade and economic engagements, offering local businesses a platform to extend their reach to a broad market.

The above-mentioned women entrepreneurs and many other women in the state have continuously fostered their private enterprises by leveraging unique local crafts and items. Their skills in their respective fields are commended and honored on both the national and international platforms. Many of their products are now valued brands in terms of quality, sustainability, and profitability. However, they continue to pursue endless research towards enhancing their ideas and knowledge. They further seek to upgrade their products by acquiring skill training and technological know-how, sometimes organized by themselves or with collaboration from agencies. These women entrepreneurs desire to reach a broader scope and are constantly collaborating with like-minded and concerned individuals or groups for support or cooperation for mutual growth.

7.3.5 Meitei Women in the Academic Sphere

The structure of the traditional Meitei society needed more scope for formal education. The people did not warmly accept the initiation of formal education under the British. Women were incredibly discouraged from getting an English education until the mid-twentieth century. Hoax news was spread to hinder the process of women's enrollment in schools. However, this does not mean that education and women remained separated like the banks of a river. By the twentieth century, we had stellar leading women figures in academics and education.

Using their royal heritage to their advantage, these princesses of the Ningthouja dynasty were the forerunners: (1) M.K. Binodini Devi is considered one of the leading women in heralding and uplifting Manipuri art and literature. She is associated with several literary figures, cinematic directors like the legendary Aribam Shyam Sharma, musical societies like Roop Raga, radio programs, paintings, and many more. As a writer, she is credited for her historical-literary novels like Bor Shaheb Ongbi Sanatombi and Maharaj Churachand-gee Emung Manung. Her short story Nunggairakta Chandramukhi is one of the best short stories in Manipur literature. Her radio drama Ashangba Nongjabi was, and still is, well-received by the people (Roy, 2018). (2) Rajkumari Tamphasana had a keen interest in education. So, in recognition of it, her father, H.H., the then Maharaja of Manipur, Churachand Singh, established the first girls' school in the state, named Tamphasana Girls' High School, in 1935. The school is now upgraded as a government-recognized higher secondary school. (3) Maharani Dhanamanjuri—She was the queen and wife of Maharaj Churachand. Her generous donation of Rs. 10,000/– (ten thousand) led to the foundation and establishment of Dhanamanjuri College on April 6, 1946. The premier college in the state has now been upgraded to a state university under the name Dhanamanjuri University (DMU) since 2018. The university was built with three Dhanamanjuri Colleges (science, arts, and commerce), Ghanapriya Women's College, and LMS Law College, Imphal (Dhanamanjuri University, n.d.). (4) Another leading figure was Nongthombam Kanan Devi, the sister of Prabhabati Devi, the then Dowager Queen of Maharaja Bodhachandra Singh. To fulfill her last wish, the Ramakrishna Mission Society established a memorial school, Kanan Devi Memorial School, in 1987 and a college of teacher education, Kanan Devi Memorial College of Education, in 1992. The first institution currently runs from nursery to class XII, affiliating with the Central Board of Secondary Education. The college is presently running, being one of the oldest colleges for teacher education in the state (Official Website of Kanan Devi Memorial School, n.d.).

Contemporary Manipuri society presents us with an image of continuing traditional norms and dynamic development with increasing academic involvement. Parents look forward to educating their daughters in the same manner as their sons. There are even positive cases of female students outperforming male students at different levels of education and in academic-related careers. Nowadays, we can witness many able women professors in the state serving academia. Not only do they discharge academic duties, but they are also involved in administrative positions at various universities in the state.

7.3.6 Gender Roles in a Challenging Time

Gender roles and functions in the Meitei community have changed over time. Increased access to education and economic opportunities has shifted men's and women's participation in diverse career paths. Women now actively engage in political and social spheres, challenging traditional gender norms. Women were forced to fulfill multiple roles, from familial sustenance to public lives. Economic contributions from both genders have become more common, reflecting a more inclusive and evolving cultural landscape. Ima Market is a center that shows the collective identity of women.

In the Meitei community of Manipur, the distinctive gender roles reflect a society where women actively engage across various spheres, setting it apart from mainstream India. Notably, women play a pivotal role in the economic landscape through their involvement in the unique Ima market, which is exclusively for women and serves as a testament to their entrepreneurial spirit and financial contributions. The active participation of women must mark the political and social landscape.

The rise of digital platforms and online connectivity has the potential to profoundly change Manipur's economy. This technological shift can create fresh business opportunities, foster entrepreneurship, and streamline global trade. The increased connectivity can amplify trade and economic engagements, offering local businesses a platform to extend their reach to a broad market.

Digital platforms are instrumental in safeguarding and promoting one's cultural heritage. Online archives, digital museums, and platforms that highlight traditional arts and crafts serve to conserve the culture. It is crucial to balance technological progress and the preservation of genuine cultural practices.

The impact of technology on gender roles is in several ways. It has the potential to offer women opportunities to engage in the digital economy, access education, and pursue employment, thereby challenging traditional gender norms. However, there is a concern about the possibility of reinforcing gender biases and inequalities in online environments. Empowering women with digital literacy and ensuring equal access to technology is imperative, which will contribute significantly to promoting more inclusive gender roles.

Digital platforms can reshape the landscape of social interactions and community dynamics. These platforms enable the exchange of ideas and collaboration by facilitating connections within and beyond regions. Nonetheless, it is crucial to acknowledge and address potential challenges, including digital divides and the spread of misinformation. Although technology has the potential for positive transformation, it challenges traditional practices. Maintaining a delicate balance is crucial, enabling technology integration without jeopardizing cultural traditions' authenticity. This entails using technology to document and share traditions while sustaining their existence in the physical realm.

The digital landscape in Manipur faces significant challenges, mainly due to extended Internet shutdowns that have severely impacted digital businesses, remote workers, and students. This restricted access not only disrupts daily life but also hampers economic progress. Nevertheless, promising initiatives are working to enhance digital literacy among women in Manipur. One standout program is the e-Nupikeither initiative, which empowers Indigenous women artisans by providing essential training in digital skills and e-commerce. The goal is to help these women preserve their traditional crafts and expand their reach into broader markets.

Another vital initiative is the Digital Literacy and Online Safety Programme offered by the National Commission for Women (NCW). This program focuses on equipping women across India, including those in Manipur, with the tools to use the Internet safely, navigate social media, and manage email effectively. These initiatives are crucial strides toward bridging the digital divide and empowering women in Manipur. Investing in these programs can pave the way for a more inclusive and prosperous digital future.

7.4 Feasibility of Connectivity in Tourism Between Manipur and Northern Thailand

7.4.1 *Observation of Tourism Between Manipur and Northern Thailand*

Based on an empirical study of the feasibility of connectivity between the two regions of northeast India and northern Thailand, the data from observation and relevant interviews are consistent with the information from the literature review.

The relationship between India and Thailand is deep-rooted, longstanding, and multidimensional, ranging from ties among people to businesses. Indian culture has played an integral role in influencing many aspects of Thailand, including religion, ceremonies, language, literature, and food—the cultural ties between India and Thailand through time.

Bilateral relations between the two countries are rooted in history, social, cultural, trade, and people contacts. According to Trading Economics, India and Thailand share maritime boundaries in the Andaman Sea, including land routes; as mentioned, Manipur was one of the trade routes between the East and the West. Thus, the land routes created trade and relations between people in early kingdoms through the exchange of goods between each other; for instance, early kingdoms in northern Thailand, such as the Lan Na kingdom, traded through Burma to the eastern border of Manipur to India and China. This route was also mainly used for migration, pilgrimage, and cargo transportation for cattle and horse traders (Pemberton, 1979). The main exchange products are raw silk, tea, paper, and fruits from China that are exchanged for forest products such as beeswax, ivory, cotton, and ponies from Manipur (Kamei, 2003).

Furthermore, tourism in India and Thailand has grown significantly over the past decade. Unfortunately, Manipur and the northeastern states of India still have few Thai tourists traveling and are not very well-known compared to other parts of India. Thai tourists prefer to travel to other states, especially cities, which relate to Buddhist beliefs and other famous tourist attractions. Despite observation in the research field, Manipur has found a safe and clean place to travel with friendly people and good hospitality. The hotels and accommodations in tourist cities are of various levels and usually provide excellent and clean services. The low turnout of tourists in this northeastern state was due to the lack of information on Manipur tourism, including tourist attractions, activities, and exclusive shopping and souvenirs for Thai tourists. The idea of promoting international tourism between the countries is that Manipur tourism organizations should provide more travel information and present a beautiful image of cultural and natural attractions, including the friendliness of the Meitei people, directly to a target group of Thai people interested in cultural tourism. This can be achieved through targeted marketing campaigns, partnerships with travel agencies, and the use of social media platforms to showcase the unique experiences that Manipur has to offer. By highlighting its rich history, vibrant festivals, and exquisite handicrafts, the state can attract more Thai tourists eager to explore its cultural heritage. Meanwhile, this also creates local income.

Nevertheless, tourism in Manipur promotes fewer local attractions regarding handicrafts and gastronomy, at least in Thailand. According to observation and interviews with locals, Imphal is famous for its textiles and the distinctive women's market known as the Ima Keithel. It is also well-known for its various local products. However, with little publicity, the Thai people are almost unaware. In addition, there are fewer souvenir shops for shopping around the tourist attractions compared to Chiang Mai, one of the famous tourist attraction cities in the north of Thailand, which attracts the world with its historical, cultural, and natural identity and serves tourists with its traditional handicrafts and local food. This can earn more income for the local people, as the Tourism and Sports Ministry of Thailand reported that in 2023, Chiang Mai welcomed 7.8 million tourists, which is 1.56 million USD (The Tourism and Sports Ministry of Thailand, 2024).

Based on an interview, the informant highlighted that academic collaboration is a crucial factor contributing to tourism growth in these regions and northern Thailand. Additionally, a cultural exchange program facilitates interaction between the two regions. Given the numerous socio-cultural resemblances in art, crafts, cultural traditions, and dancing styles, it would serve as an additional means of strengthening cultural linkages. Furthermore, it analyzes the demand and supply dynamics of the tourism market. Furthermore, to develop a collaboration with Thailand to find similarities. Conferences or cultural exchange programs can be utilized to exhibit the culture and promote local products through showcasing and advertising.

It was found that tourism and sub-sectors have similar aspects in Manipur and northern Thailand. The feasible possibility is cultural and spiritual tourism; the local informant in Manipur mentioned that visiting the religious sites in both areas can be linked up. It needs to be explored through nonliterary sources that can bring a kind of interface and connectivity. For instance, Buddhist meditation, incorporated into the education system, incorporated all the traditional values, which makes education more contextual and cultural.

7.4.2 *Feasibility of Tourism in Manipur*

The World Economic Outlook (WEO) reported economic growth worldwide, establishing a correlation between performance and crucial industries, such as tourism. Economies with substantial travel and tourist sectors demonstrate significant economic resilience and high levels of economic activity. Countries heavily dependent on tourism for their GDP have experienced a quicker recovery from the epidemic's effects, unlike nations where tourism is not a significant industry. As shown in the report made by the United Nations Tourism, in 2023, tourism was one of the most important industries in the world, producing a total of about 1.5 trillion USD, and the World Travel and Tourism Council reported that tourism created 21.6 million new jobs, bringing the total to 295 million and set to grow by 23.3%, reaching 9.2% of the global economy.

Cultural tourism is gaining popularity faster than other tourism segments and worldwide growth. The UNWTO has recently confirmed that cultural tourism is a significant component of international tourism, representing more than 39% of tourism arrivals. The reference highlights the considerable influence of historical and cultural assets on the advancement of tourism. The historical and cultural heritage presents a genuine economic opportunity for governments to generate cash. The connection between culture, heritage, and tourism is crucial. Therefore, it is necessary to establish stronger connections between cultural heritage, which encompasses the customs, practices, places, objects, artistic expressions, and values developed and transmitted by a community across generations, and tourism, which has the potential to serve as a significant tool for sustainability.

1. *Cultural Tourism.*

The feasibility of tourism that affects tourism between Manipur and northern Thailand is the basis of the historical and cultural factors that have led to tourists' decision-making in cultural tourism. Manipur boasts a rich cultural heritage with archeological sites, museums, and ancient architecture. One notable example is the Kangla Fort, or Kangla Palace, symbolizing Meitei architecture's historical magnificence since ancient times. Additionally, Manipur boasts unique Hindu temples that preserve Hindu art and the local population's devotion to worship. Notable examples include the Shri Govindaji Temple, an exquisite temple situated in East Imphal, exclusively dedicated to Lord Krishna, and the Radha Krishnachandra ISKCON Temple, a renowned pilgrimage site in Manipur that draws tourists from around the world to visit. Moreover, the Sanamahi temple is renowned for its distinctive octagonal design and pyramidal roofing. It draws tourists throughout the year owing to its veneration of Lord Sanamahi and is regarded as one of the prominent temples in Manipur.

The Meitei people in Manipur have a unique lifestyle, including distinctive women's clothing characterized by beautiful cotton weaving patterns like those worn by traditional Lan Na women. Additionally, their local cuisine is less spicy and flavorful, with ingredients that resemble those used in the northern region of Thailand. These aspects of the Meitei culture are attractive to the Thai people compared to individuals from other parts of India. Manipur's essence aligns with the concept of cultural tourism, catering to the interests of tourists seeking to acquire knowledge about the region's culture and historical heritage, explore its architecture, and immerse themselves in the local way of life.

One of the festivals that is significant for promoting cultural tourism in Manipur. The Sangai Festival is the most anticipated event in Manipur. The Tourism Festival, organized by the Manipur Tourism Department annually, is named after the Sangai to promote the brow-antlered deer endemic to Manipur. The festival, initiated in 2010, seeks to highlight Manipur's extensive cultural heritage and indigenous traditions that remain largely unexplored by many. The festival, celebrated on a large scale, attracts individuals from across India and globally. The festival celebrates

Manipur's culture, handlooms, handicrafts, indigenous sports, music, adventure sports, and cuisine. The festival showcases the cultural richness of Manipur through music, dance, and diverse indigenous art forms associated with its tribes and communities. Ras Leela, a classical Manipuri dance form, is a well-known and significant aspect of the region's folk dance repertoire. The festival allows artisans and weavers to exhibit their creativity and skill in indigenous handicrafts and handlooms. The themed huts displayed at the heritage park provide insights into the lifestyles of various tribes. Local participants demonstrate their proficiency in traditional sports, including the renowned Meitei martial art Thang Ta, which integrates spear and sword techniques alongside other athletic activities (Fig. 7.3).

According to the study, promoting tourism between Manipur and Thailand could contribute to cultural tourism. Cultural tourism involves managing cultural heritage sites by focusing on local priorities for protection and preservation. It also respects local culture and maintains local resources in response to socio-economic needs and cultural aesthetics that can sustain cultural identity.

Besides, the climatic conditions are perfect for tourists. The then Prime Minister of India, Jawaharlal Nehru, described her as the "Switzerland of the East." The invading Japanese forces also called her "a flower on the lofty height" during WWII. The scenic landscape and natural topography can be utilized as tourist attractions. The state is full of lakes, hills, and wildlife sanctuaries. It has been found that Manipur is blessed with salubrious landscapes and a pleasurable climate.

Fig. 7.3 Sangai festival. *Note*. From *Manipur Sangai festival* [photograph], by The Ministry of Tourism of India, 2024, https://utsav.gov.in/view-event/manipur-sangai-festival-1

Loktak Lake is the largest freshwater lake in northeast India. The state has high prospects for ecotourism and blessed with endemic species. Loktak Lake is developing into an ecotourism hotspot. There are homestays built on the floating phumdis[24] of the lake, where tourists can stay to enjoy the scenic beauty of the lake (Manipur Tourism, n.d.) (Fig. 7.4). Keibul Lamjao National Park, the only floating national park in the world, is the home of the Sangai (Cervus Eldi Eldi). Sangai is endemic only to the Keibul Lamjao National Park. The lake and the park cannot be missed from tourists' itineraries. The Shirui Hills is the home for Shirui Lily (Lilium Mackliniae). The lily is also endemic only to the Shirui Hills. There is a vast prospect of developing the hill into a tourist hotspot.

2. *Spiritual tourism.*

Spiritual tourism as a form of cultural tourism is expanding as individuals increasingly seek personal spiritual growth and explore the spirituality of different cultures. In 2007, the UNWTO recognized spiritual tourism as the fastest-growing segment, becoming a trend—a newer form of pilgrimage tourism (Bone, 2013). According to the government of India's tourism policy, spiritual, and religious

Fig. 7.4 Loktak lake with its floating Phumdis, the unique floating islands made of decaying vegetation, soil, and organic matter. *Note*. From Loktak lake, Manipur: complete 2024 travel guide to India's only floating island [photograph], by unseen Northeast India, 2024, https://unseennortheastindia.com/loktak-lake-manipur/

[24] Phumdis are floating accumulations of flora covering an area of 40 square kilometers. They take on various geometric forms as they float on the lake's surface. The floating islands are unique to Loktak Lake in Manipur, northeastern India.

tourism has emerged as the predominant form of tourism, attracting domestic and international visitors amidst various other trends. (Ministry of Tourism of India, 2020). Tourism development in the northeastern region is a prime area of focus for the Ministry of Tourism of India, which is carrying out various activities to develop and promote tourism in the region. This includes the Swadesh Darshan and Pilgrimage Rejuvenation and Spirituality Augmentation Drive (PRASAD). Manipur is enlisted in the Swadesh Darshan Scheme for the spiritual circuit and spiritual tourism (Ministry of Tourism of India, 2018).

Thus, the feasibility of tourism in Manipur is to promote spiritual tourism. The Meitei have high spiritual values. A large section of the people is still influenced by the Vaishnavism traditions, which were rooted during the reign of Maharaj Bhagyachandra in the late eighteenth century. The revivalism of Sanamahism in the mid-twentieth century also profoundly changed the spiritual beliefs of the Meiteis. Spiritual tourism is a common practice in the community, where older members go on pilgrimages to sacred lands associated with Hindu gods and goddesses. This tradition reflects a deep spiritual connection and a profound reverence for the divine. The pilgrimage journeys are not merely physical travels but spiritual quests fostering a sense of devotion and cultural connectivity within the community. There are a lot of spiritual tourist centers, such as Moirang Thangjing, Hiyangthang Lairembi, and Sanamahi Temple at Haying Khongbal Tolong Yumpham, Govindajee, Kaina, etc. Though there is a lot of tourism potential in such places, there is a need for proper utilization and execution.

The Meitei community of Manipur has several storytelling traditions and overall literature, including folktales, myths, and legends passed down through generations. In addition, a longstanding practice exists of creating pottery and handicrafts, such as weaving, basketry, and wood carving. Also, various social customs and rituals, such as marriage ceremonies, funeral rites, and other traditional practices, play a significant role in Meitei society. The Meitei performed several cultural and religious festivals. People use different types of attire for various occasions.

By showcasing the richness of the Meitei community, visitors are invited to experience the vibrant festivals, cultural performances, and historical sites that contribute to the local economy. Also, by empowering the local artisans and craftsmen to create traditional handicrafts and highlighting the local cuisine by promoting traditional recipes, organizing food events, and offering cooking classes, the flavors of different communities can be shared. These initiatives create economic value while celebrating the culinary heritage.

The Meiteis perform several religious festivals in a year. The most important is the Lai Haraoba festival, which sheds light on the early religion of Manipur. Imoinu Eratpa is celebrated on the 12th day of the Meitei lunar month of Wakching and generally falls in January. On this day, people serve traditional dishes. People regarded Imoinu as a goddess of health, prosperity, and abundance in Manipur. In the month of Lamta in March, a spring festival is marked by traditional sports, folk dances, and the Thabal Chongba dance, in which men and women participate. Gender roles may be more fluid during such events, allowing for inclusive participation. Cheithaba, a formal announcement at the end of a year, was established during the reign of King Kyamba (r. 1467–1508), which is still practiced today. Cheiraoba, the new year of the Meiteis, is celebrated on the first day of Sajibu in April with rituals, feasts, and community gatherings. It involves cleaning houses and preparing and organizing these activities, which may include contributions from both genders. People arrange traditional cuisine feasts for their families, relatives, and friends. First, they offer food to their local deities and serve their ancestors at their home entrances. After enjoying the meal, people ascended nearby hills to offer prayer. The presentation of gifts by married ladies to their fathers and brothers is also an essential feature of this festival.

Kang Festival is celebrated to honor Lord Jagannath, Balabhadra, and Subhadra. This festival includes the offering of khechri, a dish made with rice and dal. Lord Jagannath leaves his temple in a traditional wooden car called "Kang" in Manipur, pulled by devotees who vie with one another for this honor. Heikru Hidongba is celebrated in September, and Heikru Hidongba is a boat race. Long, narrow boats are used to accommodate many rowers. An idol of Lord Vishnu is installed before the commencement of the race. Ningol Chakkouba is a remarkable social festival of the Meiteis celebrated on the second day of the Hiyanggei month in November. Married daughters of the family come to their ancestral house to celebrate with a feast. Gifts are presented by the male relatives to their sisters and daughters. It is a form of family reunion.

Apopka Khurumba, or the worship of ancestors, is another critical ritual by the Meitei: respecting their ancestors by offering eatables. The Ukai Phalang Kappa ceremony is held in the month of Phairen in February. Mera Men Tongba, also known as Mera Shantuba, is a tremendous ceremonial festival attended by the king and his subjects, both from the hill and plain, to promote integration and goodness among them. Kongba Leithong Phatpa (digging the hole at Kongba) is a ritual executed by Amaiba Loishang, or the priest department at Kongba River, by offering edible and inedible things to know the good and bad luck of the forthcoming year (Parratt, 2013, p. 51). During Langban in September, people offer edible fruits to their ancestors. Ancestral worship is one of the essential characteristics and religious features of the pre-Hindu society in Manipur. Different clans worship their ancestors in different months with offerings, separate fruits, eatables, and clan swords. With the coming of Hinduism in Manipur, tremendous changes occurred in religious ceremonies and cultural activities. With time, religious syncretism took place. However, most pre-Vaishnavism ceremonies are retained with the institution of maiba. The sixteen Samskaras (sacraments) of Hinduism reflect the profound philosophy of the religious purification rites and ceremonies for sanctifying the body, mind, and intellect of an individual. The rites-de-passage referring to pokpa (birth), Luhongba (marriage), and korou nonggaba (death sacraments) of the Meiteis reflect the religious philosophy of Sanamahism.

7.5 Conclusion

Since the pre-modern days, the women of Manipur were not passive players in society. The societal setup does not exhibit many gender biases. Although it cannot be denied that there were specific tasks assigned to each gender, the limitations were not rigid. Women took an active role in earning a livelihood for the family. In those days, women were primarily occupied in vocational trades like handloom and handicraft, such as weaving, pottery making, basketry, and fishing.

The limited access to education did not prove to be a hindrance to women's empowerment. State institutions like the lallup system had beneficial impacts on women. The system provided a scope for women's participation in the state's economic activities. They independently ran markets and bazaars. The Ima market at the heart of Imphal, the capital of Manipur, is a living testimony. The state machinery also had separate departments concerning women. This allowed women to participate in state policy during that time.

Unfortunately, women's involvement in the entrepreneurial and tourism sectors remained unsatisfactory. Despite efforts, the number of educated women and entrepreneurs remained low. The involvement of women in state machinery was also very much limited. The condition has seen significant changes since the beginning of the twentieth century. Women of royal heritage and well-to-do families became actively involved in education. They can be considered a pioneer woman entrepreneur. Their long-term vision facilitated an environment where women's roles in academia and entrepreneurship strengthened. The growing education sector, digital literacy, and financial literacy have played their part in the ever-growing space occupied by women in various trades and institutions. The impact of technology on gender roles is in several ways. It has the potential to offer women opportunities to engage in the digital economy, access education, and pursue employment, thereby challenging traditional gender norms. However, there is a concern about the possibility of reinforcing gender biases and inequalities in online environments. Empowering women with digital literacy and ensuring equal access to technology is imperative, which will contribute significantly to promoting more inclusive gender roles.

A discussion has been held regarding the challenges encountered in tourism between Manipur and Thailand. Empirical studies of cultural and natural tourist attractions in Manipur indicate that data obtained from observations and pertinent interviews align with findings from the literature review. Over the past 10 years, the tourism industries in India and Thailand have seen significant growth, with the peak travel season occurring from November to February each year, when the weather is favorable and the temperatures are moderate. Manipur remains a state with limited Thai tourist visitation and needs recognition relative to other regions of India. The limited information regarding Manipur tourism, encompassing tourist attractions, activities, and unique shopping and souvenirs for Thai tourists, contributed to this situation. Facilitating international tourism among the nations in Manipur Tourism organizations should enhance the dissemination of travel information and effectively showcase the appealing aspects of cultural and natural attractions, including the hospitality of the Meitei people, specifically targeting Thai individuals interested in artistic and spiritual tourism. It also generates local income.

Furthermore, Manipur presents significant potential for ecotourism development. Four significant river basins characterize Manipur. Manipur is referred to as the "Land of Rivers." The state features diverse lakes, hills, caves, rivers, and wildlife sanctuaries. This location offers diverse activities suitable for adventure tourism, including rock climbing, water sports, caving, trekking, and hiking. Opportunities for ecotourism and adventure tourism are increasing. Tourism faces fundamental challenges, such as transportation, tourist facilities, hygienic food, accommodation, traveler safety, ethnic clashes, tourist information systems, and brand image. Consequently, the tourism department needs to establish effective policies to address the state's limitations and foster ecotourism's growth.

Tourism between Manipur and Thailand should be organized through a structured travel system that assesses tourism's impact, maintains sustainability standards, and implements security measures for tourists. These elements are critical in influencing travel decisions. It is essential to consider social and cultural resources by fostering community and tourist awareness aimed at protecting and preserving cultural heritage while also ensuring respect for the intellectual property of communities. The approach involves not only the development of tourist destinations, landscapes, and cultural heritage sites but also emphasizes the importance of fostering lifelong learning opportunities for locals and tourists rather than concentrating solely on tourist attractions. Enhancing tourism research in Northeast India in collaboration with Thailand is advisable to explore the potential for developing an appropriate area to support cultural and sustainable tourism.

References

Arambam, S. (2021). Women's empowerment in Manipur. *Journal of Social and Political Sciences., 4*(1), 25–39.

Bharucha, R. (1991). Politics of indigenous theatre: Kanhailal in Manipur. *Economic and Political Weekly, 26*(11/12), 747–754.

Bone, K. (2013). Spiritual retreat Tourism in New Zealand. *Tourism Recreation Research, 38*(3), 295–309.

Cady, J. F. (2009). *Southeast Asia*. Surjeet Publication.

Devi, N. P. (2011). *The cultural history of Manipur*. Times Publication House.

Dhanamanjuri University. (n.d.). *About us*. (n.d.). Retrieved 20 July 2024, from https://dmu.ac.in/about_us

Emilia C. (2023). *Chitwan national park: A wildlife paradise worth exploring*. Retrieved 23 July 2024, from https://kidneysolutionreview.com/chitwan-national-park.html

Government of India. (2012). *The North-Eastern areas (re-organization) and other related laws (amendment) ACT, 2012*.Retrieved 3 October 2024, from https://www.india.gov.in/north-eastern-areas-reorganisation-amendment-act-2012?page=1

Hodson, T. C. (1908). *The Meitheis*. Akansha Publishing House.

Kabui, G. (2003). *History of Manipur: Precolonial period*. National Publishing House.

Kamei, G. (2003). *History of Manipur–pre-colonial period (second revised and enlarged).* Akansha Publishing House.
Lukram, M. (2019). *Manipuri women leaders, their storyline.* Akansha Publishing House.
Manipur Tourism. (n.d.). *Loktak Aquamarine Homestay.* Retrieved 20 July 2024, from https://manipurtourism.gov.in/loktak-aquamarine-homestay/
Ministry of Tourism of India. (2020). *Home/Ministry of Tourism Government of India.* Retrieved 5 August 2024, fromwww.tourism.gov.in
Ministry of Tourism of India. (2018). *Press Information Bureau Government of India: First project under the Swadesh Darshan Scheme, 'North East Circuit: Imphal & Khongjom' in Manipur Inaugurated Today.* Retrieved 5 August 2024, from https://pib.gov.in/Pressreleaseshare.aspx?PRID=1542990
Ministry of Tourism of India. (2024). *Manipur Sangai Festival* [Photograph]. Retrieved 11 November 2024, from https://utsav.gov.in/view-event/manipur-sangai-festival-1
National Commission for Scheduled Tribes. (2023). *State/Union Territory-wise List of Scheduled Tribes in India.* Retrieved 3 October 2024, from https://ncst.nic.in/content/other-information-relating-tribals
Northeast Bullet. (2023a). *Nupi Lan Movement: Fighting Spirit of Manipuri Women* [Photograph]. Retrieved 11 November 2024 from https://www.northeastbullet.com/nupi-lan/
Northeast Bullet. (2023b). *Meira Paibi: Spirit of Manipuri Women* [Photograph]. Retrieved 11 November 2024 from https://www.northeastbullet.com/meira-paibi/
Official Website of Kanan Devi Memorial School. (n.d.). *Life Sketch of Shrimati Kanan Devi.* Retrieved 20 July 2024, from https://kdmschool.com/
Pemberton, R. B. (1979). *The eastern frontier of India.* Mittal Publication.
Parratt, S. N. (1989). Garib Niwaz: Wars and religious policy in 18th century Manipur. *Internationales Asienforum, 20*(3–4), 295–302.
Parratt, S. N. (2013). *The religion of Manipur.* Spectrum Publications.
Roy, L.S. (2018). *Binodini's women: The three strong characters of my son, My precious.* Retrieved 20 July 2024, from https://indianexpress.com/article/gender/binodinis-women-the-three-strong-characters-of-my-son-my-precious-5091936/
Salam, B. (2023). The Meira Paibis of Manipus: Locating agency. *Indonesia Law Review, 13*(3), 1–10.
Sebastian, R. (2021). Refashioning kingship in Manipur in the 18th century: The politico-religious projects of Garibniwaz and Bhagyacandra. *Religions, 12*(1041), 1–27.
Singh, N. K. (1975). Manipuri language, status, and importance.: N. Tombi Raj Singh.
Singh, N. I. (1976). *The Manipur administration (1709–1907).* Friends and Co.
Singh, K. (1995). Naothingkhong Phambal Kaba. Shri Ningthoujam Khelchandra Singh.
Sudhir, H. (2013). Market and state: Women's political participation in Manipur in historical perspective.: Sunmarg Publishers & Distributors.
Singh, N. (2014). *Geography of Manipur.* Rajesh Publications.
Singh, N. J. (2015). Emergence of Manipur as a nation-state.: Akansha Publication House.
Singha, M. (2016). Marriage diplomacy between the states of Manipur and Burma, 18th to 19th centuries. *Proceedings of the Indian History Congress., 77*(2016), 874–879.
Singh & Singh. (2023). Lallup: A socio-economic and political organisation. *Journal of Emerging Technologies and Innovative Research (JETIR)., 10*(1), 757–759.
Sudhir, H. (2012). *Social history of Manipur.* Sunmarg Publishers and Distributors.
The Tourism and Sports Ministry of Thailand. (2024). *Tourism Statistic 2023.* Retrieved 31 July 2024, from https://www.mots.go.th/more_news_new.php?cid=411
Unseen Northeast India. (2024). *Loktak Lake, Manipur: Complete 2024 Travel Guide to India's Only Floating Island* [Photograph]. Retrieved 11 November 2024, from https://unseennortheastindia.com/loktak-lake-manipur/

Chapter 8
Cultural Commoditization and the Empowerment of Ethnic Women in Community-Based Ecotourism

Ariya Svetamra

Abstract This paper uses a feminist political ecology approach to look at how gender and tourism are connected, focusing on how gender roles are shaped by environmental issues in the context of globalization and the commercialization of tourism. The study provides an intersectional analysis of tourism, gender, power, and the environment. Data were collected through documentary research, field observations, and in-depth interviews with women who operate homestays in Northern Thailand and Nagaland, Northeast India. This paper argues that the cultural commoditization process might reinforce rather than diminish local cultural themes and values. Additionally, emerging forms of alternative tourism provide opportunities to challenge gender stereotypes by promoting economic independence and empowerment among women.

Keywords Cultural commoditization · Ethnic women · Women's empowerment · Community-based ecotourism · Nagaland

8.1 Introduction

Tourism in Thailand has experienced rapid development, resulting in positive and negative impacts. Initially, sustainability was not a priority, but it has gained importance. The Tourism Authority of Thailand (TAT) now emphasizes sustainable tourism by providing training for industry stakeholders. The rise of community-based tourism (CBT) projects reflects the growing demand for sustainable tourism practices from locals and tourists. CBT is particularly popular in northern Thailand, especially in Chiang Mai and Chiang Rai, where activities like hill tribe homestays

A. Svetamra (✉)
Department of Women's Studies, Faculty of Social Sciences, Chiang Mai University, Chiang Mai, Thailand

Y. Santasombat (ed.), *Bio-Cultural Diversity, Ethnicity, and Local Knowledge*,
https://doi.org/10.1007/978-981-95-4242-0_8

and trekking blend nature and culture. Organizations such as the Community-Based Tourism Institute (CBT-I) and the Northern Thailand Community-Based Tourism Network (CBT-N) support these initiatives by providing research, training, and resources to empower communities and promote sustainable tourism development (Breugel, 2013).

Nagaland, located in Northeast India, is characterized by its mountainous terrain, rich biodiversity, and diverse tribal cultures. The state's traditional practices, such as community forest conservation, are being adapted to address the challenges of Jhum cultivation. Since 1981, tourism development in Nagaland has been managed by the Department of Tourism, which views tourism as a tool for cultural and socio-economic growth. Recognizing tourism as a "Priority Sector" under the State Industrial Policy of Nagaland-2000, the state aims to promote ecotourism and sustainable development, particularly in rural areas where nearly 80% of the population resides.

With 88% of Nagaland's forests under private ownership and the limitations of traditional jhum cultivation, there is a pressing need for sustainable alternatives. Nongovernmental organizations (NGOs) play a crucial role in promoting ecotourism and sustainable practices. Integrating traditional values with modern education among the younger Naga population offers promising prospects for advancing sustainable tourism development while preserving the state's cultural and environmental heritage.

The state department of tourism, in collaboration with other relevant departments and NGOs, has been organizing the annual Hornbill Festival at Kisama Heritage Village in the first 7 days of December, and since 2013, it has been extended to the 10th of December. It showcases the cultural heritage of Nagaland's tribes and attracts both domestic and international tourists. The state promotes various forms of tourism, including cultural, rural, historical, adventure, and ecotourism as a recent addition. The tourism department encourages community participation by supporting the development of homestays in villages like Khonoma, Kigwema, and Jakhama (Kinny & Lanusosang, 2016).

In tourism, cultural commoditization can be seen as a subtle facet of capitalist development in international tourist markets. However, it can also be argued that the commoditization process may reinforce, rather than undermine, local cultural themes and values. Furthermore, emerging forms of alternative tourism offer opportunities to challenge gender stereotypes by fostering economic independence and empowerment among women. These initiatives enable women to develop skills in marketing, business management, and ecosystem conservation. Women are increasingly moving beyond traditional domestic roles, as demonstrated by their successful establishment of tourist accommodations for nature enthusiasts. These ventures enhance their respect within the community and promote local culture and environmental sustainability. The economic benefits are widely distributed due to community involvement, utilizing local resources such as materials, labor, food, crafts, performances, and tour guides (Dunn, 2007).

This study aims to examine how gender norms are negotiated within environmental struggles amidst globalization and the cultural commoditization of tourism.

The study utilizes both secondary and primary data. Secondary data are sourced from books, journals, government statistics, newspapers, and online resources. Primary data are collected through field studies, in-depth interviews, and participant and nonparticipant observations.

8.2 Cultural Commoditization of Tourism

The academic debate surrounding tourism's impact on local cultures emphasizes the negative effects of tourism, arguing that it leads to the degradation of cultural practices by transforming them into consumable commodities. This process of commodification, according to this perspective, strips cultural expressions of their deeper spiritual significance and repackages them for the consumption of tourists who are largely motivated by a desire to possess cultural artifacts rather than engage with their true meaning (Mathieson & Wall, 1982, cited in Shepherd, 2002).

An alternative viewpoint is that cultural involution, which suggests that tourism can stimulate a renewed interest in traditional cultural forms, reinforces local identity and fosters economic benefits. McKean (1989) proposes that cultural tourism can act as a mechanism for preserving and even revitalizing local traditions by generating cultural pride and material gains for the community. However, this perspective relies on the assumption that local actors can clearly distinguish between sacred cultural elements, which should remain untouched by tourism, and more "profane" elements that are open to commodification (Picard, 1996, 1997, cited in Shepherd, 2002). The issue, as the text highlights, lies in the difficulty of maintaining such a distinction, particularly as tourism fosters increased interaction between locals and tourists, often blurring these boundaries.

Further complicating this discussion is the concept of "authenticity." Many critics argue that tourism, by its very nature, undermines the authenticity of cultural practices. As the text notes, scholars, like Greenwood (1989), have explored how the commodification of local festivals and traditions transforms them from intimate, meaningful practices into spectacles tailored for external consumption. This process, in turn, diminishes their significance for the local community. It can similarly be argued that the transformation of cultural heritage into a tourist product redefines its value, shifting from a "cultural value" to a "commercial value," which often results in the reinvention of the past in ways that cater to tourist expectations rather than historical or cultural accuracy (Lanfant, 1995, cited in Shepherd, 2002).

However, challenging the static notion of "authenticity" and the tendency to romanticize an idealized, unspoiled past, scholars such as Said (1978), cited in Longkumer, 2015 argued that culture is inherently hybrid, dynamic, and fluid. The idea of pinpointing a singular "original" moment in time when a culture was pure or authentic is problematic. If culture evolves over time, then what is deemed "authentic" may change. Thus, today's commodified cultural practice might well become tomorrow's recognized tradition. The focus should therefore shift from a lamentation of what has been "lost" to an exploration of how authenticity is constructed in the contemporary context of tourism.

Hence, moving beyond the binarism of commodification versus authenticity is critical. Instead of just pointing out what has been commodified and made inauthentic, it is important to investigate how different social forces shape the construction of competing authenticities in the tourism industry. These competing authenticities are the products of ongoing cultural (re)invention, shaped by the forces of consumption, local identity, and global tourism markets. In doing so, it encourages a deeper understanding of how tourism functions not just as an economic or cultural exchange but as a site where meanings are contested and negotiated.

Ultimately, the study of tourism and culture must avoid simplistic dichotomies of authenticity versus commodification. Accepting commodification as a given does not require one to either embrace global capitalism or romanticize resistance against it; instead, it opens up space to ask more critical questions about how cultural meanings are produced, consumed, and transformed within the context of tourism. This approach allows for a more dynamic understanding of how people engage with culture in an increasingly globalized and commodified world (Shepherd, 2002).

According to Cohen (1988), most of the literature recognizes three essential elements: demonstration, authenticity, and commoditization. Even if cultural commoditization and authenticity are frequently apparent, showcasing culture in its purest state can have significant effects, both favorable and adverse. On the flip side, exhibiting culture can result in modifications that cater to the demands of tourists, including using light and sound displays to tell historical stories, combining traditional and fusion dance styles, or using martial arts as tourist attractions. On the other hand, the benefits include the preservation of performing arts, the involvement of younger generations in traditional arts as a means of livelihood, and a decrease in migration for commercial employment (Cohen, 1988, cited in Maheshwari, 2022).

Cultural commoditization can be viewed as a subtle aspect of capitalist development within international tourist markets. Greenwood (1989) argues that the appropriation of a culture's rituals and symbols for external consumption can inadvertently deprive local people of the meanings by which they understand and organize their lives. However, it can also be argued that the commoditization process may reinforce rather than undermine local cultural themes and values (Jarvenpa, 1994).

The growing influence of the tourism industry has significantly impacted the authenticity of cultures. Cohen (1988) argues that tourism, as an economic activity, has been implicated in the commodification of cultures. Objects and performances that were originally created for local consumption are increasingly tailored to the tourism market, leading to their exploitation, debasement, and trivialization (Smith & Robinson, 2006, p. 177). The Hornbill Festival, for example, has evolved into a cultural industry. While this form of tourism may be seen as a politically strategic move by the government, it appears that the aim is to globalize the festival by attracting sponsors, promoters, and various stakeholders, each with their own interests.

However, the process of commodifying the past for tourist consumption can involve (dis)inventing traditions. To make a commodity marketable, the government assigns brand names such as "Nagaland: Land of Festivals" or "Hornbill Festival: Festival of Festivals." These slogans may attract tourists and help the Hornbill Festival gain global recognition, thus establishing its identity within the cultural industry. Nevertheless, this commodification raises questions about whether the government of Nagaland is transforming the Hornbill Festival into a commercial "state" festival and effectively selling Nagaland to the outside world.

The festival also commodifies culture through souvenirs. As Smith and Robinson (2006, p. 191) note, "souvenirs are the most easily available commodity of culture and heritage," with both souvenirs and experiences often being perceived as authentic (cited in Tatongkala, 2016). The Hornbill Festival represents a negotiated balance between local and national/global influences and between an imagined past and enforced modernity. For the state, it signifies that peace is essential for development, while for the Naga groups, it offers a platform to showcase their revived identities. Critically, it illustrates the dual nature of heritage—it serves both political and economic functions. Thus, while the Hornbill Festival may be a sacred venue for enacting Naga identities, it also functions as a marketplace for cultural commodities (Tatongkala, 2016).

Most literature on the Nagas focuses on the impact of Western colonialism, often portraying a one-way process that diminishes traditional culture. Some argue that only remote areas retain authentic practices (Stirn & van Ham, 2000, cited in Longkumer, 2015). Despite efforts to revitalize culture through events like the Hornbill Festival, these initiatives are sometimes perceived as superficial and disconnected from historical traditions (von Stockhausen & Wettstein, 2008, cited in Longkumer, 2015). However, it can be argued that the Nagas are not merely passive observers, for they are active participants in crafting a distinct national image, using colonial-era images of "primitivism" to both promote and revitalize traditional culture. This process involves a reverse gaze on primitivism and the performance of identity and cultural hybridity. These strategies help the Nagas position themselves within the global discourse of indigeneity, turning formerly primitive images into symbols of a distinct national culture.

The Hornbill Festival, held annually in Kisama, Nagaland, markets the state as the "land of festivals." It showcases Naga culture through a variety of events and activities, providing a platform for both local and international audiences. The festival represents the 16 official tribes and serves as a space for local and national unity. With nearly 95% of Nagas identifying as Christians, the role of Christianity in the festival is significant. Although some churches believe that certain features of the celebration, like rice beer, are incompatible with Christian principles, the festival does not aim to revive every facet of pre-Christian society. In 2012, a cross was erected at Kisama to symbolize the site as a Christian location (Longkumer, 2015).

The Hornbill Festival holds multiple meanings for the Nagas. It serves to educate younger generations about their traditions, fosters a sense of Naga community among diverse ethnic groups, and represents a state-managed cultural event that transcends local variations. The festival also features national performances and attracts international attention. Besides the uniting of communities and tribes of Nagaland, which is home to over 17 tribes and sub-tribes, the 2023 edition of the festival is a testimony of the state government's focus on promoting the festival on a global scale. It focuses on integration and cooperation, with the USA, Germany, and Colombia joining the festival as "Country Partners" and Assam as the official "State Partner." This year, the Hornbill Festival will also have musicians from Bolivia, Korea, Germany, and Israel performing at the festival apart from the local artists (Sharma, 2023). According to Longkumer (2015), the festival attempts to project a distinct Naga identity that is aligned with indigenous rhetoric and self-determination. The festival has become a significant cultural and economic event, attracting diverse visitors and contributing to a broader sense of Naga identity. While some tourists seek "authentic" experiences, the festival's hybrid nature highlights the complex interplay between tradition and modernity, positioning the Nagas within the global discourse of indigenous culture (Longkumer, 2015) (Figs. 8.1, 8.2, and 8.3).

Thailand is renowned for its rich culture, traditions, and natural beauty, establishing itself as a top global destination for travelers. Given that tourism is a major revenue source for Thailand, it is crucial for the country to prioritize the economic benefits of this sector. Tourism is the country's second-largest revenue source, significantly contributing to income distribution across the nation. However, it has been noted that most of this revenue is concentrated in major cities (Japan International Cooperation Agency, 2020).

Currently, community-based tourism offers an economic alternative for communities impacted by government policies on forest and natural resource conservation,

Fig. 8.1 Two men dressed in traditional attire are preparing for their performance at the Hornbill Festival

Fig. 8.2 Ethnic women in traditional dress at the Hornbill Festival

Fig. 8.3 Colombian ambassador to India, USA ambassador to India, chief minister of Nagaland, UK high commissioner to India) in Hornbill Festival 2023. Credit: @abumetha on X. Source: https://nagalandgk.com/nagaland-current-affairs-digest-oct-dec-2023/#google_vignette

which have restricted land use in forested areas. The central government perceives the hill tribes as a source of problems related to deforestation and soil erosion, growing opium poppies, being an unstable political factor, and often not considering themselves to be Thai nationals (Cooper, 1979, cited in Dearden, 1991). The government's push to promote tourism as a key revenue source has affected ethnic communities rich in biodiversity, ecosystems, and cultural heritage. These communities have had to adapt by transforming their natural and cultural resources into economic assets. Penetration of the highlands, even by the Thai government, was achieved only recently, although missionaries have been active there for many years. Change is now fueled by a variety of factors, some of which are supported by Thai government agencies, particularly following the prohibition of opium cultivation in 1959. The market pressures that now infiltrate the highlands, frequently without the programs' knowledge, may have gained more traction than any of these institutional influences. The growth of tourism in the highlands constitutes another example of the unpredictable effects of unleashed market forces (Dearden, 1991).

A key approach to developing tourism in a way that benefits local communities is through "community-based tourism" (CBT). According to the ASEAN Community-Based Tourism Standard, community-based tourism is defined as "tourism activity that is community-owned and operated, managed or coordinated at the community level, and contributes to the well-being of communities by supporting sustainable livelihoods and protecting valued socio-cultural traditions and natural and cultural heritage resources" (Japan International Cooperation Agency, 2020).

The Thailand Community-Based Tourism Institute (CBT-I), established in 2006 under the Thailand Research Fund (TRF) Regional Office in Chiang Mai, has been instrumental in supporting Thai communities to develop small-scale tourism programs suited to their cultures and environments. CBT is designed to include and benefit local communities, especially indigenous people and villagers in developing countries. CBT is a socially sustainable form of tourism, typically initiated and operated by local and indigenous people. It emphasizes shared leadership, community well-being over individual profit, and the preservation of traditional culture and the environment. CBT offers travelers unique opportunities to engage with local communities directly, providing an alternative to unsustainable development and offering rural and impoverished communities additional income sources (Wirudchawong, 2012).

Community-based tourism (CBT) has become a vital component of the villagers' social fabric, enhancing their confidence in representing their culture as a form of self-empowerment and forging a new identity. While CBT's primary goals include generating income, reducing poverty, and preserving both culture and nature, it also plays a significant role in strengthening community cohesion and fostering a sense of pride and ownership among the villagers (Promburom, 2021). The highland people in the villages are creating the image of "forest and natural environment protectors" against the image of "the destroyers of the forest," which might be perceived or portrayed by others. The villagers' efforts to project themselves as protectors might involve highlighting their role in conservation efforts, traditional ecological knowledge, or community-based initiatives, such as community-based tourism that focuses on preserving the environment. They might also be working to challenge stereotypes or misconceptions that portray them negatively.

8.3 Feminist Political Ecology

Feminist political ecology is an interdisciplinary framework that examines the intersections of gender, power, and the environment. It explores how social structures, like patriarchy and capitalism, shape environmental policies and practices and how these factors impact women and marginalized communities. This approach emphasizes that environmental issues must be understood through the lens of gender and power dynamics, highlighting inequalities in resource access, environmental degradation, and climate change impacts. It advocates for more inclusive and equitable environmental justice.

Feminist political ecology provides insights into how intersecting social categories affect environmental outcomes. It considers gender as a crucial factor in shaping resource access and control. This includes interacting with class, caste, race, culture, and ethnicity to influence ecological processes and the struggle to maintain environmental sustainability. This framework emphasizes gendered knowledge in creating and sustaining healthy environments, addressing gendered environmental rights and responsibilities—such as property, resources, space, and legal and customary rights—and exploring gendered environmental politics.

Feminist political ecologists critically examine the impacts of development and globalization, focusing on how these processes perpetuate gender inequalities and environmental degradation. There is an ongoing debate about whether mainstream development models effectively address gender disparities and environmental sustainability. Besides, debates over epistemological and methodological approaches in environmental research are engaged. Furthermore, feminist approaches that emphasize experiential knowledge and the lived experiences of marginalized groups are advocated, challenging traditional scientific methods.

A key issue in this field is the inclusion of local and indigenous knowledge in environmental management. Feminist political ecologists argue for recognizing and valuing the environmental insights of women and indigenous peoples, who often possess crucial knowledge about sustainable practices. Overall, feminist political ecology enhances our understanding of environmental issues by integrating gender analysis and challenging existing power structures, advocating for more inclusive and equitable environmental justice and policy solutions (Rocheleau et al., 1996).

According to feminist political ecology principles, which emphasize well-being, sustainable livelihoods, and the interdependence between people and ecosystems, livelihoods should foster relationships of mutual support within communities and with the environment. Thus, livelihoods extend beyond mere economic fulfillment and income generation to include daily interactions, emotions, and attitudes toward the ecosystem (Harcourt, 2017).

8.4 Community-Based Ecotourism and Ethnic Women Empowerment: Thailand

Jabo Village, located in Pang Mapha District, Mae Hong Son Province, is a small ethnic Lahu community with a total population of 273 people and 95 households. The community has experienced multiple relocations, with the current Lahu residents having moved from nearby villages due to an outbreak of disease. The community, named "Jabo" after the village leader, was officially recognized as a village in 1989. Almost all members of the community are relatives, and they continue to use the Lahu language and traditional clothing. The main occupations in the community are farming (corn, melons, and red beans) and livestock raising (pigs and chickens), with secondary occupations including gathering forest products for sale and general labor. Many villagers are involved in trading, acting as middlemen who purchase agricultural products from various villages to resell to city merchants. Most of the income comes from selling rice and corn, and nearly every household has debts from borrowing to invest in fertilizers, pesticides, and seeds.

The community is located within a national forest reserve and wildlife sanctuary that enforces strict regulations for nature and environmental conservation. Although there is no official land title, the land has traditionally been divided among families from generation to generation. Currently, the government has set a limit of 25 rai (about 10 acres) of land per person, with landholders required to be at least 18 years old. Landholders have also divided their land among their descendants. Government agencies, including the National Land Policy Committee, conducted a land survey in the community over the last 2 years.

Jabo Village has seen an increase in tourism and now operates as a homestay, allowing visitors to experience Lahu life and culture. It is a successful example of a community-based ecotourism village in Mae Hong Son Province and is well-known on social media. Thai tourists particularly visit during the winter to view the sea of mist, learn about the culture, go trekking in the community forest, and understand rotational farming. The village also has prehistoric tourist sites, such as the Phi Man Cave.

The community tourism group was established in 2001 with support from the "Revitalizing Life and Culture Project" and the "Tourism for Life and Nature Project." It has since evolved into the "Community-Based Tourism Institute" (CBT-I) with research support from the Thai Research Fund. The community aims to provide tourists with insights into Lahu cultural practices and conservation efforts while generating supplementary income. Initially, all 30 households in Jabo Village joined the tourism group, but some have since withdrawn. Currently, the tourism group includes 34 households, with 24 operating homestays and 10 others contributing by cooking and guiding treks. The group has sought certification for Thai homestay standards from the Department of Tourism and Sports, and they have received their first certification since 2009. Currently, the homestay company welcomes over 2000 guests a year, of whom 70% are Thai and 30% are from other countries. Tourism generates significant annual income for households, with some relying on it as their primary source of income during the tourist season.

The tourist activities include homestay accommodation, cultural learning experiences (such as weaving, Lahu music, and making traditional bamboo instruments), and local tours to nearby attractions. These tours include natural and historical sites. For instance, the community forests (formerly rotational farms, which have become lush with large trees) are preserved for hiking activities that cover 10 km, lasting 2–3 h. Other sites are the rotational farms and pigpens of the village, Phi Man Cave, Phu Pha Mok, and Mae Lana Cave. Transportation is also provided to and from these points of interest (Figs. 8.4 and 8.5).

Historically, tourism did not offer opportunities for women, but it has since empowered women and challenged gender stereotypes. Initially, men were more involved in tourism management due to the need for coordination with external agencies. Today, women play a more significant role in tourism operations, as they now predominantly manage tourism activities. Currently, Jabo Village has a woman who has been elected as the chairperson of the tourism group.

Nana, the current female chairperson of the tourism group in Jabo Village, expressed her views on the group's cohesion, contrasting with other villagers who have experienced conflict due to leadership and family ties:

> Jabo is unique in that it remains a strong and unified village. We only have one central gathering area. One reason for this unity is that we work together and communicate openly.

Fig. 8.4 The Lahu people of Ban Ja Bo welcome tourists. Source: Facebook community-based ecotourism by Ban Ja Bo Community

Fig. 8.5 Community trekking tours and rotational farming activities

It's not about boasting; tourism is similar. We are all siblings, so we slowly talk things through and remind each other that we share the same blood and come from the same group. In other villages, conflicts arise, and people cannot collaborate. Here, the village head and the ritual leaders are all relatives, so everyone is connected. For example, the village head's wife is his sister-in-law. When issues arise, we resolve them. If the discussion is still unclear, we give it some time for everyone to calm down before coming together again. The Lahu people help each other, gradually improving the situation through cooperation.

In the past, when electing the village head 10 years ago, there were nearly divided factions. It depended on the ritual leaders, who, if divided, would lead to split factions and resist cooperation. Some villages even split their ceremonial grounds into two factions. Here, the ritual leaders did not support such divisions. Our tourism group helps the ritual leaders by providing support, such as money from the communal fund for their rituals, which builds goodwill.

There are both tourism and non-tourism groups in the village. Non-tourism groups have no income, while the tourism group, which deducts a percentage for the communal fund, helps support them. For instance, if a house is damaged or in times of need, the group assists. During festivals, we contribute to purchasing a pig for the community. They see the benefits of the tourism group. Since the tourism group's formation, there have been four chairpersons, three of whom were men.

When comparing male and female leaders, Nana noted that women handle conflicts better. While men might respond to harsh words with aggression, women tend to remain calm and address issues more diplomatically. Women arrange meetings to discuss and reach consensus, avoiding division. Previously, the group handled all media and tourist scheduling, but now individuals handle these tasks independently. Despite this, there's a collective monthly deduction to support village development.

The tourism group is working to include everyone in a single tourism group and ensure standards. They conduct evaluations every 3 to 4 years to maintain certification, with some places being demolished if they fail. The village is moving towards a unified tourism group model, ensuring everyone contributes a percentage, and tourists find comfort, such as having hot water in winter. The group also focuses on environmental conservation, such as planting trees and cleaning up the forest. They also manage communal pigpens to reduce odors and maintain firebreaks annually. Funds are used for various needs, including signage and community events.

> During New Year celebrations, they contribute to slaughtering pigs and hosting traditional events, supporting the community's cultural practices. They also collaborate with the neighboring Lisu community, collecting small contributions from each household to support their activities. This cooperation fosters mutual support and helps sustain traditional practices (Nana, personal communication, May 3, 2022).

In addition to income from homestays and tourism, some women in the community benefit from tourism in other ways. For example, during the winter season when tourist numbers peak, some have opened souvenir shops and noodle stalls. An interview with the owner of a popular noodle shop, well-known on social media and frequented by many tourists daily, revealed that this business significantly contributes to the village's income. This has helped raise the village's profile and generate a considerable amount of revenue from tourism (Fig. 8.6).

As women's economic roles in tourism have increased, they have gained more influence both in households and in public. This shift allows them to negotiate gender roles more effectively, such as asking their husbands to assist with household chores and childcare during the tourist season. Women can also negotiate the use of earnings to ease household and agricultural burdens, for instance, by purchasing appliances like washing machines or vehicles for farming.

However, income earned by women from homestay work is not entirely autonomous; it often requires approval from their husbands before it can be utilized. In some cases, women use earnings from selling souvenirs to buy personal items, such as televisions and mobile phones, as this money is considered their own personal earnings.

Namee, a woman who manages a homestay with her family and has two children, explained her decision to end the relationship and her confidence in the ability to support herself and her children independently:

> I have five siblings, comprising four sisters and one brother. I am the third. This year, I have been helping to clean the homestay and the campsite above the village. Yesterday, we had three rooms occupied by tourists. We just started operating this year. Last year, my younger sister managed it, but after she miscarried, I took over. I have to clean and change the bed

Fig. 8.6 Souvenir shop in Jabo village. Source: Facebook community-based ecotourism by Ban Ja Bo Community

> linens. My sister's homestay has two rooms, my brothers have five rooms, and mine is still under construction. My brother is building it alone; he's investing his own money and labor in it. Currently, it's incomplete, with only the bathroom left to finish. I have to purchase materials like steel, and my brother does the work without charging me. My grandmother provided the land for my homestay because I didn't have any. I also have a one-rai (approximately 0.4 acres) plot for farming, which is under my father's name. My father said we don't need to divide the land and should help each other. I focused solely on my studies, and since graduation, I have had two children. I didn't work outside; I stayed in the same school dormitory from Grade 2 to high school. I got married in my early twenties and lived together for 15–20 years in his village. I have only recently returned to living here with my two children. I decided to separate from him because he was involved with another woman. After my mother advised me to end the relationship, I made the decision to move forward. I am confident that I can support myself and my children independently (Namee, personal communication, May 10, 2022).

However, the absence of a suitable site for homestay lodgings and the lack of financial means to invest in such endeavors provide difficulties for women who do not profit much from tourism. Compared to women who gain more from tourism, they have less negotiating power within their houses. Due to differences in location and financial resources, these women are less able to take advantage of tourism-related income, which limits their ability to bargain and better their living conditions.

Naha, a woman who experienced fewer benefits from her homestay business, provided her insights:

> I'm currently 32 years old. I got married when I was 18, but I later got divorced. My 14-year-old daughter, who is in the third year of secondary school, lives with her grandmother in a nearby village. I moved here to live with my partner, and we have a nine-year-old son. My partner is saving money to build a house, which he has now completed. We have two land parcels and a farm that his uncle gave us. We grow rice for consumption and rear pigs. We also raise corn, where we make a little profit. During the tourist season, I can't sell any goods because I don't know how, and I can't make any adjustments because I didn't go to school. My parents didn't send me to school. I'm currently studying through a non-formal education program, but I don't have a calculator or a mobile phone.
>
> I also want to run a homestay, but my location is not ideal as it lacks a scenic view of the mountains, which tourists prefer. Even after more than a decade of living here, I still lack the land and amenities necessary for it, such as a decent bathroom. I resort to using a friend's bathroom. When tourists arrive, I work as a delivery person for barbecue meals. I earn 50 baht per trip, and if there are tourists, I make around 200–300 baht per night. My partner sells wild products like honey once a year. If he didn't support me, I would borrow money from relatives. After selling corn, I repay the debt every year. My mother can't work due to knee pain, and my brother sends her some money every month. My father works in construction but drinks alcohol every day. My partner spends his days drinking and playing games on his mobile phone, and we often argue. When we first got married, he didn't drink or have any of these habits, but he has changed after making many friends who have influenced him (Naha, personal communication, May 15, 2022).

Within and outside of the community, women use group networks to negotiate changes to gender norms in public. They engage in community networks and community-based tourism networks, and they use their spiritual beliefs to negotiate both inside and outside of the community. They do this by referencing customary laws, community rights, and their spiritual beliefs.

A female spiritual leader advised against the violation of Lahu culture, as it could bring misfortune to any documentary crew filming the ritual dance performance. Consequently, the crew was prohibited from taking photos. Additionally, a trekking tour to a sacred spring behind the school was canceled because it disrupted the village's sacred shrine. This was described as follows:

> There was once a team of documentary crew who came to film the performance of a ritual dance on the dancing stage. They set up lights, and the team's host was walking in and out during the ritual. People went to consult with a female spiritual leader in the village. She said that the "dancing ritual" was sacred and converting it into a performance for tourism was considered a violation of Lahu culture. If we still wanted to make the ritual a performance, then we should do it outside the village. They said that going against the spirits could bring misfortune or disasters to the village. Consequently, it turned out to be true. There was an outbreak of disease, causing the death of almost all the pigs in the village. Besides, due to days of continuous rain, a landslide occurred, affecting many people. We have our own beliefs, so we said that this should not be done. We spoke directly to them, no matter who they were. After that, they were forbidden from taking photos. Additionally, there used to be a trekking tour program that led visitors to a sacred spring behind the school. However, the tour management canceled this program because it involved taking tourists into the forest, requiring them to pass by the village's sacred shrine, which might disturb the village's sacred elements (Weera, personal communication, May 5, 2022) (Fig. 8.7).

In Thailand, women who are involved in tourism have gained increased prominence and acceptance in various roles. They have capitalized on their participation in tourism to enhance their decision-making power, impacting gender dynamics both within their households and in public life, both locally and beyond their

Fig. 8.7 Lahu's ritual dance, Ban Ja Bo

communities. By leveraging tourism networks and spiritual beliefs, these women are able to negotiate and assert their influence effectively. Additionally, they use their kinship connections to manage and resolve conflicts. They actively work to prevent political divisions and maintain community cohesion, a challenge faced by other communities. Empowered by tourism, they play a significant role in both personal and communal decision-making processes.

8.5 Community-Based Ecotourism and Ethnic Women Empowerment: Nagaland

Nagaland, the "Land of Festivals," is renowned for its diverse tribal traditions and rituals. Each tribe in the state celebrates its festivals throughout the year, contributing to a vibrant cultural landscape. The Hornbill Festival, organized by the state government, exemplifies this rich tradition. The festival, held annually at the Kisama Naga Heritage Village, aims to preserve and showcase Nagaland's cultural heritage. Located approximately 10 km from Kohima, the Kisama Naga Heritage Village serves as the main venue for the Hornbill Festival. This heritage complex features 16 traditional tribal houses, a World War II Museum, a Bamboo Heritage Hall, a Bamboo Pavilion, and a stadium for live performances. The village is designed to protect and promote Nagaland's ethnic and cultural heritage through a common management approach and comprehensive preservation methods.

Tourism has become a catalyst for economic growth at the grassroots level, providing opportunities for individuals, especially women, to start businesses like managing homestays or selling homemade food products. The rise of homestays around Kohima, particularly in Kigwema, demonstrates tourism's role in economic diversification and sustainability. Many small rural communities, like Kigwema, leverage social media for connectivity and are seeing increased tourism through homestays, guest houses, and lodges.

Kigwema village has seen success with homestays, particularly in the newly developed Kipfüzha area. The village, situated near Mt. Shürho and other attractions, has around 13 homestays that thrive during peak seasons, like the Hornbill Festival and spring (Morung Express News Kohima, 2018).

Kigwema, a historic village in southern Nagaland, lies 15 km south of Kohima. Nestled at the foothills of Mount Japfü, Kigwema is one of the oldest Angami Naga villages. The village gained historical significance during World War II as Japanese troops were stationed here. It is a popular trekking destination, particularly for those aiming to climb Mount Japfü, home to the world's tallest rhododendron tree. As of the 2011 Census, Kigwema had a population of 3872, predominantly Christian. The village, surrounded by the Barail Range and Dzüko Valley, offers natural beauty and cultural richness.

Rural communities involved in tourism are calling for increased government support for training, infrastructure improvements, and promotional activities. Due to limited financial and human resources, these communities heavily depend on central government programs. However, such reliance may result in them being politically influenced.

A woman who runs a homestay highlighted the challenges faced due to insufficient government support:

> What is happening in all these homestays? They started with nothing. But over the years, with the demand for heaters, geysers, or Western toilets, they're improving with the little they get. Unlike Sikkim, Meghalaya, or Arunachal, which receive more government support, we've had minimal help here. Our infrastructure issues include unreliable electricity, lack of street lighting, inadequate water supply, and poor drainage. We've been managing ourselves, but government support would be invaluable (a woman, personal communication, 30 November 2023).

> Meno, a pioneer in the homestay business in Kigwema, Nagaland, launched GreenVilla Homestay in 2010 in her father's house. With prior experience working in small hotels in England, where she got married and had a daughter, Meno returned to Nagaland to start her venture. She has since trained local villagers, including single mothers and youth, in homestay management, significantly contributing to community income and development (Morung Express News Kohima, 2018).[1]

> I trained women, especially the youth, in homestay management by inviting them to my house for hands-on training. They learned various skills, such as house arrangement, room preparation, and guest communication, before implementing these practices on their own. Currently, my homestay employs two single mothers from the village who are gaining experience and learning the ropes of the homestay business alongside me (Meno, personal communication, November 29, 2024).

> The experience that I'm trying to give people has a unique and personal touch. Whenever someone visits a place, the hosts will show the guest around. They are proud to do so and are delighted to tell the guest everything they can because they call the place their home. To them, it isn't just a job; it's an interaction.
>
> When I have guests in my house, I look after them. That's the same idea behind promoting rural homestays. The locals absolutely love to have these guests. The women play an active role in interacting with them, making them comfortable. It makes them amazingly happy. To add to it, they benefit because they earn money. So, when they have guests sent to them, not only do they prosper financially, but they also interact with the guests and learn from them. They take pride in their culture while showing it to an outsider. And so, organically, the local community becomes socially and economically stronger.
>
> I've always loved interacting with travelers and people from different regions. I've traveled all over India and abroad over the years, and I've learned a lot and gained a lot of experience. My love for travel and working holidays took me to many places in the Himalayan region as well as abroad (where I've worked in the hospitality sector). But deep inside, I've always meant to come home and promote Nagaland and the Northeast. Once I'd returned and started this, I realized that this was what I really wanted to do. I wanted to be my own boss and help people have life experiences.
>
> My vision is to let the local community benefit and take pride in their surroundings, environment, and culture. When locals show others their home, they're proud and delighted.

---

[1] Morung Express News Kohima. (2018). The thriving business of Homestays in Nagaland. https://morungexpress.com/thriving-business-homestays-nagaland

> Not only does this interaction financially support them, but it also strengthens their cultural pride (Explore Nagaland, 2015).

Meno's personal journey reflects her commitment to promoting Nagaland and supporting local communities through tourism. She believes in showcasing the region's beauty and fostering positive, memorable interactions with visitors. The following is what she said:

> The Northeast is not a dangerous place, but it is a destination filled with beautiful people and tribes. You should experience it for yourself, for these are real human interactions that you will never forget (Explore Nagaland, 2015).

In 2024, Kigwema's village leaders and participants from YouthNet's[2] Homestay training program visited Shillong and Mawlynnong (Asia's cleanest village in 2003) for exposure. They gained insights into local markets, homestay management, and cleanliness practices. Inspired by their observations, they aspire to transform Kigwema into a model homestay village in Nagaland.

I have met several women and youth, as well as some men running homestays with their mothers and sisters, who received advice from Meno. They are now successfully operating homestays in Kigwema village and the Kipfüzha area, which is located near Kisama Heritage Village, the venue for the Hornbill Festival.

Avino, a 42-year-old married woman, received advice from Meno on running a homestay. She and her husband were introduced to the concept when they moved to Kipfüzha, Kigwema, in 2012. They initially started by offering three rooms for tourists. She said:

> It has been seven years now. We have eight rooms available for tourists, including dormitories. Through this homestay business, we were able to build and upgrade our house, meeting our needs and more.
>
> This house is our own. Initially, we lived downstairs and had a dormitory there. However, we have now renovated and expanded the area upstairs. This year, we have four rooms with attached bathrooms and a new building with two additional rooms, making a total of seven rooms with attached bathrooms. My husband occasionally runs a small business as a carpenter, making tables, seats, doors, and beds, and even constructing parts of the house. Over the past 12 years, we've saved a lot of money in our savings account.
>
> Actually, I don't go outside much. As a mother of four children, I used to stay at home all day and night. I've learned a lot from the stories shared by outsiders. I feel very fortunate to have this opportunity, thanks to Meno, who helped me immensely. She was the first one to assist me. Initially, I lived in the village, but after moving here, she gave me valuable ideas and stories. Her help has been crucial, and because of her support, I'm seeing significant growth.
>
> It's been really helpful for me and my children. Not only have I benefited, but my kids have too. Their world has become more open to the tourists who visit us. Otherwise, they would have to venture elsewhere to learn about the world. The tourists bring to us a wealth

[2]YouthNet is a nonprofit organization dedicated to empowering youth by helping them acquire knowledge, develop life skills, and cultivate attitudes that enable them to become self-directed, positive, productive, responsible, and contributing members of society. Established on February 1, 2006, by a group of young Naga professionals. Since its inception, it has become one of the leading youth organizations in Northeast India, focusing on addressing issues related to education, unemployment, entrepreneurship, and livelihood (https://www.youthnet.org.in/)

> of knowledge and experiences. My children can now help me in many ways as they have grown older, making things a bit easier for me. My two daughters, one in Class 10 and the other in her second semester of college education, can clean rooms and cook. They are confident in speaking English with the tourists and have gained broader experiences. They've learned to respect and cooperate with others. As a mother, I've learned a lot about cleanliness, arranging bedrooms, and maintaining the house. My cousins and sisters help me, and they also learn valuable skills if they ever want to run a homestay themselves.
>
> I've learned to use social media platforms like WhatsApp and Instagram to send pictures. Some family members help me with this, and my daughter also assists me. I'm gradually improving my skills. I take pictures of the guests and post them online.
>
> Sometimes, if I need help, I will ask my husband. However, I prefer not to burden him with too many homestay duties, such as cooking and cleaning. My two sons are still young, so I handle most of the tasks when tourists arrive. I ask them what they need and serve them myself. For cooking, I often seek help from my sisters. I keep the homestay income separate for household expenses and occasionally for the children's education. While my husband contributes financially, having my personal income allows me to spend freely. I consider myself a lucky woman with a wonderful family and great children. I feel blessed that God has given me such a supportive family (Avino, personal communication, November 30, 2023).

Asa, a 45-year-old married woman, narrated her experiences as follows:

> I have never worked outside the home. The homestay business is considered well. We have a total of 10–11 rooms and some tents in the garden, accommodating around 20–30 people. Guests often tell us that they only need a place to sleep, and if they request accommodation, we don't want to turn them away. Currently, we host 10–15 guests on weekends. Our homestay is simple and not standardized, but it has a convenient location near the highway, offers delicious food, and has ample parking spaces. I handle cleaning, cooking, and gardening. I enjoy cooking and preparing both Naga dishes and Indian food. These are skills that I learned from childhood.
>
> I'm not very active on Facebook or Instagram. Most bookings come through phone calls. We work with agents from Assam, Bangalore, and West Bengal who send groups our way, primarily from India but occasionally from abroad.
>
> I have a 14-year-old daughter, and our family of nine lives together in this house. The children are also learning to help. The income from the homestay contributes to the children's education. My daughter assists me with cleaning vegetables, tidying the house, and washing clothes. I don't feel tired even when we have many guests; I'm happy doing it.
>
> We consider our guests as friends. They come and go, always introducing new experiences. We learn from them just as they learn from us (Asa, personal communication, November 30, 2023).

In addition to the married women, there are also young women running homestays in this area.

Mona, a 32-year-old woman, related her experiences running a homestay in Kigwema village and the impact it has had on her life in the following account:

> Before I became a homestay owner, I worked as a guide in this village. I was eager to learn new things, but as a guide, the time spent with guests was limited. I wanted to engage more deeply with people, spend more time with them, and potentially earn money at the same time. I often discussed with my dad the idea of creating a space where people could come together, chat, and learn from one another. That was the vision I had.
>
> It's been two years since we started the homestay. We began in December of the previous year with three rooms. This year, due to increased guest demand, we converted the seating area into an additional bedroom, bringing the total to four rooms. The land belongs

to my father, and with the encouragement from him and my brother, they built the homestay with their bare hands. Our neighbors, who are also relatives, assisted with the construction. It's my own business now.

To reach customers, we use Instagram, Facebook, and Google Maps, thanks to my brother, who set these up since I'm not very tech-savvy. Having been a guest house owner previously for two years, I have developed a regular clientele.

Running the homestay has profoundly changed my perspective on people. Initially, I made assumptions about others, but I've learned that my views were quite narrow. For example, I realized that homestays are not common everywhere; while hotels and guesthouses are widespread, not everyone welcomes strangers into their homes. I thought that welcoming guests was universal, but I discovered otherwise. Additionally, I used to think that organic farming and local produce were simple, but I now appreciate their significance and benefits more deeply. I've learned that commercially produced vegetables, often treated with pesticides, aren't as healthy as the local, organic options we have. This has changed my food habits; I'm moving away from fast food and increasingly valuing local and wild produce.

During my studies—pursuing a BA in Economics, an MA in Sociology, and an LLB (Bachelor of Laws)—I ran a guesthouse and helped with my cousin's homestay. After completing my LLB, I became an advocate and continued with the homestay. I don't plan to work in the city; I love the peaceful village life and the close-knit community here. I'm content with my current work and the people I meet through it. There's no regret in my choices.

Our homestay rates are modest, reflecting our basic facilities. We charge 1200 rupees per person for a bed and breakfast, with lunch and dinner available on request with a two-hour notice. We provide tea and snacks for free. We accommodate dietary preferences and allergies, sourcing ingredients like fish, which is not readily available locally. Our vegetables are mostly locally grown and organically produced from our garden. We also cultivate rice and rely on government rations whenever it's necessary. While I enjoy cooking, I admit that my skills are still developing.

I learned about running a homestay from Meno, who owns the first homestay in our village. Even as a guide, I took training under her. I consulted her when I started my guesthouse, and I continue to seek her advice. I've also completed a six-month training course with YouthNet, which provided valuable insights into homestay management, customer interaction, and operational practices.

Managing the homestay alongside household chores doubles my workload, but I'm fortunate to have a supportive mother who helps with the house, and my younger brother takes care of it when I'm not around.

In the future, I plan to improve the homestay by building a better house with an attached bathroom and addressing parking issues. Although the current setup is sufficient to maintain our income, I believe that more improvements are needed to ensure better financial viability.

The local community benefits from the homestay as well. For instance, if I don't have bananas from my garden, I buy them from neighbors. This way, they also earn from selling their produce. I also support local artisans by introducing guests to their products, such as handmade shawls and kitchen towels. This mutual support enriches both the guests' experience and the local economy (Mona, personal communication, November 30, 2023) (Fig. 8.8).

Aze, a 24-year-old single woman, described her experiences as follows:

I've been running my homestay for six years. I learned how to manage it from Meno, my aunt. We have three rooms and two vintage-style non-attached bathrooms: one room for four people, one for three, and another for six. During the peak season from October to

Fig. 8.8 A stall selling local produce in front of the house to tourists

January, we are fully booked. In the off-season, between June and July, we still receive some guests.

We get to know the people and their cultures. They ask about our culture, and we learn about theirs. It's a mutual exchange of knowledge. I stay at home and don't go out much.

I'm currently finishing my final year of college with a major in sociology and planning to pursue an MA. In the past, everyone was seeking government jobs, but now those positions are saturated. Thus, having alternative career options is important. Sometimes, students tend to give up when they can't find a job right away. I have an elder brother who graduated this year and now works as a tour guide in the area. He helps with welcoming and interacting with tourists, while my mom and I handle cleaning and cooking. My dad, who works in Kohima, helps us when he gets home early. Our guests enjoy the food, and we cook according to their preferences.

Mom manages the homestay income. We use it to buy things for the house in preparation for the next tourist season, especially around Christmas. We also use it to buy gifts for family members. The primary income comes from my father's regular job, and the homestay provides additional funds for improvements.

We sometimes get stressed when there are no guests. Occasionally, tourists book but then don't show up, leading to cancellations. We're now starting to ask for advance payments to mitigate this issue (Aze, personal communication, November 28, 2023).

Amo, a 22-year-old student who helps her family manage the homestay, provided the following account:

We started in 2018 but had to close it down for a couple of years. We reopened last year and are still going strong. My family manages the homestay. We have three rooms with twelve bunk beds and one dormitory, which can accommodate around 20 people. Annually, we host 200–300 guests, especially during the Hornbill Festival in December. Besides the festival, we also see numerous visitors during Christmas and New Year holidays. Most of our guests are young people, and our setup is ideal for large groups.

I handle hosting and management, while my mom does most of the cooking and cleaning. During peak festival seasons, we hire additional help to manage large groups, as it becomes too much for just us. For dormitories, we charge 1500 rupees per night, and for rooms with attached bathrooms, the rate is 4,000 rupees per night. During the off-season, rates drop to 1200 rupees for rooms and 500 rupees for dormitories. For now, I handle most of the responsibilities. We receive bookings through Airbnb as well as personal referrals, so we operate year-round.

I completed my bachelor's and master's degrees in English this year. I initially planned to study outside Nagaland, but given my schedule, I was wondering whether I should embark on my studies here instead. I took a gap year to reassess my plans. I was uncertain about whether to pursue further education, but now I'm confident and eager to continue. I balance my studies with managing the homestay. My uncle helps with bookings and phone calls. If I find better opportunities elsewhere, I might move, but if not, I will continue with my family to run the homestay. I love writing and hope to develop my skills further, possibly becoming a writer in the future. I enjoy spending time with friends, especially in the mornings. I love sunrise treks and often go with friends and neighbors. We meet up in the morning, hike up the mountains, and enjoy breakfast with a beautiful sunrise view.

I've gained a lot from running the homestay. Meeting people from diverse backgrounds allows us to learn about different cultures, professions, and fields of study. For example, guests from IT, engineering, and other fields share their knowledge, which enriches my understanding. I've also grown personally. I wasn't very talkative before, but interacting with guests has helped me become more engaged and open. We exchange knowledge and learn about each other's lifestyles.

One of the biggest challenges is meeting the expectations of some guests, who often expect luxury comparable to five-star hotels. It can be tough because we cannot always meet such high demands. Unlike hotels, our homestay is a shared space with limited amenities. I have to communicate clearly with the guests by explaining that while we may not offer the same level of luxury as hotels, we do our best to provide a comfortable experience with our resources. I stress that respect for the shared nature of the space is important, and I encourage guests to make the most of their stay despite the limitations (Amo, personal communication, November 29, 2023).

In Kigwema, several young men run homestays with their mothers, sisters, or young wives. While most are educated, they choose to live in a community setting. Some of them have prior experience with youth training programs.

Aja, a 35-year-old single man, runs the homestay with his sister and mother. He explained his experiences as follows:

There are around 40 homestays in this area, but Meno was the pioneer. We learned a great deal from her about hosting and interacting with guests. She inspired us when we started Rooftop Café.

Earlier, my sister was the host, but after she got married and moved to a different house, I took over. It's been six years since I started hosting. Before this, I ran a hostel and engaged in other business activities. Our homestay is seasonal, with guests primarily coming for Dzukou Valley trekking from October to February. During the Indian Durga Puja[3] holiday

[3] Durga Puja, a major festival of Hinduism, celebrates the victory of the goddess Durga over the demon king Mahishasura. It begins on the same day as Navratri, a nine-night festival in many northern and western states that more broadly celebrates the divine feminine (shakti) (https://www.britannica.com/topic/Durga-Puja)

> in October and throughout the trekking season, we host 10–15 people each month. Guests come from both India and abroad. My role involves managing everything, like a traffic officer, ensuring that the market runs smoothly, checking room supplies like toilet paper and soap, and preparing Naga food for guests. I learned to cook while studying in Delhi. For cooking ingredients, we visit the market and use produce from our own garden, where we grow fruits, vegetables, corn, ginger, and more. This year, we also grew rice in our paddy field.
>
> We have one dormitory, four double rooms, and two single rooms with attached bathrooms. The dormitory can accommodate six or seven beds, depending on the group size. We provide hot water and heating. Guests often gather in the kitchen to relax, where we make a chimney fire and serve meals. They typically have breakfast and dinner with us and spend the rest of their time in the dormitory.
>
> We use Instagram, Facebook, and Google Maps for bookings. Many of our previous guests recommended us to their friends and relatives. Most of our bookings are made in advance. However, we occasionally get last-minute guests when we're not fully prepared. I inform guests beforehand that they are coming to a village on the outskirts of town and should not expect city amenities. It's important to set realistic expectations for the tourists to avoid culture shock.
>
> We enjoy interacting with guests and learning from their diverse backgrounds. For us, the most appealing aspect lies in the conversations and cultural exchange. Hosting allows us to remain in our home and enjoy the comfort of familiar surroundings, unlike managing a hotel in the city, which can be lonely and stressful.
>
> I have no plans to move to the city. While the city has its unique characteristics, I appreciate the natural environment here, despite some ecological issues during the hornbill season when many visitors come to the area. My family—comprising my mother, father, nephew, brother, and grandmother—lives here. We have a small house nearby for my grandmother, who prefers her personal kitchen and undisturbed independence (Aja, personal communication, December 1, 2023).

His sister, who came later, revealed her personal experiences as follows:

> "We started the homestay together, with me initially taking the lead. My brother has since taken over, but we share similar experiences. Our mom played a crucial role in encouraging us when we were hesitant. We're grateful for her support.
>
> For women, running a homestay often means more freedom, independence, and income while staying at home. It's a convenient way to earn without having to leave home.
>
> I've learned a lot from hosting, such as how to manage temper and handle different personalities. Guests might complain initially, but they usually become more understanding and appreciative after a day or two. They particularly enjoy the food, which often leads to a positive change in their attitude. It is, as the saying goes, "The way to a man's heart is through his stomach."
>
> Running a homestay means you're in your own home, surrounded by comfort and familiar things, unlike managing a city hotel, where you deal with strangers and face loneliness and stress. This is one of the greatest advantages of homestaying (Aja's sister, personal communication, December 1, 2023).

Rovy, a 27-year-old married man who runs a homestay with his young wife, stated the following:

> I started the homestay in 2018. Meno is considered the mother of homestays in our area. When we first settled here, she came over to help us, even cleaning the room and a toilet together with our family. That's how we got started.
>
> Recently, we offered six rooms, one dormitory, and two cottages on the top floor. We added four new rooms, each with three beds and eight bathrooms. We have three staff mem-

> bers. We provide a bed and breakfast and offer lunch and dinner upon request. Our busiest times are during the Hornbill Festival, when guests stay with us for two to ten days. We can accommodate around 20 guests at full capacity. Throughout the year, we also host visitors, primarily from Kolkata and Assam, who come for trekking in the famous Dzukou Valley. The trek usually takes two and a half hours, but some people take four to five hours due to the steep climb. We are active on Instagram and Google Maps and registered with the Nagaland Tourism Department.
>
> We aim to accommodate our guests and get to know them better. Initially, we were shy and not very open. But as we hosted more guests, we learned to interact with them and build our confidence. We've had many lovely guests, but we faced some challenges last year. For instance, some guests complained about Wi-Fi or requested changes that were difficult to accommodate. Some even asked for outdoor photos or video calls before booking. We found that managing these expectations can be tough. We don't offer the same amenities as a hotel, and we are not professionally trained. Nonetheless, we strive to offer genuine hospitality.
>
> I've changed my mindset significantly since starting homestay. Previously, we were quite reserved and focused solely on our community. Through interacting with guests from around the world, we've gained a new perspective and passion. It's been wonderful to meet people from diverse backgrounds, and I've learned a lot from these interactions (Rovy, personal communication, November 30, 2023).

In conclusion, homestay owners find personal fulfillment and financial benefit in their roles, driven by a passion for hospitality and cultural exchange. They often receive strong support from family and community, which helps in both the establishment and management of their businesses. Challenges include balancing work with personal life, managing seasonal fluctuations, and meeting diverse guest expectations. Marketing relies on social media and traditional methods, with varying levels of tech-savviness among hosts. Financially, modest rates and income from homestays support household needs and improvements. Continuous learning and future plans for enhancements are common among owners, reflecting their commitment to providing better guest experiences.

The women who run homestays often revealed a rich tapestry of experiences, challenges, and successes. Firstly, homestays provide a crucial source of income for many women. This aspect can be particularly significant in areas where other economic opportunities are limited. The income from homestays can support not only the host family but also contribute to local economies. Secondly, many women who run homestays are driven by a deep passion for hospitality and sharing their culture. They enjoy meeting people from diverse backgrounds and providing personalized experiences. Thirdly, running a homestay offers a unique opportunity for cultural exchange. Hosts often highlight the enriching experiences they gain from interacting with guests and learning about different cultures. Fourthly, support from the local community can be vital. Many women benefit from local networks, advice, and resources that help them navigate the complexities of running a homestay. Lastly, running a homestay often leads to personal growth. Hosts frequently describe how the experience has helped them develop skills in communication, problem-solving, and adaptability.

8.6 Comparative Analysis Between Thailand and Nagaland

A comparison between Thailand and Nagaland reveals differences in geography, socio-economic context, gender norms, and resource challenges. In Thailand, the Lahu ethnic community lives within a national forest reserve with strict conservation regulations and no official land titles, with land traditionally passed down among families. In Nagaland, the Angami ethnic community follows customary land ownership laws, with much land privately owned and shifting cultivation common. While both regions face pressures from land use, Nagaland plans to adopt a new forest policy focused on conservation, organic farming, and ecotourism to promote sustainable development and preserve cultural heritage (Department of Environment, Forest and Climate Change, Government of Nagaland, n.d.).

In Thailand, homestay operations are part of a community-based tourism model, where local communities offer cultural experiences like weaving, music, and visits to preserved forests and farms. Supported by organizations like CBT-I and CBT-N, this model empowers locals to manage tourism and benefit from it, with regular evaluations to maintain standards. In Nagaland, homestays cater to trekkers and visitors to the Hornbill Festival, with a focus on local culture. A pioneer in Kigwema has trained women and youth in homestay management, promoting sustainable tourism and cultural preservation. Both regions use homestays to support sustainable tourism and cultural experiences, with Thailand having a more organized, community-driven model, while Nagaland is still developing its homestay sector.

In both Thailand and Nagaland, women are becoming more involved in tourism, particularly through managing homestays, but their experiences differ. In Thailand, women have gained more agency in negotiating gender roles, such as asking husbands to help with household chores during tourist seasons. While they can use their earnings to ease household burdens, they often require their husbands' approval to spend this income. In some cases, this economic empowerment leads to significant personal changes, like women becoming financially independent and ending relationships with their husbands. In Nagaland, women, traditionally confined to the home, have found more independence through managing homestays and interacting with tourists, often via social media. This independence allows them greater control over finances, particularly for their children's education, and offers the freedom to earn without leaving home. Thai women still face some limitations in using their earnings independently, while Naga women enjoy greater financial autonomy. In Thailand, women are negotiating gender roles within the household. In both situations, women's roles shift from being confined to the home to being business owners.

Women who own homestays benefit differently from the business compared to those who do not. Homestay owners typically have advantages, such as a prime tourist location, investment resources, human capital (skills and knowledge to attract tourists), and support from husbands, kinship networks (land, investment, and labor), and community-based tourism networks. In contrast, women without homestays or with lower incomes from the business lack these resources. In Thailand, these differences influence the negotiation of gender roles, both domestically and publicly. Women who manage homestays often face the challenge of balancing both household chores and the demands of the tourism business. In Thailand, women can negotiate by requesting their husbands' assistance with domestic and production tasks, particularly when tourists are present. However, this support is often temporary, as husbands typically refuse to help when there are no tourists. This situation reflects a gendered division of labor where support is conditional on the business's demands, and the women's responsibilities remain largely unrelieved outside those times. In Nagaland, women also seek help from their husbands but tend to avoid burdening them with tasks like cooking and cleaning. Instead, they rely on other family members, such as sisters or daughters, for support. This suggests a more consistent reliance on family networks, rather than on the husband alone, indicating a different approach to sharing domestic responsibilities. Overall, while both groups of women manage the dual pressures of homestay business and domestic responsibilities, the nature and consistency of support differ significantly between the two regions.

According to the resource struggles, the highland people in Northern Thailand are actively reshaping their image from "destroyers of the forest" to "protectors of the forest and natural environment." This effort includes highlighting their conservation work, such as community-based tourism, which focuses on environmental preservation. Villagers engage in activities like tree planting, forest clean-ups, managing communal pigpens, and maintaining firebreaks. The tourism group also integrates Lahu cultural practices and conservation efforts, aiming to educate tourists while generating supplementary income. They often use spiritual beliefs to negotiate both within their community and with outsiders to reinforce their conservation initiatives. In Nagaland, young women running homestays benefit from training provided by the government and nonprofit organizations like YouthNet. These programs cover local market knowledge, homestay management, cleanliness practices, and creating plastic-free communities. The training supports the use of indigenous knowledge, including organic farming, traditional food, and local wild produce, empowering the women to share these practices with others in their communities and neighboring areas. While Northern Thailand focuses on community-based tourism and environmental conservation to reshape its image and educate tourists, blending cultural and ecological preservation, Nagaland emphasizes youth training in sustainable practices and homestay management, using indigenous knowledge to promote environmental sustainability and community welfare. Both regions are using their cultural heritage and environmental efforts to empower their communities, with a strong focus on sustainability, cultural preservation, and economic empowerment through tourism.

8.7 Conclusion

There are differing views on tourism's impact on local cultures. Some scholars argue that tourism leads to cultural degradation, commodification, and a loss of authenticity. This perspective suggests that tourism reduces cultural products and traditions to mere items for sale, often disconnected from their original meanings. For example, festivals and traditions may be transformed into public spectacles for tourists, losing their original significance and meaning to locals.

On the other hand, proponents of cultural involution argue that tourism can revive interest in traditional practices, benefiting local communities both culturally and economically. However, this view assumes locals can easily differentiate between sacred and commodifiable cultural elements. It also critiques the idea of authenticity, suggesting that culture is dynamic and fluid, making the notion of an "original" moment in time problematic.

Given the complexity of cultural commodification, a more thorough examination of how authenticity is constructed and negotiated in the tourism industry is necessary, rather than concentrating only on the loss of authenticity. This shift moves beyond lamenting cultural loss to examining how people navigate and create meaning within the context of tourism (Shepherd, 2002). The process of cultural commoditization can reinforce local cultural themes and values rather than undermine them (Jarvenpa, 1994). Emerging forms of alternative tourism, such as community-based tourism, offer opportunities to challenge gender stereotypes by promoting economic independence and empowerment among women. This type of tourism provides women with more freedom, independence, and income while allowing them to stay in their communities. Through homestay businesses, women have been able to improve their homes, support their children's education, and meet various needs. As some women have noted, "Having my personal income allows me to spend freely." Interactions with guests have broadened their worldviews, benefiting not only women but also their children, who gain exposure to new cultures and lifestyles. Additionally, women have learned to use social media platforms and engage in the globalized world, extending their experiences beyond their local communities.

Women's engagement in community-based tourism enhances their decision-making power, influencing gender dynamics both within their households and in public life. They leverage networks, including kinship connections, women's networks, and government and NGO support, to empower themselves and manage conflicts, prevent political divisions, and maintain community cohesion. This empowerment through tourism enables women to play significant roles in personal and communal decision-making processes.

From the perspective of feminist political ecology, development and globalization often perpetuate gender inequalities and environmental degradation. Feminist political ecology examines how intersecting social categories—such as gender, class, caste, race, culture, and ethnicity—affect environmental outcomes and influence efforts to maintain environmental sustainability. This framework highlights the importance of recognizing and valuing the environmental insights of women and indigenous peoples, who often possess crucial knowledge about sustainable practices (Rocheleau et al., 1996).

Community-based ecotourism supports the development of small-scale tourism programs that are culturally and environmentally appropriate. It emphasizes shared leadership, community well-being over individual profit, and the preservation of traditional culture and the environment (Wirudchawong, 2012). This form of tourism provides an alternative to unsustainable development and offers additional income sources for rural and impoverished communities. However, livelihoods encompass more than economic gains; they also include daily interactions, emotions, and attitudes toward the ecosystem (Harcourt, 2017). Women involved in community-based tourism, like Meno and other homestay operators in Nagaland, emphasize the value of these interactions. They said, "The local community benefits and takes pride in their surroundings, environment, and culture. When locals show you their home, they're proud and happy. This interaction not only supports them financially but also strengthens their cultural pride."

A key issue in this field is integrating local and indigenous knowledge into ecotourism while simultaneously addressing gender dynamics and promoting environmental management. Indigenous communities, particularly women, often possess valuable insights into sustainable practices and resource management. On the other hand, gender inequality can impact decision-making and gender role negotiation by influencing access to and control over resources. Additionally, ecotourism presents an opportunity to showcase local knowledge, support community-based conservation efforts, and provide economic benefits. By incorporating both local knowledge and gender perspectives, along with promoting sustainable ecotourism, we can create more inclusive, effective, and equitable strategies for managing natural resources and addressing environmental challenges.

References

Breugel, L. v. (2013). *Community-based tourism: Local participation and perceived impacts: A comparative study between two communities in Thailand*. Master Thesis, Radboud University Nijmegen, The Netherlands. Retrieved August 10, 2024, from http://tourismlibrary.tat.or.th/medias/RU0232/RU0232_fulltext.pdf

Cohen, E. (1988). Authenticity and commoditization in tourism. *Annals of Tourism Research, 15*(3), 371–386.

Cooper, R. G. (1979). Tribal minorities of northern Thailand: Problems and prospects. *Southeast Asian Affairs, 14*, 323–332.

Dearden, P. (1991). Tourism and sustainable development in northern Thailand. *Geographical Review, 81*(4), 400–413. https://doi.org/10.2307/215607

Department of Environment, Forest and Climate Change, Government of Nagaland. (n.d.). *Nagaland environment & forest policy*. Retrieved December 15, 2024, from https://nagaland.gov.in/storage/PostFiles/State_Forest_&_Environment_Policy_2015.pdf

Dunn, S. (2007). *Toward empowerment: Women and community-based tourism in Thailand*. MA. Thesis, University of Oregon.

Explore Nagaland. (2015). *Redefining the tourism industry*. Retrieved August 10, 2024, from https://www.explorenagaland.com/about-explore-nagaland/

Greenwood, D. (1989). Culture by the pound: An anthropological perspective on tourism as cultural commoditization. In V. Smith (Ed.), *Hosts and guests: The anthropology of tourism* (pp. 171–186). University of Pennsylvania Press. [orig.pub.1977].

Harcourt, W. (2017). Gender and sustainable livelihoods: Linking gendered experiences of environment, community and self. *Agriculture and Human Values, 34*(4), 1007–1019.

Japan International Cooperation Agency. (2020). *Survey on community-based tourism market in Thailand.* Retrieved September 20, 2024, from https://www.jica.go.jp/Resource/project/thailand/025/materials/ku57pq00003u68o4-att/survey_202003.pdf

Jarvenpa, R. (1994). Commoditization versus cultural integration: Tourism and image building in the Klondike. *Arctic Anthropology, 31*(1), 26–46.

Kinny, A., & Lanusosang, T. (2016). Exploring the potentials for ecotourism and sustainable tourism development in Nagaland India. *International Journal of Applied Research, 2*(9), 156–160.

Lanfant, M.-F. (1995). International tourism, internationalization and the challenge to identity. In M.-F. Lanfant, J. Allcock, & E. Bruner (Eds.), *International tourism: Identity and change* (pp. 24–43). Sage.

Longkumer, A. (2015). As our ancestors once lived': Representation, performance and constructing a National Culture amongst the Nagas of India, Himalaya, 35(1,10), 51–64. Retrieved August 5, 2024, from http://digitalcommons.macalester.edu/himalaya/vol35/iss1/10/

Maheshwari, K. (2022). Analysing cultural commoditization in relation with tourism in India. *International Journal of Advanced Research in Commerce, Management & Social Science, 5*(1), 170–176.

Mathieson, A., & Wall, G. (1982). *Tourism: Economic, physical and social impacts*. Longman.

McKean, P. (1989). Toward a theoretical analysis of tourism: Economic dualism and cultural involution in Bali. In V. Smith (Ed.), *Hosts and guests: The anthropology of tourism* (pp. 119–136). University of Pennsylvania Press. [orig.pub.1977].

Morung Express News Kohima. (2018). *The thriving business of homestays in Nagaland.* Retrieved August 25, 2024, from https://morungexpress.com/thriving-business-homestays-nagaland

Picard, Michel. (1996). Bali: Cultural tourism and touristic culture (trans. Diana Darling). : Archipelago Press. [orig.pub.1992].

Picard, M. (1997). Cultural tourism, nation-building, and regional culture: The making of a Balinese identity. In Picard & Wood (Eds.), *Tourism, ethnicity, and the state in Asian and Pacific societies* (pp. 181–214). University of Hawai'i Press.

Promburom, T. (2021). *Performing gender roles in homestay tourism: Place-making and social change for Lahu women in Northern Thailand.* Ph.D. Dissertation (unpublished), Chiang Mai University.

Rocheleau, D., Thomas-Slayter, B., & Wangari, E. (Eds.). (1996). *Feminist political ecology: Global issues and local experience*. Routledge.

Said, E. (1978). *Orientalism.* London: Routledge.

Sharma, D. (2023). *Hornbill festival 2023: Redefining local cooperation and global integration.* Retrieved August 4, 2024, from https://morungexpress.com/hornbill-festival-2023-redefining-local-cooperation-and-global-integration

Shepherd, R. (2002). Commodification, culture and tourism. *Tourist Studies, 2*(2), 183–201.

Smith, M. K., & Robinson, M. (Eds.). (2006). *Cultural tourism in a changing world: Politics, participation and (re)presentation*. Channel View Publications.

Stirn, A., & van Ham, P. (2000). *The seven sisters of India: Tribal worlds between Tibet and Burma*. Prestel.

Tatongkala. (2016). Orchestrated Celebration: Performing reconstructed identity in the hornbill festival of Nagaland. *The International Journal of Humanities & Social Studies, 4*(4), 147–150.

Von Stockhausen, A., & Wettstein, M. (2008). 'Cultural extravagance' and the search for identity in present-day Nagaland. In R. Kunz & V. Joshi (Eds.), *Naga: A Forgotten Mountain region rediscovered* (pp. 180–187). Christoph Merian Verlag and Museum der Kulturen Basel.

Wirudchawong, N. (2012). Policy on community tourism development in Thailand. *Journal of Tourism, 6*(2), 13–26.

Chapter 9
Segregation and Struggling Ethnic Women in the Tea Industry in Northeast India and Northern Thailand

Siya Uthai

Abstract Ethnic women are the main labor force in the tea industry of Northeast India, where more than 50% of India's tea production occurs. The study aims to examine the factors that constrain the occupational mobility of ethnic women within the tea industries of Northeast India and Northern Thailand and explore women's adaptation to these constraints. Data were collected between May 2022 to March 2024 through related documents, in-depth interviews, group interviews, and observation techniques. Findings reveal distinct patterns of occupational segregation. In Northeast India, limited access to education and systemic rights deprivation confine ethnic women primarily to tea-plucking roles. In Northern Thailand, while ethnic women are also overrepresented in lower-skilled, informal labor within the tea garden, they exhibit greater adaptability through skill development and gain more occupational options. Meanwhile, female workers in Northeast India faced difficulty in changing their livelihoods. Consequently, their primary strategy for improving their circumstances focuses on facilitating access to education for the next generation to pursue a better future beyond the tea garden.

Keywords Ethnic women · Tea plucking · Tea garden · Labor segregation

9.1 Introduction

Before industrialization, men and women were mainly involved in the agricultural sector for household food production. In shifting cultivation that relied on basic tools, such as hoes and digging sticks, women could participate as food producers like men (Sorgner, 2021). Both Northeast India and Northern Thailand relied heavily on agriculture, with women participating alongside men in subsistence farming. The declining number of women in the labor force is related to economic

S. Uthai (✉)
Department of Social Science and Development, Chiang Mai University, Chiang Mai, Thailand

Y. Santasombat (ed.), *Bio-Cultural Diversity, Ethnicity, and Local Knowledge*,
https://doi.org/10.1007/978-981-95-4242-0_9

development. Plowing equipment changed traditional agriculture to mass production for the market economy. Due to the plow requiring "significant upper body strength, grip strength, and burst of power" (Alesina et al., 2011, p. 2), women's labor forces on farmland are reduced and limited to work at home. Industrial development changed the form of labor forces and pushed people from agriculture to the industrial sector, including women's labor forces. Women, especially poor women, who are limited to working at home and childcare, are motivated to get paid jobs, like factory workers. However, women are the first group to be terminated during an economic recession (Parwez, 2019).

In modern society, due to industrialization, women are employed in various kinds of jobs, both as high-skilled and low-skilled workers. However, the status of women is not yet equal to men in many countries, particularly India. The issues of an unwanted daughter, dowry, poor education, and healthcare, as well as harassment at the workplace, are usually found (Saikia, 2020). In Northeast India, the status of women is better than in other parts of India, as the rate of women entrepreneurs is about 20%. This rate is higher than the country average of 10% because women in many hill tribes dominated the market. Moreover, no dowry system oppressed women in marriage among indigenous cultures and Assamese society (Saikia, 2020). However, there is unequal employment between men and women in Northeast India in terms of the rate of wages and the form of employment. The complex dynamics of women's experiences in Northeast India are essential to consider the interplay of historical, cultural, political, and economic factors.

Assam State, situated in Northeast India, is the largest state by population at 31,205,576 (Assam State Portal, 2021). Its social structure is woven from diverse ethnicities, languages, and beliefs. Assam has a rich history rooted in tribal cultures. The population is divided into three categories: hill tribes, plain tribes, and the non-tribal population. The tribes of the state are identified as "indigenous" and "non-indigenous," and 23 indigenous groups received 'Scheduled Tribes (STs)' status (Sharma, 2018). Some ethnic groups, like the Ahom, Chutia, Moran, Motok, and Koch, are integrated into Sanskritization, but they retain strong tribal traditions and customs (AspireIAS, 2020). Hinduism is less influential than in other parts of India, but most Assamese people are Hindu. However, people in rural areas are not strict with caste norms (Assam State Portal, 2021). The Bodo tribe is one of Assam's largest indigenous groups, constituting about 5% of Assam's population. They fought for statehood and were granted special autonomous district status in 2003. However, Bodo has conflicts with Bengali-speaking Muslims who settled in the overlapping area before the British established the state of Bangladesh in 1947. The campaign against illegal Bangladeshis was started in 1979 by the All Assam Student's Union and demanded the removal of illegal immigrants from the electoral rolls. The Indian central government ignored the union's request, and it led to violence in 1983 with massive deaths. The Assam Accord addressed the issue of illegal immigration by categorizing immigrants based on their arrival dates. People who arrived before 1966 are considered citizens. However, minor tension still exists between the Bodo tribe and the Muslim group (Bhattacharyya, 2012).

Recently, regarded as migrant and non-tribal (non-ST) groups, the Bengali-speaking groups were marginalized. They included the tea tribes, which form the ethnic group that migrated from various parts of India during colonial times. Since the British occupied, the workers in the tea estates have been regarded as non-indigenous groups. Tea plantations and tea tribes are British legacies to Assam at present. Tea tribes sought Scheduled Tribe (ST) status because they lost various social welfare and opportunities, like educational scholarships, housing subsidies, and healthcare facilities (Deccan Herald, 2024).

According to Assam State Portal (2021), "Assam has the single largest tea-growing area in the world, constituting around one-seventh of the global tea production. It has its own variety, Camellia Assamica. The state accounts for over 50 percent of the country's overall tea production." By this information, tea estates have been extended to become large plantations for tea production, which the global markets label as "Assam Tea." The colonizers are gone now, but tea tribe workers still become the primary labor force in Assam tea plantations. The population of "tea tribes" in Assam is approximately 5 million, and they do not receive ST status (Sharma, 2018). ST status in Assam is related to politics and the proportion of seats in the parliament (Sinha, 2022). However, the impacts are on tea tribe workers, particularly the lack of social welfare in education, employment, and health care. Many of them are still stuck in the tea estate, which is a low-paid job that only costs two USD/day (Elks, 2018).

In comparison to the status of women in northern Thailand, which can be called in the former term "Lanna," the social structure of Lanna society in the past was matriarchal kinship. In marriage, a man will be moved to a woman's family, and the family ancestors are counted only as cousins of women (Kitsutorn & Eiamrerai, 2023). The Lanna women in a high-status family received significant roles related to religious supporters, such as a chairperson in temple building. Furthermore, women's leadership role as rulers of the empire is found in the Lanna legend. The women dominated the marketplace as merchants and sellers (Ungkawanichakul, 2017). In Northern Thailand, rural communities depended on agriculture for their livelihoods. The Buddhist temple serves as a religious and social center where the community activities revolve. Tai ethnic groups are the dominant group in Northern Thailand (UNESCO World Heritage Center, 2015). However, many ethnic groups settle along the mountainous land, such as the Hmong, Karen, Akha, Lahu, and Lisu. For Lisu communities that settled along the highlands of Chiang Rai Province, northern Thailand, women are regarded as honored when working on farmland like their husbands (Klein-Hutheesing, 1990).

In Chiang Rai, the tea plantations have been extended for decades because of a high demand in domestic markets. Chiang Rai occupies the largest tea plantation in Thailand (Thai PBS, 2020). Tea plants, specifically Camellia sinensis, thrive in the cool, misty climate of Northern Thailand's highland forests. Local people use fermented tea leaves to eat after meals instead of beverages (Bouphun et al., 2017). They often ferment tea leaves into a unique product known as "Miang," a post-meal digestive aid. In recent years, the commercial cultivation of tea has expanded, with a focus on producing high-quality black, green, and oolong teas. These teas are sold domestically, and a small-scale export is made to international markets, particularly China and Taiwan. Large tea enterprises in Chiang Rai are often owned by Thai-Chinese families and employ local ethnic communities, particularly women, to work in their tea gardens, primarily for the labor-intensive task of tea plucking.

The study aims to understand the production processes in the tea industries between Northeast India and Northern Thailand, which are finally involved with employment segregation. Assam was selected as an exemplary case in Northeast India, and Chiang Rai was selected in Northern Thailand because it has the highest rate of tea production in each region. The article attempts to analyze the factors and circumstances of segregation and how ethnic women seek to move on from the constraint. Qualitative research methods are applied through in-depth interviews, group interviews, and observation techniques for data collection from May 2022 to March 2024. Key informants were selected for in-depth interviews, and group interviews comprising 20 female tea workers, a Jorhat tea company manager, and two small tea garden owners in Assam were conducted. In Chiang Rai, northern Thailand, 18 female tea workers, three tea company managers from three districts, a female marketing manager, and a small enterprise's female owner were selected for interviews.

9.2 Concepts and Literature

9.2.1 Gender Segregation

Women's employment is the way to put them into paid jobs. Women shift from home to the market and enter the formal economy within the occupational structure through gender stratification (Cohen, 2004). Therefore, gender inequality in the workplace is caused by the unequal structure, which occurs in both vertical and horizontal dimensions. Women's labor force can be economic empowerment indicators, but employment segregation limits women to unpaid, informal, and low-return jobs (Carranza et al., 2023).

In developing countries, the female labor force is increasing in parallel with the educational opportunities. However, the cases in South Asia differ as women are employed in a few sectors and occupations (Sahoo & Klasen, 2018). Customs and cultural practices in India are significant causes of gender inequality. The gender pay gap and different working hours are the leading indicators of inequality. ILO informed that women choose to work in part-time jobs because of domestic responsibilities (Dewi et al., 2024). As informal workers, they are at risk during a crisis, as the women will encounter. Women need to find a way to sustain themselves in the constrained structure.

9.2.2 Social Adaptation Strategy

In psychology, social adaptation is related to self-esteem, which is the process that involves confronting, compromising, and cooperating with the environment, depending on coping skills and interpersonal relationships (Neerly-Prado et al., 2021). Many studies on migrants and social dynamics have applied social adaptation to explain individual and group interactions. Wu (2022) studied female workers in the food delivery sector in Shenzhen, China, which is a male-dominated job. Women adapted themselves in three aspects—economic, occupational, and psychological—to gain income freedom, but they were also confronted with sleeplessness and a hurried work state.

Liu (2023) studied the Chinese diaspora in Japan by examining the Chinese female workers who graduated from Japanese colleges. Three factors are involved in their adaptation: individual rational choices, mutual pressure of in-group and out-group interaction, and national policy between Japan and China. Besides, Irtasyah (2023) conducted a study on the social adaptation strategies used by migrant populations to cope with a new culture. The first common strategy is an adjustment to following social norms, rules, and cultural practices. Learning the host language is another strategy to get closer to local people. However, they strategize to maintain their identity, culture, and values. To feel connected to the new environment, they engage in traditional celebrations and traditional food cooking.

9.2.3 Tea Industry in Northeast India

During the 1820s, British travelers explored the forests of northern Assam and discovered tea plants in the forests, which could be proved by scientific knowledge. Previously, the locals in Assam did not use tea leaves as a beverage, unlike in China. In 1823, the former Governor of Bengal, which covered the area of present-day Assam, became a trader after his retirement and made an agreement with the local ruler to cultivate tea in upper Assam (Sharma, 2009).

Assam is India's largest tea production region, accounting for more than 50% of the total tea production (Goswami, 2006). Recently, Assam's tea industry has been occupied by Indian and Assamese capitalists who have their own tea estates and have the option of sourcing tea leaves from over 60,000 smallholders in Assam. Female workers are also found to be highly involved in the tea production process (Das & Mishra, 2018). Women have been pushed to be workers in the tea industry since colonial times. They were not locals in Assam but were imported from other parts of India, such as Bihar, Orissa, West Bengal, Andhra Pradesh, and Madhya Pradesh (Sarmah, 2020).

9.2.4 *Women's Role in Industrial Workforces*

The rapid economic growth and industrialization across Asia in recent decades have significantly increased women's participation in the industrial workforce. This trend can be traced back to colonial times when the demand for tropical commodities and the subsequent integration of Asian economies into the global market spurred increased migration and economic activity (Talbot, 2002). In recent decades, India has experienced rapid economic growth, urbanization, and social changes. While women's education and fertility rates have improved, female labor force participation remains low. On the supply side, societal norms and expectations often restrict women's ability to work, particularly in rural and wealthier households. On the demand side, legal, normative, and economic constraints limit women's employment opportunities (Fletcher et al., 2017).

In Thailand, women are a significant part of economic progress. The gender wage gap in Thailand has narrowed over the past few decades, primarily due to women's educational attainment. However, discrimination against women persists in the Thai labor market. Young women are often paid less than men, even when they are more productive. The gender intensity of industries, or the proportion of male and female workers, significantly influences gender wage gaps (Thi Bui & Permpoonwiwat, 2015). Thai women in workplaces are at a higher rate than in other countries in Southeast Asia, but gender inequality in the job position remains a crucial issue.

9.3 Ethnic Women in the Tea Industry in Northeast India

Assam has been the area of tea production since colonial times. During the control of British companies, labor forces were imported from other parts of India because of the resistance from Indigenous groups (Sarmah, 2020). Since the 1850s, Assam has become the leading tea-producing region in the world through British rule, which transformed Indian subsistence peasants into industrial farmers whose economic self-sufficiency was reduced into market dependence. The British transformed India's economy, focusing on urban areas and plantation agriculture for export but diminishing the rural and tribal economies (Karmakar, 2015). Assamese labor forces were limited due to their reluctance to work in the plantations under British labor obligations. The workforce in tea plantations came from various regions of India through the "Sardari system," which involved sending experienced workers back to their home regions to persuade others to join. In 1859, the act enforced this system to make it more efficient and cost-effective for tea plantation owners. The colonial time in India exposed the complex and often exploitative practices surrounding women, particularly in terms of marriage, labor, and sexual exploitation; for example, the case of a girl sold in Orissa by their father or husband (Sen, 2010). Therefore, workers in the tea estates worked like slaves, as they could not move out.

By 1889, almost 50% of workers in Assam tea plantations were from Chotanagpur, where Adivasi and Sadan (Mulwasi) ethnic groups received Scheduled Tribes (ST) and Scheduled Castes in the Indian Constitution but not in Assam (Sentinel Digital Desk, 2023). Nowadays, Adivasi and non-Adivasi workers are called "tea tribes," and their descendants continue working and living in the tea industry. Some of them got married to villagers in communities nearby and moved out from the tea estate, but they still worked as tea pluckers for small tea growers. Adas (interview, August 3, 2023), a 45-year-old female worker who called herself an "Adivasi," got married and lived with her husband in the community near the large tea estate. Although she did not work in the tea estate, she still worked as a tea plucker in the small tea garden of locals for more than 20 years. Moreover, her daughter, Adi (Interview, August 3, 2023), has worked in the same tea garden, even though she used to study in primary school. She narrated, *"I did get a chance to attend a school, but now I am married. My husband has gone to Chennai for work, and I also dream of leaving Assam someday to build a better life."* The young generation, who experienced new skills and environments apart from tea gardens, attempted to move out.

In small tea gardens, female workers are from nearby villages, mostly Adivasi. Some of them identify as Bengali, the descendants of tea workers since colonial times. When asked about the treatment difference between genders at the workplace, both Adivasi and non-Adivasi workers stated that only women work on this task (interview, August 3, 2023). Furthermore, they receive payment in cash every weekend from the garden owners. According to the small tea garden growers (August 2, 2023), they hired workers from the village after they had informed them of the period of work, number of workers, and payment methods. Many people accepted the work. They have two main jobs: tea plucking and spraying. Women will work on tea plucking; old men will sometimes join, albeit only a few. Mostly, men in the village work on their farmland, and women help men work on farmland during harvest season (Interview, a small tea grower, August 2, 2023). There are two cases of female workers who work in the small tea garden whose husbands work in the factory of a nearby tea company. First, a 47-year-old worker stated, *"My husband works in a nearby tea garden in the factory* (August 3, 2023)." Next, a 55-year-old worker who is from the same community also mentioned the same of her husband. Their husbands have chances to work at the tea factory, but they are not hired. Thus, they have to work with small tea growers.

Due to Assam's disapproval of the Adivasi group's ST designation, the Adivasi ethnic women residents of the village lack the right to own land (Behal, 2016). However, they fight by using the legal system and collective action (Sentinel Digital Desk, 2023). Therefore, many cannot go anywhere outside the tea estates, and nowadays, many women marry natives in the community nearby. Women who married the same Adivasi workers had to follow their husbands to work in the tea estate. According to an interview with a temporary female worker at the tea company in Jorhat (August 1, 2023), she moved with her husband to work at the tea company, but her husband passed away. She stated, *"I have been working for 20 years now, and I do not know how to work in other places because I know nothing apart from the tea garden."* Lack of education and skill development compels female workers to adapt.

Traditional Indian families are considered patriarchal and collectivistic, and they join families that include three or four living generations. Women are expected to accept a position as subservient and subordinate to males and the needs of others (Chadda & Deb, 2013). Women's status is related to motherhood functions, which, by having sons, will ensure their position as the mother-in-law to have significant power in households. During motherhood time, women can be safer from domestic violence (Reed, 2021). In tea plantations, women are responsible for housework, childcare, and family care. Moreover, women are employed as workers for daily wages outside the home. They have a "double burden" that limits their time and opportunity for educational pursuits and career advancement. A permanent female worker at a tea company (August 2, 2023) said, *"I have to wake up early to finish household tasks before going out to work for a whole day in the tea garden."* Also, temporary female workers at the company replied with the same response, *"We wake up early in the morning at around 4 am to finish the household work and then come here to work"* (August 1, 2023).

The labor structures between small tea gardens and tea companies are different in the production processes. Most small tea garden owners do not have factories because of the high cost of machinery. Therefore, they sell fresh tea leaves to the company and run their business only in the agricultural sector. The tea company occupied all tea production processes, from the tea garden to wholesale and retail products for sale (interview, a tea company's manager, July 30, 2023). Workers in the tea company are divided into two types: permanent and temporary workers because of the high demand for manpower during the rainy season when the young tea buds grow faster. Both permanent and temporary workers are hired to work in the gardens and the factory at the same daily wage rate, according to Assam State's regulation (interview, a tea company's manager, July 30, 2023). Outside the rainy season, the labor force is reduced by half. At this time, the temporary workers will not be hired to work, and they will not be paid by the company (interview, a temporary female worker, August 1, 2023).

The permanent workers, both men and women, were born and grew up in the tea company. Another type of worker in the tea plantation is the temporary worker, who gets the same wage as the permanent worker, but the period of work is limited to 6 months during the rapid growth of tea leaf buds. The manager of the tea company (interview, July 30, 2023) employed men to work at the factory since many of the jobs there required strength to operate large machinery. Women in tea plucking are faster than men, according to their manager and female employees. A female worker can collect tea leaves of about 50–60 kilograms/day during the rainy season, which is double that of a male worker. Therefore, the tea company allocated tea plucking to women for efficiency and productivity.

Both in tea companies and small tea gardens, women are segregated to work only in tea plucking, even though there are other jobs in the garden, for example, spraying and cutting grass. Male workers are allocated to work with equipment that requires strength (Alesina et al., 2011, p. 2), while women must work with their hands without any new skill training. According to a respondent, she said, "I learned the way of plucking from my parents when I was young" (Interview, a permanent female worker, July 31, 2023). Tea plucking is only a skill that women in the tea industry have. As a result, they cannot consider working outside of the tea garden. Furthermore, their civil rights to live in safety are not protected by their ST status.

9.4 Ethnic Women in the Tea Industry in Northern Thailand

Chiang Rai Province, located in Northern Thailand, occupies the largest area of tea production in Thailand, approximately 13,600 hectares (Thai PBS, 2020). Residents of Northern Thailand's upland forests grow tea plants, and they utilize the fermented leaves of the Camellia sinensis species, which resemble Assam tea, to make "Miang," a local delicacy (Tea and Coffee Institute, Mae Fah Luang University, n.d.). Tea plantations in Chiang Rai started with experimental plots before Chinese-Thai entrepreneurs expanded their business by importing and breeding tea trees from Taiwan (Pongchababnapa, 2015). Recently, Chiang Rai became a large area of tea plantations, including Assam tea plants and Chinese tea plants in Thailand.

The suitable location of tea plantations in Chiang Rai is mountainous, where highlanders from various ethnic groups settle down. The Chinese military of Kuomintang from the Southern Region, one of the settling ethnic groups, did not move to Taiwan but are scattered around the Thailand–Myanmar border (Interview, an old tea company owner, May 13, 2022). To sell tea leaves to a Chinese caravan from Yunnan, China, the young tea company owner's father and his group discovered tea plants in the forest of Chiang Rai. They also hired the Lahu, an ethnic group that resides in the tea forest area, to gather tea leaves from the forest (Interview, May 14, 2022).

When the Chinese group started their tea business upon their arrival, they hired locals to collect tea leaves instead of working by themselves. Therefore, ethnic group communities, such as the Akha, Lisu, and Lahu (Thai Rath, 2019), near the tea forests became labor forces for the Chinese tea business until recently. Nowadays, tea businesses face labor shortage conditions because many of them attempt to be workers abroad. For instance, a female worker lamented, "*If I have a chance to work in South Korea or Israel like my husband, I will go even if it is risky in Israel, I will have much money for my family*" (Interview, a female worker in Mar Fah Luang, March 14, 2024). Tea owners increasingly imported Tai people from Shan State to work in the tea industry (Mano & Pianluprasidh, 2023).

According to the interview with managers of tea plantations in Wiang Pa Pao District, Mae Suai District, and Mae Fah Luang District, Chiang Rai Province, they hired workers from Shan State to work legally in their tea enterprises. Tea entrepreneurs allocated male workers to work in the factory sector. They get paid monthly as permanent workers. While working in the tea garden, the entrepreneurs hired workers for part-time jobs to get paid for the weight of tea leaves that they collected each day. Most of the people who collect the tea leaves are women. The factory's tasks are routine processes that they have to do the same thing every day, and it is managed as shift work because some processes have to be completed at night. Hence, male workers are the suitable choice for the enterprise.

Even though the local majority in Northern Thailand operated on matriarchal kinships, the ethnic minorities involved with the tea industry in Chiang Rai are both entrepreneurs and workers. Chinese entrepreneurs run the Chinese family system in business models in which family members are major shareholders in corporate equity. Chinese family businesses are comprised of family members who are in executive positions. Therefore, it is not easy to separate family and business responsibilities (Tsai et al., 2013). Chinese traditional patriarchal attitudes limit women in families from taking care of family affairs. Women also have to assist in the family business, but it is still uncommon for women to be considered potential successors, as men in the family are often prioritized. Nowadays, Chinese women with higher education can also be promoted to significant positions in the family business (Chen et al., 2018). Although women accessed high positions in the business field, they remained in the family, taking care of the main responsibilities. According to the tea company sales manager in Mae Suai during an interview, she expressed, "*I have to see my child for a while, for it seems like he has awoken* (May 14, 2022)." This was similar to a situation in Mae Fa Luang's tea company when she managed the shop while taking care of her children after coming back from school (Interview, marketing managers in Mae Fa Luang, March 14, 2024).

At the worker level, they are ethnic minorities who work in the tea plantation. Mostly, Tai Yai, or Tai people, who originated in Shan State, Myanmar, have become the main workers in many tea companies in Chiang Rai, both male and female. Tai Yai families are usually in small households that comprise a father, mother, children, and sometimes grandparents. A male-dominated family is a customary practice in the Tai Yai ethnic group. Following the Buddhist religion, men can be ordained as Buddhist monks, and parents also get merit from their sons. While women cannot do it, the status of men in Tai society is higher. Therefore, women have to work for a whole day on farm and household affairs (Santasombat, 2008). Akha people are another ethnic group that works on tea plantations. Traditionally, Akha men have superior roles to control women. After marriage, women become men's assets, and they must obey their husbands' mothers to do all the housework and eat after all the other family members. A son will receive all family assets in the next generation, but a daughter is obligated be an asset to other families. Akha men consider the marriage to a woman by her farming performance because she also needs to help her husband in farming apart from housework (Wongnuch et al., 2023).

Many tea garden workers are ethnic women who switched from working on their family farms to tea picking for the tea corporations to increase their income. However, men are also capable of picking tea in the tea gardens; this task is not just for women. There is a lack of men in tea plucking because they can collect tea leaves in small amounts daily (Interview, a manager in Mae Suai District, May 14, 2022). As a result, their revenue will be lower because the payment is based on the quantity of tea leaves rather than the daily rate. Many women in the tea industry in Chiang Rai Province are hired to pick fresh tea leaves for business owners without being formal employees (Interview, an old tea company owner, May 13, 2022). They are ethnic women from nearby communities and Tai Yai who married tea factory workers of tea enterprises. Women who come to work as tea pluckers, even though they do not enter the formal system, are satisfied with a flexible work environment. This is because the women's priority role in families still involves taking care of household chores and raising children. A Tai woman accepted her working conditions: *"Working without a contract like this is better for me because when my family has any urgent affairs, I can freely avoid working on that day"* (Interview, a female worker in Mae Suai, October 20, 2022).

Moreover, a group of migrant women entered via the Thai–Myanmar border, many of whom did not have work permit documents; even their husbands had those documents issued by the tea enterprise owners. Therefore, especially for married women with children, working apart from the tea plantations is impossible. In Northern Thailand's tea sector, women play a role in marketing in addition to plucking tea. The Chiang Rai manufacturing company's tea supply chain markets itself by reaching out to customers directly on wholesale and retail channels. The primary responsibilities in the domestic retail market are carried out by a woman who is the daughter-in-law of a tea firm owner through a Chinese family business (interview, marketing managers in Mae Fa Luang, October 22, 2022).

There is a female entrepreneur in Wiang Pa Pao who sells tea leaves at her shops and online platforms (Interview, female entrepreneurs in Wiang Pa Pao, September 23, 2022). It is because her Chinese father, who relocated his family from Bangkok to Chiang Rai, left her an inheritance of a tea estate. Following that, she and her spouse increased the tea plantation's size to 16 hectares. To gather new tea leaves, she used laborers from neighboring settlements. She and her family members manage all the subsequent procedures until the final product is ready for sale. Educated women have more access to various tea supply chain operations owing to the Northern Thai tea sector. However, it is still limited to the ethnic group with more social and economic capital.

9.5 Discussion

In the Assam tea industry, women's roles are limited only to tea pluckers in the tea garden using their hands at work. The circumstances did not only occur in the big tea estates; even in the small tea gardens of locals, women were segregated to work manually in agriculture. The segregation resulted from unequal opportunities, which are not only due to gender issues. In a male-dominated society, women get fewer opportunities in the educational system because of their responsibilities in household affairs. Hence, female workers lack new skills to find other jobs, in comparison to men who have more opportunities in education and social engagement (Chadda & Deb, 2013). Sahoo and Klasen (2018) state that education opportunities are significant to women's employment. Many "tea tribe" workers did not invest in schooling; they could not think of anything apart from tea gardens. Moving outside the tea estates becomes risky for them, especially women. Cultural practices in India confined women to housework and childcare responsibilities (Dewi et al., 2024). Moreover, ethnic inequality disadvantaged "tea tribe" people through civil rights deprivation and lack of social welfare. Tea tribe female workers have been discriminated against through gender and ethnic bias.

Apart from the Hindu society in Assam, female education is not encouraged in the tribal society of Northeast India. They are also forced to perform domestic tasks and marriage (Kumar & Shobana, 2022). Therefore, the skill of tea plucking is only the knowledge that they learned from their parents when they were young (a permanent female worker, July 31, 2023). Some women who marry locals and live in nearby villages can use their skills in tea plucking to earn a part-time job after they finish household affairs. However, during harvest season, women also help their husbands work on the farmland. The circumstances of female workers are more complicated than those of wealthy families. In Northeast India, the caste system is not obligated intensively, especially in rural areas. However, people of different class statuses were burdened unequally. Educated women in well-off families have to take care of household affairs after marriage, but they can concentrate only on domestic affairs without professional work (Biswas & Mukhopadhyay, 2018).

In Assam, the segregation of ethnic women by industrial influences and social norms has limited the roles of women. Hinduism and tribal customs confine women to domestic responsibilities. Women are obligated to marry and belong to their husbands. Therefore, girls have limited access to education. Finally, their career options are restricted, and they are dependent on their husband's family. Meanwhile, ethnic women who are in poor families have a "double burden" of both household chores and childcare, together with being low-skilled workers in the tea gardens to gain more income. Even though there are in different wages based on gender factors for labor workers in tea plantations due to state regulations, the hiring opportunities of women for permanent jobs are lower than for male workers. Especially in the factory, employers hire only male workers to work because of their physical strength and shift-work adaptability. These circumstances limited women from learning new skills, while men could gain experience in machinery working skills, which brought the latter more opportunities to work in other factory jobs. Moreover, ethnic inequality in Assam through legal systems deprived civil rights and basic social welfare of 'tea tribe' workers by limiting their access to resources.

However, ethnic women struggled with the constraints of earning from part-time jobs as tea pluckers in small tea gardens. Lifting the financial situation allowed them to bring more opportunities to their children's education, which they recognized as a pathway to a better future beyond tea gardens. Optimistically, small tea growers open job opportunities for women to do part-time jobs that are more flexible than working in tea companies. They can balance their family affairs and social engagement with other women in their communities. Women are probably set on collective actions to empower each other beyond their limitations.

In Northern Thailand, ethnic women in the tea industry are also working part-time jobs in tea plucking. To incentivize employees to gather as much tea as possible to increase their earnings, Chiang Rai tea entrepreneurs employed workers to gather tea leaves in their tea gardens and weighed the leaves. Furthermore, the quantity of tea buds varies, especially during the dry season when none are present; yet, tea entrepreneurs do not incur any costs related to employee responsibilities. Women were segregated from other processes after collecting tea leaves as informal employment (Carranza et al., 2023) because the primary responsibilities were related to childcare and domestic tasks. They are limited to working in the tea factory, doing routine tasks, and working the night shift. However, women who have more capital, particularly those with education opportunities, are able to apply their knowledge and skills to work in the marketing sector and develop themselves into small entrepreneurs.

Ethnic women in Chiang Rai attempt to adapt themselves to earn more income while caring for domestic tasks. When there are no jobs in tea gardens during the tourist season, they acquire new skills in sewing, weaving, and tailoring to create goods to sell in the local marketplaces. Many women realize the limits of women's status in traditional society. Thus, they send their daughters to a school similar to their sons, incurring a higher expense. These conditions drove women to do many informal jobs to afford their children's expenses because they hoped their daughters could find a good job in the city with a higher education (Interview, female workers in Mae Fah Luang, March 14, 2024) to gain more income freedom (Wu, 2022). Meanwhile, women with greater financial resources can adjust to working with modern technologies as a tactic to deal with the new company culture (Irtasyah, 2023). E-commerce provides a new strategy for women to position themselves in working with equipment that does not require to use body strength (Alesina et al., 2011), like working on farmland and factories.

Ethnic women in Chiang Rai and Assam are engaged as low-skilled workers in the tea industry, and traditional practices have constrained them to domestic responsibilities. Women are automatically pushed to professional work outside to gain additional income for their families, but they cannot share household chores with their husbands. They faced double burdens from the traditional responsibilities and feminization of labor in industrial sectors. However, ethnic women in Chiang Rai demonstrated more possibilities for livelihood adaptation. They engaged in other income-generating jobs, such as weaving and selling their craftworks to tourists. In contrast, ethnic women in Assam have limited skills and live in a closed area that fewer outsiders can access. A few economic activities can be found apart from tea plantations.

Education is a significant factor in adaptation strategies. Women in Chiang Rai prioritize their children's education as a means of livelihood diversification. In Thailand, people's access to the education system is broader and easier compared to the Assam case. Particularly, Adivasi or tea tribe workers who have rights deprivation face challenges to accessing public services, including the educational system. Moreover, traditional practices give priority to men rather than women because they have to get married and become their husband's assets. Parents thus prefer to support their boys. Lack of education constrains tea tribe women from depending on their husbands. Nowadays, some tea companies allow the government to open the primary school in their territory. Many girls from tea tribes are sent to school like boys because female workers want their kids to have an education for their choice of occupation outside the tea garden (interview, a temporary female worker, August 1, 2023). Adi (interview, August 3, 2023), a young woman who used to be in a school, stated that in the future, she would like to go out of the tea garden and Assam as the adaptation strategy (Irtasyah, 2023). She believed that she would have a better life in a place where Adivasi status was legal. Because she had a chance to study, she could learn more Assamese and Hindi languages and other things that she could not learn in the tea garden. Thus, she is finding a better way for her life (Interview, Adi, August 3, 2023).

9.6 Conclusion

The tea industry in India was established during colonial times, and India's economy has inherited the legacy of colonial exploitation in recent times. Women in tea gardens have been the commercial image of tea products from India for decades, as they are tea pluckers in the tea gardens who use their hands instead of equipment. More than 50% of India's tea production is from Northeast India, particularly in Assam. Despite being the backbone of the Indian tea industry, "tea tribe" women in Assam face significant constraints. This includes their limited access to education and restricted employment opportunities, confining them to tea plucking with poor living conditions. Their lack of Scheduled Tribe status generated their vulnerability and confined them to tea estates.

Ethnic inequality exacerbates these challenges, as "tea tribe" women experience civil rights deprivation and inadequate social welfare. Opportunities for advancement are scarce, as male workers dominate factory tasks and skill-building opportunities, leaving women at a disadvantage. Lacking education and training from employers, women are segregated from the work that they have been familiar with since they were young. The roles of tea tribe women in Assam are limited to tea plucking. However, women struggled for their resilience by engaging in part-time tea-plucking jobs in small tea gardens, a flexible job with the potential to uplift their families' financial circumstances.

In contrast, Chiang Rai's tea entrepreneurs find tea pluckers from locals who belong to ethnic minority groups, with some of them having migrated from Myanmar through the Thai-Myanmar border. Ethnic women in Chiang Rai also face challenges, such as informal employment and limited access to social safety nets, and they demonstrate greater adaptation due to diverse economic opportunities within their communities. Tea entrepreneurs paid only the wage based on the amount of tea leaves they collected. Ethnic women were relegated to informal workers because of their domestic responsibilities and the inevitable shift-work process of tea companies. However, the presence of tourism allows them to pursue alternative livelihoods, such as home-based businesses and seasonal work.

Furthermore, Chiang Rai's social circumstances offer both official and informal platforms for learning new skills and finding various informal occupations to work from home; thus, increased educational options translate into greater economic flexibility. Additionally, women who have access to education can adjust to the high position in the tea sector that has historically been held by men. Meanwhile, tea tribe women faced many challenges in accessing educational systems because of traditional practices and domestic responsibilities. Despite having the chance to marry locals and moving out to live in the area, women who have no other skills continue to work at low-paying jobs like picking tea. In their responsibilities as mothers, women find it difficult to support their kids' education, which they see as a way to escape the cycle of dependency and provide a brighter future.

The findings highlight the profound impact of social, economic, and cultural contexts on shaping women's roles and experiences within the tea industry. Although gender-based segregation and limited upward mobility are present in both locations, there are notable differences in the level of limitation and the adaptation of livelihoods for empowerment. In Assam, the challenges faced by "tea tribe" women are deeply entrenched in historical, systemic inequities. Both industrial practices and sociocultural norms have marginalized these women. The study underscores the pressing need for comprehensive policy interventions to redress these historical injustices and foster gender equity. Policy interventions in Assam are necessary to secure their rights, improve access to education and skills training, and promote equitable labor practices within the tea industry.

References

Alesina, A., Giuliano, P., & Nunn, N. (2011). *On the origins of gender roles: Women and the plough*. Working Paper Series 17098. National Bureau of Economic Research. https://doi.org/10.3386/w17098

AspireIAS. (2020, January 6). Non-tribal communities can never claim to be indigenous: AAKS. Aspire IAS. https://www.aspireias.com/daily-news-analysis-current-affairs/Non-tribal-communities-can-never-claim-to-be-indigenous-AAKS

Assam State Portal. (2021). Assam at a glance. https://assam.gov.in/about-us/393

Behal, M. (2016). Marginalization: A narrative of the Adivasi community in Assam. North East Network, India. https://northeastnetwork.org/wp-content/uploads/2021/07/MARGINALIZATION-A-NARRATIVE-OF-THE-ADIVASI-COMMUNITY-IN-ASSAM.pdf.

Bhattacharyya, A. (2012, August 29). Understanding the historical conflicts behind today's violence in Assam. Center for American Progress. https://www.americanprogress.org/article/understanding-the-historical-conflicts-behind-todays-violence-in-assam/

Biswas, C. S., & Mukhopadhyay, I. (2018). Marital status and women empowerment in India. *Sociology International Journal, 2*(1), 29–37. https://doi.org/10.15406/sij.2018.02.00030

Bouphun, T., Wei, X., Yihe, J., & Qi, Z. (2017). Miang (Thai fermented tea) and its cuisine. *Journal of Tea Science, 37*(6), 645–652.

Carranza, E., Das, S., & Kotikula, A. (2023). *Gender-based employment segregation: Understanding causes and policy interventions*. Jobs Working Paper No. 26. International Bank for Reconstruction and Development.

Chadda, R. K., & Deb, K. S. (2013). Indian family systems, collectivistic society and psychotherapy. *Indian Journal of Psychiatry, 55*(Suppl 2), S299–S309. https://doi.org/10.4103/0019-5545.105555

Chen, S., Fang, H. C., & MacKenzie, N. G. (2018). Female leadership in contemporary Chinese family firms. *Asia Pacific Journal of Management, 35*, 181–211. https://doi.org/10.1007/s10490-017-9515-2

Cohen, P. N. (2004). The gender division of labor: "keeping house" and occupational segregation in the United States. *Gender & Society, 18*(2), 239–252. https://doi.org/10.1177/0891243203262037
Das, K. S., & Mishra, D. K. (2018). Woman and work in rural Assam: Pattern and determinants. *Journal of Social and Economic Development, 20*(2), 213–239.
Dewi, N. N. C., Al Putra, F. A., & Wijanarka, T. (2024). Gender segregation in the labour market in the Indo-Pacific region: A case study of Indonesia, Vietnam, and India. *Insignia Journal of International Relations, 11*(1), 70–92.
Elks, S. (2018, August 13). *Twinings reveals suppliers amid drive for slave-free Indian tea.* Thomson Reuters Foundation. https://cms.trust.org/item/20180813161648-b1xzc
Fletcher, E., Pande, R., & Troyer Moore, C. (2017). *Women and work in India: Descriptive evidence and a review of potential policies.* CID Faculty Working Paper No. 339. Harvard University.
Goswami, M. (2006). Emerging of entrepreneurship in tea plantation: A study of small tea growers of Assam. [PhD Thesis, Guwahati University, Assam].
Herald, D. (2024, September 26). Tea tribes marginalised in Assam despite contribution to economy: Soren writes to Himanta. Deccan Herald. https://www.deccanherald.com/india/jharkhand/tea-tribes-marginalised-in-assam-despite-contribution-to-economy-soren-writes-to-himanta-3206930
Hutheesing, O. K. (1990). Emerging sexual inequality among the Lisu of northern Thailand: the waning of dog and elephant repute. Leiden: E.J. Brill, 1990.
Irtasyah, M. R. (2023). Social adaptation strategy for migrant communities in a new culture. June 21, 2023. https://doi.org/10.31219/osf.io/g38av
Karmakar, K. G. (2015). Colonial rule and its effects on India's rural economy. *Journal of South Asian Studies, 3*, 277–288.
Kitsutorn, T., & Eiamrerai, T. (2023). Power and factors affecting the existence of Lanna women in ritual media space to worship ancestral spirits (Phi Pu Ya). *Journal of MCU Peace Studies, 11*(2).
Kumar, J. S., & Shobana, M. D. (2022). Status of women education in north-East India. *International Journal of Multidisciplinary Educational Research, 11*(3), 50–56.
Liu, J., & Chen, S. (2023). Embedded coexistence: Social adaptation of Chinese female white collar workers in Japan. Sustainability 2023, 15(2), 1294, https://doi.org/10.3390/su15021294
Mano, P., & Pianluprasidh, P. (2023, April-June). Approach to overcome the workforce shortage of the northern Thai tea plantation. *Journal of Business, Economics and Communications, 18*(2).
Neerly-Prado, A., van Elk, M., Navarrete, G., Hola, F., & Huepe, D. (2021). Social adaptation in context: The differential role of religiosity and self-esteem in vulnerable vs. non-vulnerable populations. *Frontiers in Psychology, 12*, 519623. https://doi.org/10.3389/fpsyg.2021.519623
Parwez, S. (2019). Feminisation and exploitation of labour in India. ILERA European Congress 2019. https://ilera2019.giraweb.de/content/feminisation-and-exploitation-labour-india#:~:text=The
Pongchababnapa, C. (2015, April–June). Geographical indication of Chiang Rai Tea. *Tea Newsletter, 5*(19). (In Thai).
Reed, M. N. (2021). Reproductive transitions and women's status in Indian households. *Population Studies, 75*(3), 325–341. https://doi.org/10.1080/00324728.2021.1904147
Sahoo, S., & Klasen, S. (2018). Gender segregation in education and its implications for labour market outcomes: Evidence from India. Discussion paper series: IZA DP No. 11660, IZA—Institute of Labor Economics, Bonn.
Saikia, R. (2020, May). Status of women in India with special reference to Assam. *International Journal of Research Cultural Society, 4*(5). https://ijrcs.org/wp-content/uploads/IJRCS202005049.pdf
Santasombat, Y. (2008). Gender roles and gender relations. In *Lak Chang: A reconstruction of tai identity in Daikong* (pp. 133–150). ANU Press. http://www.jstor.org/stable/j.ctt24h3gf.11
Sarmah, P. (2020). Tea tourism of Assam: Potential and challenges. *Journal of Critical Reviews, 7*(13) ISSN-2394-5125.
Sen, S. (2010). Commercial recruiting and informal intermediation: Debate over Sardari System in Assam tea plantation, 1860-1900. *Modern Asian Studies, 44*(1), 3–28.

Sentinel Digital Desk. (2023, September 28). Adivasis and non-adivasis in the Tea estates of Assam. The Sentinel. https://www.sentinelassam.com/more-news/editorial/adivasis-and-non-adivasis-in-the-tea-estates-of-assam-668787

Sharma, K. (2009). *Globalization and plantation workers in North-East India*. Kalpaz Publications.

Sharma, I. (2018). Tea tribes of Assam: Identity politics and search for liberation. *Economic & Political Weekly, 53*(9), 74–78.

Sinha, R. N. (2022, October 17). Politics behind denial of ST status to 6 Assam OBCs. NewsClick. https://www.newsclick.in/Politics-Behind-Denial-ST-Status-6-Assam-OBC

Sorgner, A. (2021). Gender and industrialization: Developments and trends in the context of developing countries. Discussion Paper Series: IZA DP No. 14160, IZA. Institute of Labor Economics, Bonn.

Talbot, J. M. (2002). Tropical commodity chains, forward integration strategies, and international inequality: Coffee, cocoa, and tea. *Review of International Political Economy, 9*(4), 701–734.

Tea and Coffee Institute, Mae Fah Luang University. (n.d.). Tea history. https://teacoffee.mfu.ac.th/tc-tea-coffeeknowledge/tc-tea/tc-teahistory.html (in Thai).

Thai PBS. (2020, October 5). Chiang Rai's tea received global awards. Thai PBS. https://www.thaipbs.or.th/news/content/297076 (In Thai).

Thai Rath. (2019). Experience the way of life of 7 Hill Tribes and Golden Sunflower at Hue Mae Kham. https://www.thairath.co.th/news/local/north/1696746 (In Thai).

Thi Bui, M.-T., & Permpoonwiwat, C. K. (2015). Gender wage inequality in Thailand: A sectoral perspective. *International Journal of Behavioral Science, 10*(2), 19–36.

Tsai, H., Wu, T., & Yeh, S. (2013). A study of Chinese Guanxi type in family business from the perspective of power-based and leadership behaviours. *South African Journal of Economic and Management Sciences, 16*(5), 102–114. https://doi.org/10.4102/sajems.v16i5.680

UNESCO World Heritage Center. (2015). Monuments, sites and cultural landscape of Chiang Mai, Capital of Lanna. https://whc.unesco.org/en/tentativelists/6003/

Ungkawanichakul, P. (2017, July–December). A study of status and way of life on northern Thai women from documents and mural paintings in Lanna. *Institute of Culture and Arts Journal, 19*(1), 82–91.

Wongnuch, P., Apidechkul, T., Srichan, P., et al. (2023). Women oppressed in the daily lives and cultural practices of the Akha people, Thailand: How can the situation change? *BMC Psychology, 11*, 311. https://doi.org/10.1186/s40359-023-01361-6

Wu, J. (2022). Research on the social adaptation of female food delivery workers in Shenzhen. International Conference on County Economic Development, Rural Revitalization and Social Sciences 2022. https://doi.org/10.2991/aebmr.k.220402.030

Chapter 10
The Cross, the Crown, and the Creature: Footprints of the Young Legacy Across Centuries

Kengkij Kitirianglarp, Pana Kantha, and Sarid Siriteerathomrong

Abstract This paper traces the multifaceted legacy of the Young family, whose influence spanned from Northeast India to Thailand across seven decades of the nineteenth and twentieth centuries. By examining their evolution from American Baptist missionaries to military operatives and finally to naturalists who founded Chiang Mai Zoo, this study reveals how colonial epistemologies persisted and transformed across changing imperial regimes in Southeast Asia. The Youngs operated within an intellectual tradition where scientific documentation served both religious conversion and imperial objectives. Their close relationships with ethnic minorities, particularly the Lahu and Wa people, proved invaluable as they navigated shifting geopolitical landscapes from British colonialism to American Cold War influence. The establishment of Chiang Mai Zoo represents the crystallization of these intertwined historical forces—where missionary zeal, colonial authority, and naturalist inquiry converged. Drawing on Aníbal Quijano's concept of "coloniality," this paper argues that institutions like zoos, often presented as neutral educational spaces, emerge from specific historical contexts that reflect and reproduce naturalist knowledge with colonial origins. By analyzing the Young family's footprints across centuries, we gain insight into how the pillars of colonial modernity—cross, crown, and creature—have fundamentally shaped our understanding of contemporary Southeast Asia.

Keywords Young family · Chiang Mai Zoo · Colonial epistemology · Missionary naturalists · Coloniality · American Baptist Missions · Southeast Asia · Thailand · Ethnic minorities · Lahu · Wa · Cold War · Natural history · Imperial knowledge systems

K. Kitirianglarp (✉) · P. Kantha · S. Siriteerathomrong
Chiang Mai University, Chiang Mai, Thailand

Y. Santasombat (ed.), *Bio-Cultural Diversity, Ethnicity, and Local Knowledge*,
https://doi.org/10.1007/978-981-95-4242-0_10

10.1 Introduction

Nestled at the foot of Doi Suthep mountain, Chiang Mai Zoo appears to be merely a recreational space where families observe wildlife and learn about biodiversity. Visitors wander through exhibits, observing taxonomically arranged specimens and reading scientifically classified information panels. Yet beneath this apparently neutral facade lies a complex institutional history. The systems of classification, display, and knowledge production embodied in such spaces have genealogies tracing back to specific historical contexts. In the case of Chiang Mai Zoo, these contexts involve missionary networks, changing imperial influences, and evolving forms of knowledge production about nature. The zoo's establishment in the mid-twentieth century emerged as part of broader historical processes connecting Southeast Asia to global networks of knowledge circulation. These processes involved numerous actors, including the Young family, American Baptist missionaries whose presence in Asia extended over seven decades, spanning territories from Northeastern India to Northern Thailand.

Julian Go (2012) notes that while European colonial power declined after World War II, this facilitated the emergence of an American empire that inherited many European colonial legacies. Anne Foster (2010) identifies the Spanish–American War as enabling growing American influence in Southeast Asia, characterized less by direct territorial control and more by economic penetration and ideological influence. During the Cold War, this influence intensified through anti-communist interventions that shaped institutional development throughout the region. Within this context, the role of American missionaries, particularly Baptists, as vehicles of Western epistemological frameworks deserves greater scholarly attention. Despite substantial research on political and military dimensions of colonialism in Southeast Asia, the contributions of missionary networks to establishing knowledge structures have been comparatively neglected.

This oversight is significant because American Baptist Missions operated in Asia since the early nineteenth century, predating direct American political involvement in the region, and often worked with British Empire support in territories including Burma, Northeastern India, and Southern China. The significance of this study lies in its potential to illuminate how colonial processes of knowledge production evolved across different imperial regimes and manifested in institutions like zoos. By focusing on the Young family, whose activities traversed British colonial territories and later areas of American influence, we can trace continuities and transformations in how knowledge about nature was collected, organized, and institutionalized. This approach corroborates Walter Mignolo's (2021) observation that colonial power extended beyond human subjects to encompass control over nonhuman beings, creating epistemological frameworks that continue to shape our understanding of the natural world.

The establishment of Chiang Mai Zoo is a revealing case study for several reasons. First, it was founded during the Cold War era, when American influence was replacing earlier European colonial structures in Southeast Asia. Second, its creation stemmed from the activities of the Young family, whose background in Baptist missionary work potentially shaped their approaches to natural history and wildlife conservation. Third, the zoo represents the institutionalization of particular ways of organizing, displaying, and understanding wildlife that may reflect broader patterns of colonial knowledge production. This research addresses a significant gap in understanding how colonial epistemologies persist and transform across time. While extensive scholarship exists on formal political and economic dimensions of colonialism in Southeast Asia, less attention has been paid to how seemingly apolitical institutions like zoos might embody colonial legacies.

Similarly, this research posits that the study scope of missionaries as agents of change should be extended to explore their role in constructing colonial knowledge systems, particularly regarding the natural world, which has not received enough attention. The Young family presents an ideal focus for examining these questions. Their involvement in the region spanned three generations and over seven decades, bridging the era of British colonialism and the period of American Cold War influence. The family members performed multiple roles, as missionaries, military personnel, and naturalists, adapting to changing geopolitical circumstances while maintaining certain continuities in their approaches to knowledge. This makes them valuable subjects for understanding how colonial knowledge systems evolve despite changes in formal political structures.

Several key questions emerge from this study: How did the transition from British colonial influence to American imperial power affect knowledge production in Southeast Asia? What role did American Baptist missionaries play in establishing colonial epistemologies? How might seemingly neutral scientific practices perpetuate colonial power relations? And specifically, how did the Young family's evolving activities contribute to developments like Chiang Mai Zoo?

This study bridges historical research on missionary activities, colonial governance, and natural history in Southeast Asia, fields often studied separately. It contributes to understanding American influence through nonmilitary channels, examines institutional development across different imperial regimes, and connects insights about coloniality and knowledge production to concrete case studies.

To explore these issues, this chapter is organized into two main sections. First, we examine the broader historical context of American Baptist activities in Asia under British colonial influence during the nineteenth and early twentieth centuries. It provides essential background information for understanding the structural conditions that shaped missionary work in the region. Second, we focus specifically on the Young family, analyzing their background and tracing the evolution of their roles over decades of engagement across multiple Asian contexts. By connecting these historical threads, we aim to develop a nuanced understanding of how Chiang Mai Zoo emerged from complex processes of colonial continuity and transformation in Southeast Asia.

10.2 American Baptist Missions in Asia: Expansion Through Colonial Networks

The discovery of tea in Assam in the early nineteenth century marked a pivotal moment in India's history, fundamentally altering the region's economic and cultural landscape while simultaneously creating conditions for American Baptist missionary expansion. This expansion was inextricably linked to colonial networks, botanical exploration, and scientific advancement, which collectively created a complex web. Connecting missionaries, merchants, and colonial officials, this web would eventually extend through the work of the Young family and contribute to developments like the Chiang Mai Zoo in Northern Thailand.

British military success in the First Anglo-Burmese War (1824–1826) laid the essential foundation for these transformations. This conflict reshaped the geopolitical boundaries of Southeast Asia and Northeast India, with the Treaty of Yandabo in February 1826 formalizing British possession of Arakan and Tennasserim, while Burma relinquished claims of sovereignty over Assam and Manipur (Peers, 2012). This territorial control created the political environment necessary for both economic exploitation and missionary activities. In 1834, Andrew Charlton confirmed the existence of naturally occurring tea plants in Assam, sending samples through colonial administrative channels that ultimately led William Bentinck, the Governor-General of Bengal, to establish a Tea Committee for investigating commercial tea cultivation possibilities. By 1838, the first export of Assam tea to England occurred, followed by the establishment of the Assam Tea Company in 1839, marking the beginning of large-scale commercial tea cultivation in India (Raman, 2021).

The expansion of American Baptist missionary activities across Asia was inseparably intertwined with these European colonial enterprises, creating a complex relationship where religious and imperial ambitions often reinforced one another. In the same vein of thought, Pynhunlang NM Shullai (2017, p. 328) made the following observation: "It is obvious that Christian mission activities and colonialism go hand in hand with each other in spite of certain misunderstandings between them. Colonialism as an instrument paves the way for Christian mission activities and at the same time helping them to establish power." This relationship was particularly evident in the botanical and scientific dimensions of missionary work, where religious conversion efforts often coincided with scientific exploration and cultural transformation.

Missionary efforts in Burma began before the expansion of British colonial control. On July 13, 1813, with support from the Baptist mission in Serampore and receiving salaries from the English Baptist Mission, Dr. Adoniram Judson and Ann Hasseltine Judson landed in Rangoon (American Baptist Foreign Mission Society, 1914, p. 5; Merriam, 1900, pp. 20–21). Their early work faced significant obstacles under the Burmese rule, with limited success and constant government suspicion. These early missionaries were inspired by figures like William Carey, whose contributions extended far beyond religious conversion. At Serampore, Carey developed an extensive five-acre garden containing over 427 plant species arranged according to the Linnean system, maintaining correspondence with botanists worldwide and exchanging specimens with Dr. William Roxburgh of the Calcutta Botanic Garden. His scientific achievements earned him recognition through plants named in his honor: *Careya saulea, Careya herbacea, and Crinum careyanum* (Srinivasan, 1961). Concurrently, his founding of the Agricultural and Horticultural Society of India in 1820 institutionalized the connection between missionary work and scientific exploration (Srinivasan, 1961). This pattern of missionaries doubling as naturalists created knowledge networks that documented local flora and fauna while advancing colonial scientific interests.

The first Anglo-Burmese War fundamentally altered these dynamics. Following the British victory, in 1826, Dr. Judson and the Wade family relocated from Yangon to Moulmein (now Mawlamyine), a territory newly established under British control. There, free from government opposition they had faced in Yangon, they established a new missionary station, complete with living quarters and a preaching hall called a "zayat" (Moffat, 1868, pp. 103–104). The American Baptist Mission established headquarters in Moulmein, significantly expanding its reach and creating a hub for missionary activities throughout the region.

Moulmein provided crucial access to the Karen people, who visited during the autumn to sell various goods. The missionaries noted the Karens' distinctive group movements and songs, but language barriers initially prevented effective communication. A breakthrough came when a Burmese Christian named Ko Shwa-ba helped free a Karen slave named Ko Thah-byu, who, after joining Dr. Judson and the Wade family, converted to Christianity. This relationship opened opportunities for the missionaries to learn the Karen language and culture, facilitating evangelization in Karen villages (Moffat, 1868, pp. 105–109). The Karen people proved particularly receptive to Christian teachings, partly because their existing legends and songs spoke of "white foreigners" who would bring them "God's book." Karen tradition held that their ancestors once possessed a lost sacred book with prophecies of its eventual return. Furthermore, Karen legends contained narratives similar to biblical stories like Adam and Eve, creating cultural bridges that facilitated religious conversion. Missionaries learned the Karen language, translated biblical texts, and taught literacy, resulting in widespread acceptance of Christianity among the Karen—a success rate notably higher than among the Burmese population (Moffat, 1868, pp. 104–113).

In Assam, colonial economic expansion coincided precisely with missionary opportunity. Revealed by a British officer's request in Sadiya in 1834, the primary driver of the invitation of missionaries to the region was colonial interests. This initiative had three main objectives: to "convert the heathen," to address rebellion and unrest among local populations, and to assist the colonial government in maintaining regional control. The British administration believed that religious conversion could help pacify "troublesome heathens" and make them more amenable to British rule, demonstrating how Christian missionaries were utilized as tools within the broader colonial administrative strategies (Robbing, 1922, pp. 140–141).

The invitation fortuitously coincided with a surplus of missionaries at the Burman mission in Moulmein. Nathan Brown and his wife, originally selected to work among the Shan people in northern Burma, redirected their efforts to Assam to work with the Khamti people, who were ethnically related to the Shan (John, 1912, p. 39). Initially referred to as the "mission to the Shan," this work was strategically positioned, as it would eventually create a western route into China (Merriam, 1900, p. 45). In 1836, Brown and his family arrived in Sadiya, Assam, receiving substantial support from Major Francis Jenkins, the commissioner of Assam, who offered 1000 rupees upon their arrival and another 1000 rupees for establishing a printing press (Cushing, 1893, pp. 9–10).

The expansion of missionary activities among the Shan people similarly illustrates the intertwining of colonial power and evangelical opportunity. Rev. J.N. Cushing made significant contributions to this effort: despite facing difficult terrain and political hostilities, he learned the Shan language and made multiple journeys into Shan territory. Cushing and his wife ventured into remote regions, including Kengtung, overcoming both geographical obstacles and resistance from Burman officials (Cushing, 1893). Their journey to Kengtung began in November 1869 and continued to Chiang Mai, where they met Presbyterian missionaries who were facing significant challenges, including the execution of local Christian converts. This meeting facilitated a valuable exchange of experiences and regional insights, exemplifying how missionary networks operated across different colonial spheres of influence (John, 1912, pp. 63–79).

Another important turning point was the annexation of Upper Burma by the British Empire in 1886. It opened new possibilities for missionary work in previously restricted territories. Prior to this, Shanland and Upper Burma's administrative regions remained largely inaccessible to missionaries due to the Burmese court's restrictive policies. British control established regional stability, allowing missionaries and traders safer travel and residence opportunities. Dr. Cushing's earlier work translating the Bible into Shan and exploring the region provided crucial foundations for subsequent missionary efforts during this more politically favorable period (Cushing, 1893).

By 1890, the mission had opened a station in Thibaw, followed by one in Moné in 1892. The establishment of these permanent camps in Shanland was a major improvement over earlier operations from British-administered territory. British administrative control made missionary work safer and more stable, eliminating pressures previously exerted by the Burmese court and military. Local populations increasingly engaged with missionary-established educational and medical institutions, facilitating Christianity's gradual acceptance among the Shan people. Despite enhanced conditions, missionary work in these remote areas still required dedicated personnel willing to undertake pioneering efforts (Cushing, 1893), a challenge that would soon be met by William Marcus Young.

Like Carey before them, these missionary pioneers often conducted botanical and zoological observations alongside their evangelical work. Their collected specimens, established gardens, and documented historical knowledge formed an important but often overlooked contribution to colonial scientific understanding of the region. This tradition of missionary-naturalists would continue throughout the

Baptist missions in Asia, creating knowledge networks that documented local flora and fauna while advancing colonial scientific interests (Daston & Galison, 2010).

This historical narrative reveals how American Baptist missionary expansion was inextricably linked to British colonial power across Northeast India and Burma. The establishment of colonial control created the political conditions necessary for missionary activities; conversely, missionaries often served colonial interests through "civilizing" indigenous populations. The missionary presence in regions like Burma, Assam, and Shanland represented the intricate interactions between religious conversion efforts and broader colonial strategies of territorial control and cultural transformation (Foster, 2010; Quijano, 2007).

As these mission networks expanded throughout the nineteenth century, they created lasting infrastructures of knowledge production and cultural influence that would persist even as formal imperial arrangements shifted. The botanical gardens, schools, printing presses, and scientific collections established by missionaries became enduring features of the cultural landscape, facilitating ongoing processes of knowledge circulation between Asia, Europe, and America. These institutional foundations would later be repurposed and reimagined as Southeast Asia moved through successive periods of colonial rule, independence movements, and Cold War realignments.

Within the expanding missionary networks across Southeast Asia, the Young family emerged as one of the most prominent missionary dynasties. Beginning with William Marcus Young, who arrived in Burma in 1882, this family would play a pivotal role in evangelization efforts among the Shan people (Young, 1902). William Young took on the challenging mission to the remote Shan territories at a time when few were willing to venture so far from established colonial centers. His pioneering work would be continued by subsequent generations of the Young family, who not only continued evangelization efforts but also engaged in scientific and cultural studies (Chase, 2014; Young, 2014). This fusion of religious mission with scientific interest would eventually lead to remarkable developments, including the establishment of the Chiang Mai Zoo in Thailand (Lawitts, 2018, pp. 305–306). Examining the history of the Young family reveals the evolving roles of missionaries from the colonial era to the Cold War period and their lasting impact on the region's religious, scientific, and cultural landscape that continues to the present day.

10.3 Who's Who in the Young Family

The Young family's influence in Southeast Asia spanned three generations, with each member playing distinct but interconnected roles in missionary work, colonial administration, intelligence operations, and cultural development across the region (see diagram below). Their collective experiences provide a remarkable case study of how missionary families adapted to the changing geopolitical landscape of the twentieth century while maintaining continuous influence in the region.

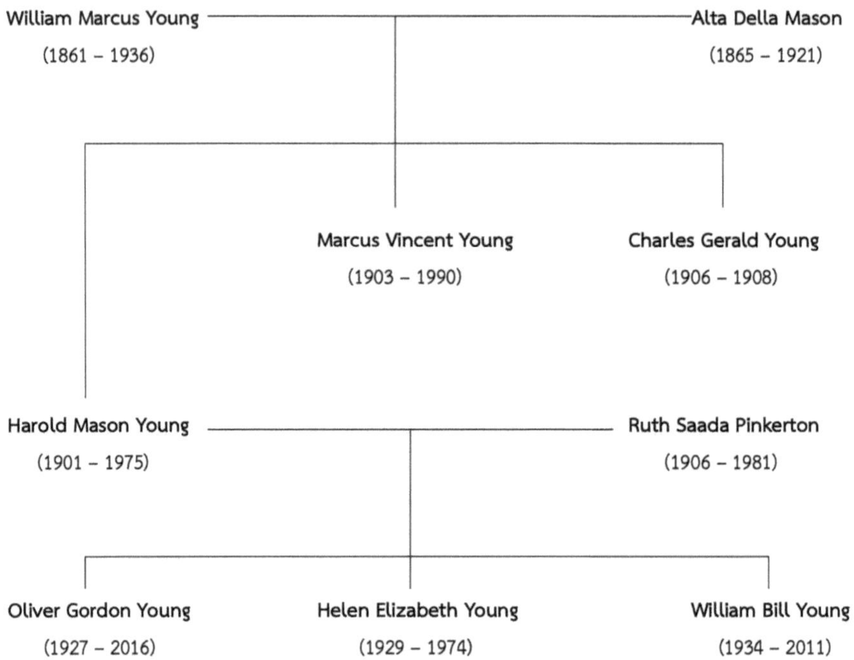

William Marcus Young was born into a Baptist family in Illinois, USA, in 1861. He grew up as a missionary. On the night of his father's death, he had a vision of Christ appearing at his bedside and ordering him to travel abroad to spread the word of God to the ends of the earth. As a result, he applied to be a member of the American Baptist Mission Society, which sent him on a mission to British Burma in 1881. He arrived in Rangoon in 1882 with four Karen assistants who traveled with him on a mission to the Shan State of Burma and was later granted permission to move to Kengtung, Shan, in 1900. During this time, he married a missionary named Della Mason on a ship. When they arrived in Kengtung, they established the Kengtung Mission House in 1901 to spread the religion to the Shan people, but it was not successful. There was no Shan convert.

William and Dell had three children: Harold, born in 1901 at the Kengtung mission house; Vincent, born in 1903; and Clarence, born in 1907. However, Clarence died young, leaving only two siblings, Harold and Vincent. After the birth of their sons, he began to travel more frequently to the Burmese-Chinese border. Harold and Vincent grew up speaking the Shan, Lahu, and Wa languages, as they had accompanied William on his missions. However, William and Dell could only speak Shan. Prior to the outbreak of World War I, the Young family was forced to return to the United States in 1907. After the war, in 1916, the American Board of Commissioners for Foreign Missions (ABCFM) approved William's request to open a new area of operations in Yunnan, and he was relocated to Banna, China, to work primarily with the Lahu people. William and his family arrived in Banna in 1920, and William and

Harold were able to build a mission station, a chapel, and a school. By 1921, Dell had died, and William sent Harold, then 20, back to the USA to study.

Later in 1926, after Harold graduated from the Bible Institute of Los Angeles, he married Ruth Saada Pinkerton. In August of the same year, Harold and Ruth arrived in Rangoon to continue their journey to Banna. The second generation of the Young family began their mission in Banna in 1927, working mainly with the Lahu people. They had a son, Oliver Gordon Young, who was raised by an Akha woman, Eva, and spoke the Lahu language as his mother tongue. In 1930, William was able to expand his mission into Wa territory. In 1931, Vincent, Harold's younger brother, married and joined the family's mission in Banna. William finally retired from the mission in 1932. Two years later, in 1934, the Young family returned to Seattle, USA. Harold and Ruth had their second son, William Bill Young, in Berkeley. In the following year, the Young family was reassigned to Burma, focusing on Wa territory and leaving Banna to the Lahu. Finally, on April 8, 1936, William, the grandfather, died, and his body was returned to California for burial.

After William died, the second generation of the Young family, brothers Harold and Vincent, took over their father's mission. They divided the mission area into two parts, with Harold responsible for the Burmese side and Vincent responsible for the Chinese side. The primary objective of the mission area's division was to attempt to enter the Wa area from both sides, since the Burmese-Chinese border separated the Wa area. Therefore, during the years from 1936 to 1940, before the outbreak of World War II, Gordon had to travel to the Burmese-Chinese-Indian area to perform his mission. During World War II, in 1942, Gordon's family had to flee to Assam, India, as their home in Taunggyi was bombed by the Japanese, who were invading Southeast Asia. Ruth, Harold, Helen (Gordon's sister), and Bill fled Burma on an evacuation plane. Harold became seriously ill and received treatment in Assam. At the age of 14, Gordon traveled to Northern India and went to school there. Vincent Young remained in Kengtung, traveling in and out of China (Banna), living on the Lahu-controlled border. Harold's enthusiasm for opening up the mission area and his expertise in working closely with ethnic groups led to Harold being employed by the British Army in late 1942 as the chief surveyor for the Ledo Road project in northeastern India and to control Shan and other ethnic forces in the Taunggyi area.

However, in 1943, having discovered that Harold was an American, the British Army discharged him. Harold then relocated to Assam to work in the Shan section of the American Office of War Information (OWI). His main duty was to provide counter-propaganda in the form of Shan language posters and airdrops until the end of the war in 1945. He went back to Burma after World War II ended and worked for the British colonialists' political department. His main duty was to help the British suppress the Wa. Later, when Burma gained independence from the British colony in 1948, the Burmese government hired Harold to oversee the conflict with the Wa. Harold's son, Gordon, served in the American Army, stationed on the Korean Peninsula until 1950. Since the political climate in Burma was unstable, the Young family left the country that same year and traveled to Chiang Rai, Chiang Mai, and Bangkok before reuniting with one another in California, U.S. Harold saw that the distinctions between Burma's ethnic groups were destroying the country's unity and degrading Westerners. For Harold, he saw that the era when the West ruled Burma was more unified. This was the end of the Young family's journey in Burma.

Following his discharge in 1950, Gordon married Peggy and pursued a degree in animal husbandry at California State Polytechnic College. Later, as the Cold War broke out, the situation regarding communism's spread throughout Southeast Asia deteriorated. The American government needed an expert to work in the anti-communist operations among ethnics groups and invited Harold to work for the CIA in 1952. Harold agreed and returned to Thailand, stationed in Chiang Mai. Harold and Ruth rented a house at Veluvan Villa in Chiang Mai. He started collecting wild animals and built a private zoo. This interest in collecting and studying animals continued the long tradition of missionary-naturalists established by figures like William Carey in the early nineteenth century. For the Youngs, as with earlier missionaries, natural history collection served multiple purposes, both as genuine scientific interest and later as an effective cover for intelligence work.

In 1953, a year after graduating from college, Gordon was contacted by the CIA. Gordon accepted the job offer and moved to Thailand. Receiving the job offer from the CIA was a turning point in Gordon's life, as he was given a secret mission to collect information in the Thai–Burma–China area, which led him to go undercover as an animal collector for a museum and zoo in 1954. While learning more about the Lahu tribe, Gordon and Peggy started gathering animals to stock the Veluvan. Eventually, the three Young family members, Harold, Ruth, and Gordon, founded the Chiang Mai Zoo in 1956. The establishment of this zoo represented a significant cultural contribution to Northern Thailand and embodied the continuing connection between missionary work and scientific exploration that had characterized American Baptist Missions in Asia since the nineteenth century. The zoo began as a private collection but would eventually grow into a major educational and conservation institution in Thailand, demonstrating how missionary legacy could be transformed into enduring cultural and scientific institutions even as formal missionary activities declined. Two years later, his younger brother, Bill Young, joined the CIA as a translator and interpreter because he could speak many ethnic languages. He also worked as an advisor to Vang Pao and the right-wing forces in Laos, based at the Udon Thani Air Force Base.

In the 1960s, the Young family gave their animals to the zoo and opened the "Chiang Mai Zoo." Harold oversaw the zoo until his passing in 1975. Another significant turning moment in Gordon's life occurred in the 1960s, when American foreign policy shifted from one of compromise to using military force against ethnic communities to stop communism from spreading throughout Southeast Asia. Gordon left the CIA as a result of his disapproval and suspicion of this shift in foreign policy. Later in 1961, he received a job offer from the US Agency for International Development (USAID), which focused more on development than military work. In addition to being an animal collector, Gordon also worked closely with ethnic groups in northern Thailand, especially the Lahu. He became an expert in the languages and traditions of Thailand's many ethnic groups because of his intimate engagement with them. In 1960, he published a significant book he wrote on ethnic groups, titled *The Hill Tribes of Northern Thailand*, by the Siam Society.

In 1975, Harold passed away at Suan Dok Hospital in Chiang Mai. The Thai government inherited Chiang Mai Zoo following his passing. Ruth returned to the USA and died in 1981. Finally, Bill continued his military career. In 1962, he moved to Northern Laos to join the Parachute Regiment and form the Secret War military units. After the war, he operated a guesthouse and sold jewelry before dying in Chiang Mai in 2011 (see also Hyun, 2023, and Lawitts, 2018).

10.4 The Baptist Mission from Kengtung to Banna

Having examined the key figures and generational timeline of the Young family, we can now explore in greater depth the specific missionary strategies, challenges, and theological perspectives that characterized their work in Kengtung and Banna. The family's approach to different ethnic groups, their reliance on colonial protection, and their theological interpretations of indigenous belief systems offer insightful information on both missionary methods and the complex interactions between Western religious expansion and indigenous spiritual traditions in Southeast Asia during this period.

According to the report to the American Baptist Missionary Union regarding the new missionary station in the Shanland at Kengtung, William Young (1902) presented a detailed ethnographic and religious analysis of the area's population. In Kengtung, Young showed a critical attitude toward the Shan people's Buddhism, characterizing it as little more than an outward manifestation of their traditional beliefs in spirit worship. He saw temples as the root of social issues and vehemently attacked the monastic establishment, charging it with a lack of morality and understanding of fundamental concepts. However, Young saw potential for spreading Christianity among the highland ethnic groups, especially the Tai Loi, Kah, and Muser (Lahu), with a total population of about 40,000. He compared these ethnic groups to the Karen people in the early days of missionary work and hypothesized that they would be more receptive to Christianity than the Buddhist Shan. Young emphasized the urgency of conducting missionary work among these ethnic groups, citing the expansion of Buddhism as a potential obstacle to future Christian evangelization. He called on American churches to support this mission, highlighting the area's strategic importance, the social conditions he viewed as deteriorating, the failure of Buddhism, and the opportunities for evangelization among highland ethnic groups whom he perceived as lacking a stable religious belief system.

10.5 Protection from the British Army

As previously mentioned, the British position in India had a direct impact on the influx of Baptists to America. In particular, the British used Baptists as an important tool to spread colonial ideas to the native peoples and to expand their influence in

Asia. In turn, Baptists relied on British protection both in their travels and in their negotiations with local authorities, especially those under British colonial rule. This protection also included military aspects since missionaries had to depend on the British Army at different locations and had to ask the British for permission before establishing stations. Additionally, the British were crucial in assisting in negotiations with local authorities to lessen the opposition the missionaries faced. In elaboration, Harold Young (2014, p. 55) said, "Then, as well as throughout the time of British administration of Burma, the only power under heaven that made the mission possible in the country was the friendly and sympathetic attitude and generous help given by the British Government to Christian missions." This relationship between missionaries and colonial powers, as described earlier, was a fundamental characteristic of Baptist missionary expansion throughout Asia, creating mutually beneficial arrangements where religious conversion efforts and imperial control reinforced one another.

In the early weeks of his arrival in Kengtung, he could not find a place to stay and start his mission, as no Shan people welcomed Westerners. William went to the British officials to ask for help, and the British sent a representative to negotiate with the Shan prince. He eventually obtained land for his mission (Young, 2014, p. 55).

The Youngs sought robust protection during the mission in Kengtung, Burma, and Banna, China, where they encountered local resistance and so-called bandits. In Kengtung, the American Baptist Missions was highly protected by the British Army. William had employed Lahu and Wa as his private guards in Banna for years. As Harold and Vincent were born in Banna, they grew up with such an army of guards surrounding their family that both were trained to use weapons like rifles since they were very young. In the Banna mission, "Harold and Vincent with their rifles, and many Lahu and Wa armed with muzzle-loaders, crossbows, and spears" (Chase, 2014, p. 87). They were able to use these weapons with ease on their subsequent missions before and during World War II due to their tight collaboration with the Wa and Lahu soldiers. To put it briefly, neither the native army nor British officials could have carried out the missions in the region without turning to weapons.

In Kengtung, where he works, William had to wait 3 years for his first conversion, but not with the nearby Shan people. In October 1904, the first Lahu convert was baptized, and this "marked the beginning of the great ingatherings among the Lahu and Wa during the years of Father's ministry to his beloved hill tribes" (Young, 2014, p. 59). The first conversion led to the arrival of the Wa, who later became a large group of converts (Young, 2014, pp. 59–63).

While William's mission primarily focused on the hill tribes, he did not abandon his mission of preaching to the Shan people in Kengtung. William's unrelenting spirit is evident in the report that he submitted to the American Baptist Missionary Union about the results of his mission. The report stated that while achieving an impressive 1636 baptisms among the hill tribes, Young did not neglect his evangelistic efforts among the Shan people in Kengtung. According to the document, he persisted in conducting regular preaching services in the bazaar, specifically for the Shan. Furthermore, Young opened an outstation five miles away from the main compound, dedicated to special work among the northern and western Shans. This outstation included a small chapel and a conducive house for native helpers, demonstrating Young's commitment to reaching the Shan population. Additionally, plans were made to start a school in the city for the benefit of Shans and Burmans, indicating a long-term strategy for engaging with these communities alongside the successful hill tribe mission (Young, n.d.).

Following the success of his mission among the hill tribes, William reflected, "This is the wonderful opportunity presented to American Baptists at Kengtung. Seldom in the history of missions has there been such an opening. The way is clear for a great advance movement into China along the frontier of Burma and Assam." (American Baptist Missionary Union, 1907, p. 8) This demonstrates his determination and desire to extend his mission into China.

10.6 Why Lahu and Wa, Not Shan?

William developed negative impressions of the Shan people after his unsuccessful attempts to convert them. Despite their self-identification as Buddhists, Kengtung was plagued by dangerous bandits during this period. William found the Shan particularly resistant to conversion compared to other groups like the Lahu or even the head-hunting Wa. Although the Shan professed adherence to Buddhist principles, in practice they combined Buddhism with animistic beliefs in an inconsistent manner. Their lack of deep religious commitment made converting them to Christianity extremely challenging (Chase, 2014, p. 82). In contrast, "the Lahu were receptive to the Gospel for many reasons and this was mostly due to the teachings of their religious leaders, songs of their bards and tribal traditions passed down from generations to expect a true religion" (Chase, 2014, p. 81). After abandoning his efforts with the Shan people, William relocated to the Banna missionary station in China in 1920 to work with the more receptive Lahu and Wa communities.

Such a negative attitude toward Buddhism among missionaries was common at the time. Similarly, in the travel logs of missionaries in Burma in the late twentieth century, it is said that "... our religion is quite different to theirs; ours is the true God, and theirs is a false god. That is true; and there is one great difference between the true religion and all false religions. The difference lies in the two words, love and fear" (Moffat, 1868, p. 20). William's failure to convert Buddhists to Christianity was not new, but it did reinforce the negative sentiment toward Buddhists and Buddhism in this area.

At first, British authorities forbade the Young from crossing the Chinese border because of the threat of headhunting from the Wa and the unstable circumstances of the Warlord Era in China. During this time, xenophobia toward Western foreigners was widespread. In 1920, William chose to begin settling their mission in Banna, China, after sailing from Rangoon to Haiphong and then traveling to Yunnan via rails and foot. According to Lawitts (2018, p. 298), "The success of this trip into China, where others had perished, helped solidify in the eyes of his followers the mythological status of William Young as a prophet under God's protection." Then, three generations of the Young family spent 22 years of their lives in the Banna station.

During the two decades in Banna, William and his sons successfully converted more than 40,000 Lahu and Wa people to Christianity. However, it wasn't William alone who accomplished this remarkable achievement; his two sons, who were fluent in the Lahu and Wa languages, played crucial roles in the Banna mission (Lawitts, 2018, pp. 295, 299). Their success was so profound that "William Young was given the Lahu title of 'Jaw Maw' or 'Man-God.' Harold and Vincent were each dubbed 'Jaw Maw Eh' or 'Son of the Man-God.' They also earned the respect and allegiance of local strongmen and military leaders. … Later, when the Young family transitioned from missionaries into covert operatives, these warriors became valuable intelligence assets and guerilla recruiters furthering American military activity in Southeast Asia" (Lawitts, 2018, p. 300). Following William's death in 1936, Harold assumed his father's position as Jaw Maw, and the mission divided into two operations: one in Burma under Harold's leadership and another in China led by Vincent.

Regarding the question of why William favored the Lahu and Wa over the Shan, he believed that the Lahu and Wa, living within an animistic cosmology, possessed spiritual innocence. Conversely, the Shan, who blended traditional animistic beliefs with Buddhism, were considered "sinful" and thus resistant to baptism. In the missionary's view, animism represented a "state of nature"—a "pre-civilized condition that preceded human sinfulness. Harold noted the impressionable nature of the Lahu, stating, "The Lahu are more easily influenced by existing conditions than most of the other hill tribes. Often, when a man becomes separated from his own environment, he will very soon attempt to adopt the ways of the others" (Young, 2013, p. 377). This susceptibility indicated that their world remained untainted by external sinfulness, making their animistic mindset more receptive to Christian teachings.

This perspective reflected contemporary Western evolutionary theories about religion. Following Edward B. Tylor's framework, animism represented an early stage of human religious evolution, followed by polytheism and ultimately monotheism. Consequently, Western monotheistic cultures were viewed as more advanced than "animistic barbarians." Buddhist societies, having established state religions, were considered too entrenched in their beliefs for conversion. The missionaries perceived Buddhism as a misguided path from which there was no return. This reasoning partly explained William's decision to abandon the Shan mission in Kengtung and focus on converting the Lahu and Wa in Banna. In essence, the missionaries viewed the animism of hill tribes as convertible, while Buddhism in the valleys, including that of the Shan, was deemed the primary obstacle to American missionary efforts.

This hierarchical view of religious evolution was common in colonial missionary thinking across Asia, as discussed previously, and shaped how missionaries like the Youngs approached different ethnic and religious groups, often predetermining which communities they deemed more "convertible" than others. Moreover, in

Lahu's folktale of the White Man and the Golden Book, "one great leader was raised up for each generation, who never taught of self but always pointed to the future when their expectations would be fulfilled. … When the appointed time is come, the messengers of the 'White Book' will follow the trails of the porcupines (small jungle paths) and sit down at the home hearth and expound the new teachings" (Young, 2014, p. 57). This was similar to Wa's belief, following Harold's notebook on the Wa followers (Young, 2014, p. 61; see also Lawitts, 2018), they were also waiting for the coming of "a White Teacher of the White Book and White Law".

In *To the Mountain Top* (Young, 2013, p. 177), Harold characterized Lahu religion as a dual system: "With most Animists, there is a strong belief in the existence of god. In fact, the form of Animism as practiced by the Lahu is equally mixed with Theism." This perspective framed their beliefs as a combination of primitive animism and theistic elements that could serve as seeds for conversion to "more progressive" beliefs such as Christianity. Harold further noted, "The teachings of these [Lahu] religious leaders laid the foundation for all the songs of the bards, as well prepared the people in a most amazing way for the coming of the Christian Gospel. They were directly responsible for the great movements of the past toward Christianity" (Ibid., p. 180). However, this interpretation of Lahu belief in a supreme deity became a subject of scholarly debate, with anthropologists specializing in Lahu culture later challenging these claims (see Walker, 1975, 2008, 2010).

The Young family's interpretation of Lahu and Wa beliefs has been subject to significant academic criticism. In Anthony Walker's critical review of Gordon's *The Hilltribes of Northern Thailand*, he identified numerous biases and errors in Gordon's portrayal. One major error was Gordon's inappropriate application of Biblical concepts like sin to Lahu religious beliefs. By emphasizing supposed connections between Lahu cosmology and Christianity, Gordon revealed his bias in favor of the Lahu over other ethnic groups he studied (Walker, 1975).

Walker refuted the claimed similarities between Christian and Lahu cosmologies that the Young family promoted: "As a matter of fact, the concept of a creator-god is a characteristic feature of Indigenous belief systems throughout the Southeast Asian world, both mainland and archipelago." However, "this creator god is usually considered a remote figure, entirely unconcerned with human kind, its creation. For their part, humans see no useful purpose in worshipping or propitiating a creator divinity that, unlike the spirits, has no impact upon their lives" (Walker, 2010, pp. 15–16). In essence, Walker argued that equating the Lahu creator divinity with the Christian God was fundamentally incorrect. The Young family's persistent overstatement of these similarities obscured the reality that their missionary work functioned as a political instrument backed by colonial military power from its inception.

Hyun (2023) emphasizes that the indigenization of Christianity was crucial to American Baptist missionary work in Asia. American missionaries in the early twentieth century commonly employed strategies such as integrating indigenous beliefs with Christian theology or reinterpreting local spirits as manifestations of God. This approach contrasted sharply with that of Portuguese Jesuits during their mission in Brazil, who harbored negative feelings toward indigenous peoples. The Jesuits perceived an "inconstancy" in indigenous mindsets that they believed hindered successful conversion to Catholicism (Viveiros de Castro, 2011). This divergence likely stemmed not from cultural differences between peoples of different continents but rather from fundamentally different missionary philosophies regarding both their mission and their understanding of indigenous peoples.

The success of the Banna mission resulted in a deep impression of Lahu among the Young family. This impression did not only concern religious reasons but also Lahu's deep understanding and love of nature. Harold and Gordon, in the later period, were highly impressed with the Lahu's systematic classification of plants and animals in the jungle. Such knowledge was very similar to the Linnaean taxonomy. Harold had a deep-seated belief that the Lahu people could be persuaded to the Baptist faith because of these similarities. Inadvertently, Harold's mission among the Lahu and Wa in Burma and China was aided by the parallels between Western and Lahu knowledge as well as their shared belief in a single supreme spirit or God.

The success of the missionary work among the Lahu and Wa in Banna laid the foundation for the Young family's continued influence in the region, which would persist even as their roles transformed dramatically during World War II. The language skills and tribal knowledge developed during their missionary endeavors, along with their deep connections with the Lahu and Wa peoples, became invaluable resources that positioned the family as key figures in subsequent military operations. This transition from missionary to military roles not only reflected the adaptability of the Young family but also illuminated the complex relationships between religious activities, political power, and conflict during this era.

10.7 Military Missions During and After World War II

The coming of WWII drastically changed the mission landscape. The axis Thai, supported by the Japanese army, occupied Kengtung in 1942. The Young brothers played a key role in commanding local ethnic armies to help protect the lands. Unfortunately, the Japanese army gained its victory, and the Young family had to flee to India and Burma by trekking. During the war, Harold joined the British Army in Assam, India, and then the Office of War Information (OWI), a part of the American Office of Strategic Services (OSS). In the former, he was a labor commander in the Ledo Road project, and in the latter, he used his proficiency in Shan and Lahu languages to help the Americans with anti-propaganda against the axis. His main operations were intelligence and subversion. Harold and his son, Gordon, served the OSS, mainly training courses in jungle survival. Relying on his close connection with Lahu and, especially, Wa, he and Gordon did well in intelligence operations. During that time, Vincent led British officers to support Chinese warlords against the Japanese army. As Harold organized Kachin guerrillas in India, Vincent helped organize Lahu and Wa in China. Finally, the Young family, including Harold, Vincent, and Gordon, worked for the OSS. According to Lawitts (2018), with their experience on the hill, even without professional training in warfare, they became valuable assets for the Allies.

After the end of WWII, Harold transferred to the military units and gained a high reputation among the British army, Shan princes, and Wa army. He commanded 350 officials of the United Military Police and controlled 7000 troops of the Shan defense force. He also helped the British suppress insurgencies in the borders between Burma and China. He had worked closely with the British until Burma's independence in 1948. Then he decided to return to the mission, but the Board of American Baptism declined his request (Lawitts, 2018, pp. 303–304).

Following its transformation from the OSS, the newly renamed Central Intelligence Agency (CIA) focused on countering communist influence. The Chinese Revolution of 1949 created an urgent need to reestablish intelligence operations in Southeast Asia, with Thailand selected as the regional headquarters. In 1951, during the early Cold War period, the CIA recruited Harold and Gordon Young, establishing their base of operations in Chiang Mai. Harold leveraged his *Jaw Maw* status to address Burmese affairs, while Gordon was initially deployed to support US Army operations in Korea. Upon Gordon's return, both brothers began recruiting Lahu and Shan individuals for CIA operations. Lawitts (2018, p. 305) notes that "Concurrent with Operation Paper in 1951, the CIA also helped Thailand create an elite unit of police Special Forces called the Border Patrol Police (BPP), which was supported by the Thai Royal Family, especially by King Bhumibol Adulyadej."

The CIA involvement extended to William (Bill) Young as well, who played a crucial role in covert operations. During the Secret War in Laos, Bill utilized his inherited divine status to "organize Lahu intelligence-gathering forays into southern China for the CIA during the 1950s." His exceptional linguistic abilities with ethnic minority languages made him invaluable to American intelligence efforts, as "he probably knew more about mountain minorities than any other American in Laos, and the CIA rightly regarded him as its 'tribal expert'" (McCoy, 2003, p. 306).

The Young family became an invaluable asset to both British colonial and American imperial military operations due to their extensive missionary experience, deep knowledge of regional ethnic groups, and strong relationships with local communities. What distinguished them from typical missionaries was their exceptional military expertise and skills. These combined capabilities enabled them to play significant roles across multiple generations and historical periods in the region.

The Young family's transition from military operations during World War II and the early Cold War to naturalist pursuits happened organically and with remarkable continuity. They effectively utilized the knowledge, skills, and tribal networks they had cultivated over decades. While continuing their involvement with CIA operations in the region, Gordon and Harold discovered a new application for their long-established relationships with the Lahu people through natural history exploration and wildlife collection. This evolution allowed them to sustain their regional influence and engagement in a transformed context, blending their strategic intelligence activities with scientific contributions that would establish their lasting legacy in Thailand.

10.8 The Naturalist Mission During the Cold War

After Gordon returned to Thailand in 1954, he re-established his relationship with the Lahu people, building on his old connections before World War II. He also wrote books recording his experience of hunting trips with the Lahu people, who were very knowledgeable about the jungle and nature. Tales involve hunting large wild animals such as tigers and wild bulls. In *Tracks of an Intruder*, first published in 1960, Gordon Young (1991, p. 23) expressed the spirit of his naturalist journey: "I was about to pick up where my father and grandfather had left off, although I would not be walking with the missionary's spirit, but rather with the spirit of the adventurer and explorer that had also possessed my forebears to an intense degree."

Gordon claimed that the family's naturalist capability was from their immense experience of living with the Lahu. According to Chase (2014, p. 102), "Harold knew many scientific terms and had an immense curiosity and love of nature; plants, reptiles, birds and mammals. His education came from long years of observation and the valuable information he learned from living among Lahu." This was the result of Lahu's naturalistic exceptionalism, which was closely similar to the Western taxonomy. As illustrated by Gordon (Young, 1991, p. 25), "While others have limitations to names of the things of the jungles, the Lahu has carried out a systematic description—taxonomy—which names to nearly all things, both large and small." And also expressed by Harold himself that "The Lahu have extremely remarkable knowledge of nature." (Young, 2014, p. 317) Moreover, as Harold mentioned the richness of Lahu's naturalist knowledge, "The Lahu are great believers in lunar control over the weather, plant life, and the habits of animals and insects. … The Lahu are masters at surviving in the jungle. During the war when I lectured to our own boys on survival techniques in the jungle, I drew almost entirely on Lahu knowledge of jungle lore" (Young, 2014, pp. 318–319).

In short, the second and third generations of the Young family, who were all born and raised in the jungles of Burma and then southern China, learned a substantial amount of knowledge about natural history from the Lahu tribe. This, combined with their Western education from their families and the time they devoted themselves to studying animals in the U.S. after World War II, resulted in them becoming rare naturalists.

Following the Chinese revolution of 1949, all Baptist missions were shut down, putting an end to European aspirations to penetrate China. In Southeast Asia alone, American Baptists believed that local Baptist churches should handle missionary activity instead of the large-scale American missionary efforts that were previously necessary. After World War II ended, the family was unable to resume their missionary work because of this.

However, they insisted on living in Thailand. Additionally, the spread of communism in this area forced the United States to carry out organized activities. One of Washington's initial decisions to conduct local operations in northern Thailand, which borders Burma and Laos, was the Young family. Harold and Gordon were also employed to help collect specimens for the entomology and herpetology departments of the American Museum of Natural History in New York. They gathered several exotic creatures from Thailand, Laos, and Burma by applying their naturalist expertise and enlisting the Lahu men to assist them in their work. Their Chiang Mai house, the Veluvan Villa, was full of wild animals, and all of them were the family's private property until Harold's death in 1975 (Lawitts, 2018, pp. 305–306).

Such commission with the museum was taken for granted in the family biography written by Gordon himself and his daughter, Debbie Young Chase. Like Gordon, she also portrayed the family's naturalistic spirit as their habitus. "As a hobby, Harold began collecting wild animals with the intent of creating a private zoo. The small zoo became very well-liked and visitors stopped by with increasing frequency. By 1954, when Harold's son, Gordon, arrived in Chiangmai, the two continued the collection of exotic animals" (Chase, 2014, p. 98). However, calling it a hobby is partly true but, at the same time, blinds us to the political forces that drive the family's practice as naturalists.

Furthermore, Harold and Gordon relied on their hunting expeditions to be supported by external forces, namely the museum, and internal forces, namely their ethnic networks and ties to the Thai government. With such support as a naturalist, Gordon determined to collect species of amphibians. During his survey trip in the 1960s, he discovered the *Ichthyophis youngorum* (Doi Suthep Caecilian or เขียดงูดอยสุเทพ in Thai) in 1960 and the *Theloderma gordoni taylor* in 1962. These species were registered in the taxonomic system and named in honor of the person who discovered them. So, to honor the Young family, who discovered them, they put "youngorum" and "gordoni" after the Latin name of that species. His name and his family names were recorded with great honor.

In addition, they also collected and kept large animals that they had hunted in their home. They built large cages in the Veluvan villa at the foot of Doi Suthep, which was a house that they had rented from the Nimmanhaemin family. Soon, the abundance of animals and the variety of exotic animals attracted visitors. In 1956, they decided to establish a private zoo by negotiating with the governor at that time to request a plot of land at the foot of Doi Suthep, Huay Kaew Road, to build a zoo. Later in 1960, the zoo was officially opened under the name "Chiang Mai Zoo," with Harold as the director until he died in 1975, after which the family gave the zoo to the government.

Harold's death coincided with the communist revolutions sweeping across Southeast Asian countries, marking the conclusion of the Vietnam War era in the region. The Young family concluded their legacy not as missionaries or military veterans, but as naturalists with a genuine passion for wildlife. Regardless of their complex roles, they will be remembered as significant ethnologists who demonstrated empathy for indigenous peoples and as the founders of Chiang Mai Zoo, Thailand's second private zoological garden. Their lasting contribution to natural history ensures that Chiang Mai Zoo and numerous amphibian species will permanently bear the Young family name.

10.9 Conclusion

> Large books have been written describing these living creatures and their wonderful habits, and still there is a great deal to tell. The works of the Lord are great, sought out of all those who have pleasure therein. These beasts and birds and insects are some of his works; and I hope you find pleasure in seeking out as much knowledge concerning them as possible. (Moffat, 1868, p. 16)

The above text, recorded by a missionary documenting American Baptists' journey to Burma in the nineteenth century, reveals how deeply intertwined religious mission and natural exploration were from the outset. By framing the study of animals and plants as approaching God's creation, missionaries established a theological justification for scientific inquiry that would characterize their work across generations.

The Young family maintained a remarkable continuity throughout seven decades across Northeast India, Burma, China, and Thailand despite shifting political landscapes. American Baptist missionary activities were deeply intertwined with British colonial projects, creating interdependent relationships where religious conversion and imperial expansion mutually supported each other. The scientific aspects of missionary work created a precedent where missionaries functioned simultaneously as naturalists, building knowledge networks that cataloged regional plant and animal species while furthering colonial scientific agendas.

This scientific activity by missionaries exemplified what Lorraine Daston and Peter Galison (2010) called the "community of letters," a system where Western explorers from various professional backgrounds contributed to shared knowledge through publication and correspondence networks. These communities were instrumental in developing systematic natural science during the eighteenth century, especially following Carl Linnaeus's taxonomic innovations. The Young family worked within this intellectual framework, where apparently neutral scientific documentation simultaneously advanced both religious conversion efforts and imperial objectives.

The Young family extended the missionary-naturalist tradition through versatile roles as religious leaders, military operatives, and scientific collectors, adapting to geopolitical shifts while maintaining consistent knowledge-gathering practices. William Young's foundational work with the Shan, Lahu, and Wa ethnic groups created pathways his sons Harold and Vincent expanded, transforming their influence from religious conversion to strategic military intelligence during World War II and Cold War operations. Their carefully cultivated relationships with indigenous communities, especially the Lahu and Wa, became strategic assets as their roles evolved.

Chiang Mai Zoo emerged as the culmination of these intersecting historical threads. The institution originated from Harold and Gordon Young's specimen collection, a genuine scientific interest that simultaneously provided effective cover for intelligence activities, before developing into a significant cultural and scientific establishment. Their scientific work, including Gordon's identification of previously undocumented amphibian species now taxonomically identified with the

Young name, illustrates how scientific practices appearing objective can simultaneously reinforce colonial power structures while advancing scientific knowledge.

The Young family's transformation, from missionaries operating under British colonial protection to CIA agents gathering biological specimens, demonstrates how colonial knowledge systems persist beyond formal imperial administrations. Their narrative compels us to understand that institutions like zoos, despite their presentation as objective educational venues, emerge from particular historical conditions shaped by missionary networks, shifting imperial powers, and evolving methodologies for producing knowledge about the natural world.

The concept of "coloniality" by Aníbal Quijano (2007) as an influence that permeates "every molecule of society" beyond formal colonial administration provides a framework for examining how the taxonomic systems, naming conventions, and exhibition methods of Chiang Mai Zoo perpetuate knowledge hierarchies with colonial origins. The evolution from amateur missionary-naturalists like the Youngs to credentialed scientists represents a continuation of colonial knowledge frameworks rather than a departure from them. Recognizing these historical underpinnings can guide more inclusive environmental education and conservation approaches that incorporate diverse knowledge traditions. Furthermore, it seeks to address what Boaventura de Sousa Santos (2014) terms "epistemicide" and advance toward more equitable validation of indigenous environmental knowledge systems.

The Young family's footprints, stretching from India to Thailand across more than a century, embody the profound convergence of cross, crown, and creature that has fundamentally shaped Southeast Asian landscapes and knowledge systems. The establishment of Chiang Mai Zoo in the late twentieth century represents not a historical accident, but the crystallization of accumulated global power structures where missionary zeal, colonial authority, and naturalist inquiry became inextricably interwoven. Their enduring legacy stands as both a remarkable cultural contribution and a compelling reminder of how these three interlinked forces continue to influence our contemporary understanding of the natural world. Without this historical trinity of power, religious conversion, imperial governance, and scientific classification, Southeast Asia would likely present an entirely different ecological and cultural reality today.

References

American Baptist Foreign Mission Society. (1914). *The Judson centennial celebrations in Burma: 1813–1913*. American Baptist Mission Press, F. D. Phinney, Supt.

American Baptist Missionary Union. (1907). *Harvest time at Kengtung*. American Baptist Missionary Union.

Chase, D. Y. (2014). My grandfather, Harold Mason Young. In *Burma headhunters*. Xlibris.

Cushing, J. N. (1893). *The Shan mission*. American Baptist Missionary Union.

Daston, L., & Galison, P. (2010). *Objectivity*. Zone Books.

de Sousa Santos, B. (2014). *Epistemologies of the south: Justice against Epistemicide*. Routledge.

Foster, A. L. (2010). *Projections of power: The United States and Europe in colonial Southeast Asia, 1919–1941*. Duke University Press.
Go, J. (2012). *Patterns of empire: The British and American empires, 1688 to the present*. Columbia University Press.
Hyun, S. (2023). Disenchanted: Thailand's indigenisation of the American cold war, seen through the experience of Gordon Young. *Journal of Southeast Asian Studies, 54*(1), 22–43.
John, W. S. T. (1912). *Josiah Nelson Cushing: Missionary and scholar Burma*. American Baptist Mission Press.
Lawitts, D. (2018). The transformation of an American Baptist missionary family into covert operatives. *Journal of Siam Society, 106*, 295–308.
McCoy, A. W. (2003). *The politics of heroin: CIA complicity in the global drug trade*. Lawrence Hill Books.
Merriam, E. F. (1900). *A history of American Baptist Mission*. American Baptist Publication Society.
Mignolo, W. (2021). *The politics of decolonial investigations*. Duke University Press.
Moffat, R. (1868). *The white foreigners from over the water: The story of the American Mission to the Burmese and the Karens*. The Religious Tract Society.
Peers, D. M. (2012). Anglo-Burmese wars (1823–1826, 1852–1853, 1885–1886). In G. Martel (Ed.), *The encyclopedia of war* (1st ed.). Blackwell Publishing Ltd.
Quijano, A. (2007). Coloniality and modernity/rationality. *Cultural Studies, 21*(2), 168–178.
Raman, A. (2021). 'Discovery' of the tea plant *Thea assamica* (now, *Camellia sinensis* var. *assamica)* in the Indian territory in the 1839s. *Indian Journal of Natural Products and Resources, 12*(1), 11–25.
Robbing, J. C. (1922). *Following to pioneers: A story of American Baptist Mission work in India and Burma*. The Judson Press.
Shullai, P. N. M. (2017). Colonialism, Christianity and Mission activities in India: A postcolonial perspectives. *International Journal of Humanities & Social Science Studies, 3*(5), 324–334.
Srinivasan, K. S. (1961). William Carey: His activities in botanical and horticultural researches in India. *Bulletin of the Botanical Survey of India, 3*(1), 1–10.
Viveiros de Castro, E. (2011). *The inconstancy of the Indian soul: The encounter of Catholics and cannibals in the 16th-century Brazil*. Prickly Paradigm.
Walker, Anthony. (1975). Review of Gordon Young's tracks of an intruder. *Journal of Siam Society*.
Walker, A. (2008). Karen and Lahu: Ethnic affiliation or Baptists' imagination. *Journal of Siam Society, 96*, 217–228.
Walker, A. (2010). The first Lahu (Muhsur) Christians: A community in northern Thailand. *Acta Orientalia Vilnensia, 11*(2), 7–28.
Young, W. M. (1902). A new station in Shan land. *The Baptist Missionary Magazine*, 179–181.
Young, G. (1991). *Tracks of an intruder*. Trasvin Publication.
Young, H. M. (2013). *To the mountain tops: Sojourn among the Lahu of Asia*. Xlibris.
Young, H. M. (2014). *Burma headhunters*. Xlibris.
Young, W. M. (n.d.). *The revival at Kengtung*. American Baptist Missionary Union.

Chapter 11
Lanna and Nagas Wood-Carvings in the Era of COVID-19 Pandemic

Wipawee Panjinda, Rongkakorn Anantasanta, and Thatchai Hongpaeng

Abstract This chapter aims to connect 3 years of continuous research to its final conclusions. Wood carving is the medium to connect common factors: the practice of art, cultures and beliefs, and the similarities and differences of the two territories, Lanna and Nagaland. Wood carving, as a form of art, unites various cultural expressions. This chapter explores how the pandemic, lockdown, and misery effects implicate and influence the art practice. The findings of the research are based on the collaboration of two regions' wood carving workshops: Lanna in 2023 and Nagaland in 2024. In these workshops, the artists and artisans expressed their perspectives about how they interpreted the pandemic state that morphed into an art form.

Keywords Wood-carving · Lanna · Nagas

11.1 Introduction

Until the sixteenth century, Lanna art and culture were very diverse due to the relationships the kingdom had with other kingdoms through its alliance with nearby areas. Such forms originated in Myanmar and the Thai Rattanakosin period (Punjabhan, 1994:14). Lanna's culture was generally related to Buddhism; its architecture was found to harmonize with the landscape, and its construction tended to be the center of the universe in concept. Sculptures, Buddha images, pottery, and tablet

W. Panjinda (✉)
Department of Visual Arts, Faculty of Fine Arts, Chiang Mai University, Chiang Mai, Thailand
e-mail: wipawee.p@cmu.ac.th

R. Anantasanta
Department of Design, Faculty of Fine Arts, Chiang Mai University, Chiang Mai, Thailand

T. Hongpaeng
The Division of Sculpture, Department of Visual Arts, Faculty of Fine Arts, Chiang Mai University, Chiang Mai, Thailand

Y. Santasombat (ed.), *Bio-Cultural Diversity, Ethnicity, and Local Knowledge*,
https://doi.org/10.1007/978-981-95-4242-0_11

frames are made of bronze, plaster, and kiln with different motifs applied to areas such as Chiang Saen, Lan Chang, or Myanmar-influenced (Punjabhan, 1994:16). Wooden Buddha images are religious objects, along with items such as candle holders. Nonreligious objects include items such as utensils for trapping animals, vehicles (ox cart and boat), and household utensils (handloom, mortar, pestles, and tray) (Punjabhan, 1994:18).

The establishment of familial ties between rulers through "marriage" was a traditional method commonly employed by Southeast Asian states to forge alliances and strengthen diplomatic relationships. This tactic was also a "strategy" used by King Chulalongkorn (Rama V) to consolidate power during the early years of his reign. He sought to marry the daughters of powerful nobles and various regional lords, including prominent leaders, to serve as concubines. One of the most significant familial connections, especially with the ruler of Chiang Mai, was through Princess Dara Rasami. The resulting marital alliance of Dara Rasami and King Chulalongkorn in 1885 (Castro, 2020:33), intended to cement the political relationship between Lanna and Siam (Central Thailand), illustrates the contingency of the historical moment produced by European colonial encroachment in the region. The context of Princess Dara Rasami refers to her role and status as one of King Chulalongkorn (Rama V)'s consorts, particularly emphasizing the significant political and strategic dimensions of her marriage. As a member of the royal family of Chiang Mai, her union with the king was not only a personal relationship but also a key element of the king's strategy to solidify alliances with regional powers. Her marriage to King Chulalongkorn was part of a broader effort to strengthen the monarchy's authority and create familial ties with influential local rulers in the early years of his reign.

Lanna temples, known for their elaborate wooden structures and intricate carvings, gained more recognition and support during this period. The royal attention likely helped preserve these artistic traditions. When Dara Rasami moved to the royal court, she introduced the richness of Lanna art, including wood carving, to Siam. This led to greater appreciation for northern craftsmanship. The increased interaction between artists from Lanna and central Thailand likely fostered the exchange of carving techniques and styles. Dara Rasami's influence may have promoted Lanna's distinctive wood carving styles beyond northern (Ratana & Chaimuangchun, 2021:65). The royal support for traditional arts enhanced the status, recognition, and longevity of Lanna wood carving traditions.

According to the Cambridge Experimental Project on the Nagas literary emergence of various Naga tribes, the records by the British, also as a ruler, and European observers, are dated to the 1900s. In the following, anthropological convention was used to describe the past (Jacob,1990/2012:10). Naga origin stories have two aspects: the mythological or supernatural origin of mankind and the actual movements from different regions in Southeast Asia (Jacob,1990/2012:13). Ritual wooden objects comprise the village gate, the house's horn, the grave-commemorative post, and the log drum. Non-ritual wood objects include wooden eating vessels and rice-pounding tables (Jacob,1990/2012:10).

The carvings often depict animals, nature, or tribal stories that hold sacred or cultural meanings; *mithun* (a semi-wild bovine) is a common motif in carvings, symbolizing wealth and tribal status. On the other hand, hornbills, sacred birds of many tribes, represent dignity, strength, and honor. Carvings have spiritual significance in rituals, ceremonies, or festivals to appease gods and spirits. These carvings are held in high regard because they maintain the spiritual fabric of local communities. Traditional wood carving is a generational skill, passed down through families or the tribe. It preserves and transmits the tribe's stories, history, symbolism, and craftsmanship. This dedication to keeping these traditions alive adds to its cultural dignity that can be seen in traditional dormitories, log drums, and wood monuments; furthermore, carved village gates are grand symbols, often featuring fierce animals or warriors to represent the tribe's power and protect the village from evil spirits.

The effects of the COVID-19 lockdowns on wood carving in Nagaland, focusing on the Dimapur and Kohima districts, were a mix of challenges and adaptations. While the traditional craft faced disruptions in production, demand, and community interactions, there were also small signs of resilience through digitization and innovative ways of craft design to sustain the practice.

The wood carving craft is often linked to tourism and export markets in Lanna, forming a major source of income for artisans in the region. Regrettably, the lockdown brought tourism to a halt, leaving artisans and businesses with fewer orders and unsold items in inventory. Besides, accessing quality wood for carving became difficult due to the lockdown, as suppliers and transport services faced restrictions. Although Lanna has a strong community system where wood carving skills are passed down through generations, often within families or through apprenticeships, the lockdown restrictions on gatherings and workshops disrupted this flow of knowledge, particularly affecting young aspiring artisans who depend on hands-on guidance from elders. Nonetheless, the Lanna community system emphasized knowledge transfer during this lockdown period, where outside activities were banned and family members received more opportunities to exchange skills and ideas.

11.2 Dots to Form

"Subject," "form," and "content" have always been the three basic components of a work of art, and they are wed together in an inseparable way. The "subject" can be considered the "what," the "form" the "how," and the "content" the "why" (Tayac, 2017: 39). The "form" of an artwork is the first thing that the viewer notices about it. The subject of visual art can be a person, an object, a theme, or an idea. Even though there are numerous and diverse ways to depict the subject, what matters is how much the artist is inspired by it. Next, the word "form" refers to the total overall arrangement or organization of an artwork. It is the outcome of applying the elements of art and the principles of organization, giving them structure and meaning.

The way an artist uses media (materials), design elements and principles, and the language of art constitutes the form of a work of art. Finally, the emotional or intellectual message of a work of art is its "content"—a statement, expression, or mood developed by the artist and interpreted by the observer. The content of a work of art is the idea or feeling that the artist is trying to communicate with the viewer.

Although the animistic custom has all but disappeared, it was once widespread and adorned the villages. For instance, villages were decorated with enormously intricate wood carvings on the village gates and *Morungs* (men's dormitories), such as the carvings of *mithun* heads and house horns that had a stunning visual appeal. Besides, additional X-shaped crosses made of large bamboo were erected when a new age group was being initiated, emphasizing the transition from one state to another in Konyak Naga custom (Jacob,1990/2012:28). Examples of geometrical designs that are typical of men include human figure outlines, elongated tiger or leopard patterns (Jacob,1990/2012:114), and a distinctive feather pattern on Konyak tattooing. It was a symbolic representation of their accomplishment of head-taking status.

Young men dressing up to be photographed using typing paper to simulate hornbill feathers (1939), Jacobs, Julian. 1990/2012:111

Realism was a European norm that attempted to dominate others. Although artistic modernism was defined in a way that was incompatible with realism, realism had little significance or prominence. Thus, the reflections of modern art were a reaction against realism. Consequently, modernism was viewed as a regional rebellion (Tangnamo, 2004:71). Moreover, realism art is related to gestures, manners, rules, and regulations. In Thailand, during the reign of King Mongkut Rama IV (1804–1868), he applied the manner of abolishing many outdated beliefs, such as abandoning superstitions. As a result, he became a model for photographers, painters, and sculptors to create royal portraits (Kunavichayanont, 2017:40). Previous beliefs held that creating a statue of himself was an inauspicious thing that would shorten one's lifespan. He accurately calculated the occurrence of solar eclipses and even observed solar eclipses without fear of the superstitions of the past.

Wooden bas-relief depicting a group of nine women figures forming the outline of an elephant, seventeenth century (Bhirasri, 1961:14)

This artwork, crafted in the seventeenth century in the Ayutthaya period, combines symbolism, mythology, and exceptional craftsmanship. Bas-relief has transcended its historical time and influenced how modern Thai artists think about form, symbolism, and storytelling. This work, which uses nine female figures to outline the contour of an elephant, conveys the peculiar, humorous side of the Thai artists. Besides, these characteristics are noticeable in the painting and lacquer designs. In Eastern countries, it is a standard practice to depict animals with a group of human figures. The subject finds the bas-relief's primitivism-inspired flavor to be quite appealing. It effectively conveys the pachyderm's form. The feminine figures exhibit certain human body aberrations, which are important to modern art. Then, it becomes clear that the idea of altering natural proportions for the composition's advantage was long held before it was thought of by older artists. To achieve unity and harmony of composition, the carver has elongated or shortened limbs or sections of the human body (Bhirasri, 1961:13–14). In Thai culture, elephants possess a sacred and auspicious significance. They symbolize strength, power, and royal authority. The elephant is also revered in Buddhist and Hindu traditions as a protective and sacred figure. The presence of nine women may represent femininity, nurturing, and unity in a shared purpose, reflecting harmony in the balance of power. Number nine (*gao* in Thai) is considered lucky and linked to prosperity, as it phonetically resembles the word for "progress." The relief may have been part of the temple's decorative structure, such as a door panel or gable, a ceremonial object or an ornamental piece within the royal court, or a funerary ornament symbolizing religious or spiritual concepts.

Door panels of the main viharn at Wat Bupharam, Muang District, Chiang Mai, with carvings of forest scenes. The upper part of the right-hand panel shows the Buddha in the midst of different animals, such as an elephant offering a lotus, and a monkey offering honeycomb surrounded by wild forest (Punjabhan, 1994: 187)

Wood-carved village gates, community halls & pillars with their rich motifs and symbols, Phek district of Nagaland (Metha et al., 2023)

(Detail) Wood-carved village gates, community halls & pillars with its rich motifs and symbols, Phek district of Nagaland (Metha et al., 2023)

Abstract art is created to record impressive memories with feelings and thoughts, to attract the attention of the viewer, and to provide personal satisfaction, so it is considered a cultural art object (Leemuangpai, 2000:58). A three-dimensional form appears to be there, yet it is an illusion. The artist's inspiration comes from life experiences, reflecting faith. During those moments, the artist had to face the truth of life, such as birth, aging, illness, death, and impermanence. Embedded in life, emptiness is the disintegration and impermanence of all things, which bears three key characteristics: *Anicca* (อนิจจัง, impermanence; transiency), *Dukkhata* (ทุกขัง, suffering or being oppressed), and *Anatta* (อนัตตา, soullessness; not-self) (Chanphosri, 2014: 9–10). In other words, the abstract art embodies the knowledge, thoughts, beliefs, and values that are concomitant to the harmony of human interactions in society (Leemuangpai, 2000:60). In contrast, the religious reforms of the West bear the characteristics of a violent overthrow of the old beliefs and a distinct transformation to modernity. However, for Thai society, the changes remain rather compromising. Buddhism in Thailand has responded to modernity in two ways: first, by using rituals as a means of meeting the material needs of people today or by turning sacred things into commercial Buddhism; and second, by interpreting Buddhism philosophy in a way that is more contemporary. The influence of Buddhism is demonstrated by the teachings' lack of reference to previous or future lives, lack of emphasis on rituals and supernatural abilities, and sole focus on achieving nirvana in this life by letting go of the self. Buddhism's teachings have been well-received by intellectuals and the well-educated middle class, partly because his teachings can be connected to modern logic better than the teachings that emphasize supernatural powers and rituals that are considered outdated (Kunavichayanont, 2017:39). Understanding abstract art and how it differs from modern art from a Thai Buddhist perspective. It is essential to comprehending modern art and the idea of modernism (Kunavichayanont, 2017:80).

Traditional wood carvings from Thailand and the Naga were analyzed for their realism and abstract art. The results showed that both cultures allowed semi-realistic or semi-abstract shapes to prevail, which illuminated their messages. The image of an elephant, which is popular in applying various gestures in Lanna wood-carving, and a *mithun's* head in Nagas wood-carving that is creatively carved in different designs are motifs that present meaning with richness and wealth. In this vein of thought, humans have fused hunting with art. Using spears and stone axes, the first cave painting techniques documented hunting with herds of animals through the patterns of colors and lines on cave walls. These techniques permeated brushes, clay, stones, wood, campfire tales, folklore, literature, books, and sound-enabled animations (Teerapongvishnupon, 2006: 56).

Left: Model of a Sema bride wearing a skirt with bead girdle and cowrie shell ear ornaments. She is carrying a basket containing carded wood, a spindle whorl, and a food dish. **Center:** Seated Konyak male figure wearing imitation hornbill feather headdress, hairpin, ear ornaments and bead necklaces, Longkhai village. **Right:** Male figure in ceremonial dress. The style of facial carving, ear ornaments, and kilt with imitation lines of cowrie shells are typically Angami. Model of a life-size grave carving. Jacobs, 1990/2012:202

11.3 Eruption of Pandemic and Its Interpretation in Lanna and Nagas Wood-Carvings

The pandemic has influenced the themes and subject matter of art, with many art practitioners exploring forms related to isolation and the global impact of the virus. This has resulted in a more introspective and socially engaged approach to art-making. The COVID-19 pandemic has had profound and far-reaching effects that have fundamentally altered daily life. For artists, particularly those working in traditional crafts like wood carving, the pandemic has posed unique challenges and disruptions to the shifts in work processes, access to materials, and broader societal changes that have influenced creative practices. Inevitably, the pandemic compelled the artists to adjust and routinize their art practices at home as a new way of life.

For many artists, the pandemic had brought a sudden halt to public events, such as exhibitions that provide essential platforms for showcasing and selling artwork. This is particularly true for wood carvers, whose work involves large-scale pieces, requiring significant time to complete. The inability to present their work to the public in physical spaces created a sense of isolation and a loss of creative momentum. Consequently, the lack of a public interface resulted in significantly fewer opportunities to sell artwork. For independent artists, this economic pressure compounded existing financial instability, as many had limited means of sustaining their livelihood during extended periods of inactivity. Furthermore, a lack of engagement with an audience diminished the immediate feedback and sense of fulfilment that comes from public appreciation. Hence, the period of seclusion accentuates the intimate connection between place and people through the visual spatial interaction.

During the pandemic, the lockdowns and social distancing measures disrupted the daily routines of artists. Many wood carvers rely on their workshops and studios, spaces that facilitate not only the creation of art but also a community of fellow artisans and craftsmen. The pandemic forced the closure of these communal spaces, leading to a sense of solitude. In some cases, the isolation occasioned creative paralysis, for regular interaction with peers or the physical presence of a supportive community is vital in maintaining inspiration and motivation.

Wood-carving is a highly tactile craft that relies on specific materials, including wood, tools, and sometimes paints or varnishes. The pandemic disrupted global supply chains, making it difficult for artists to access the materials they depend on. In particular, the interruption of shipping and the temporary closure of many suppliers resulted in shortages of quality wood and carving tools. This not only delayed projects but also forced many woodcarvers to adapt to using alternative materials or modifying their creative processes to work with what was available. The lack of access to quality materials also affected the overall quality of work, forcing woodcarvers to become more resourceful. This task opened artists and artisans to connect with an online purchasing and shipping process to enable an uninterrupted flow of creativity. As a result, some artists turned to digital platforms, attempting to shift their sales online. For some, the digitalization process meant adjusting their techniques or experimenting with different carving styles. However, the challenge of presenting their works effectively through online channels was further compounded as the artists faced competition with a wider range of digital art forms and mass-produced items that were flooding online marketplaces. Notwithstanding the challenges, the adjustments also allowed the artists to reach new audiences, expanding their global reach in ways that were not possible in pre-pandemic times.

The COVID-19 pandemic radically challenged and transformed the traditional forms of collaboration. Artists explored innovative ways to work together through online platforms, leading to a new work nature. Additionally, this pandemic has raised awareness of the importance of mental health and well-being. It is impossible to overstate the importance of artists' general well-being because of the intimate relationship that exists between the artist, the art practice, and the art content. Although art could be used as a means of coping with stressful moments through expressing inner experiential feelings, this research examines how the artists and the related institutions have adapted to the challenging period during the pandemic. It includes researching the financial difficulties of the artists, the evolution of art education, the availability of information sources, and the Internet-observed space that was typically utilized for public gatherings. The use of invisible cables has inadvertently reduced the value of face-to-face interactions and physical mobilities.

Overall, the pandemic has prompted a re-evaluation of art practice and research writing, encouraging innovation, flexibility, and a deeper consideration of the role of art in society. The streams of human art have evolved for a long time, although not as long as the evolution of violence in our genes. Art has diverse alternatives and paths. Art always serves as a mirror to reflect contemporary events. The long history of mankind shows that war and art often reflect each other. War has influenced the direction of art and literature. The availability of weapons has made hunting more successful. Millions of years have passed; hunting has become a part of life. We hunted to survive until we became modern humans, settled in cities, and learned to produce our own food. The hunter-gatherer nature still lives in us. Violence in the genes and the perception of beauty in the brain, the two seemingly contradictory things, have been translated into beauty in art (Teerapongvishnupon, 2006:58–59).

The carvings of elephants in Chiang Mai are popular. Local artisans created their distinguished masterpieces by meticulously carving elephants from wood, showcasing their expressive and magnificent imagination (Tourism Authority of Thailand, Undated). Each elephant carving was elaborately carved into different gestures and expressions. The unique characteristics, patterns, and imaginative carvings may pique the customers' interest and trends. In Hindu mythology, a well-carved elephant is symbolic of skilled art practices. Notably, as a royal person's transporter, the elephant is symbolic of high status. The humoristic side of the elephant is its defining sarcasm as a tax collector's tool, approaching where it goes; this was told by a Thai research scholar in 2023 (verbal communication with author on September 16, 2023).

Wood Sculpture by Ms. Xin Chaikul, Title: "*Hope*", Size: 15 × 15 × 35 cm., Year: 2022

"The Appearance of Life during the COVID-19 Pandemic," a wood carving art competition for students and independent artists, was conducted in 2022 during the second lockdown in Chiang Mai. Ms. Xin Chaikul's abstract artwork, "Hope," projected the COVID-19 pandemic's condition of hope and inspired optimism for the situation. Using a modular design, the creation's chosen shape is a metaphor for the COVID-19 virus to give it a contemporary appearance. This technique involves producing artwork that ranges in size from huge with a thorny surface to small with a smooth surface. With each wave of the evolving situation, the steps' design produces mobility, adaptation, and optimism for the artist, who aspires to advance toward the light, a better route out of the dilemma in the future.

Wood sculpture by Ms. Tanatcha Chairin, Title: "*Hope*", Size: 25 × 38 × 48 cm., Year: 2022

The other artwork, titled "Hope," the creation of Ms. Tanatcha Chairin, has taken inspiration from the symbolic flower in its wood carving form. Due to the spread of the COVID-19 virus, the general public is concerned about the situation. Despite the dire circumstances, the few medical professionals are making every effort to organize their forces to vaccinate the general public against the virus, which is perceived as a hope in preventing the pandemic's catastrophe. The creation of the piece "Hope" inspires all citizens and medical professionals. "Hope" provides encouragement for medical personnel and all citizens. The *Dok Pib,* or Kasalong, is a Lanna word that means a tree that provides shade for life. It also represents the role of public health professionals in fostering growth and development, such as remaining steadfast, providing care, and inspiring hope in the face of adversity.

In the three components of art, "content" may be the most difficult to identify, because the audience, without direct communication with the artist, must decipher the artist's thoughts by observing the work's subject and form. Ideally, the viewer's interpretation is synchronized with the artist's intention. However, the viewer's diversity of experiences can affect the communication between artist and viewer. For many people, content is determined by their familiarity with the subject; they are confined to feelings aroused by objects or ideas they know. A much broader and ultimately more meaningful content is not utterly reliant on the image but is reinforced by the form. This is especially so in more abstract works, in which the viewer may not recognize the images as a known object and must, therefore, interpret meaning from shapes and other elements. Occasionally, artists may be unaware of what motivates them to make certain choices of image or form. For them, the content of the piece may be subconscious instead of deliberate.

In 2023, a collaborative wood carving workshop involving Thai and Naga artists took place in Chiang Mai. There were six Naga artists from communities: Ao, Angami, and Rengma from Nagaland, and twelve Thai artists shared their experiences and ideas through communication. Additionally, they exchanged sketching and practical skills with one another. The Thai artists' wood-carvings are delicate and rich in details, with intended smooth and rough textures and designs in their presentation. On the other hand, the Naga artists' creations are rich in mass, solidity, fewer decorative details, and natural texture finishing. Finally, a "relationship" is forged, resulting from a blended interaction that strengthens each other's cultures and characteristics.

Wood sculpture in collaborations of Mr. Petch Wiriya (Thailand) and Mr.Vikuolelie Keditsu (Nagaland). Title: "*Hassadeeling Bird*", size: 60 × 50 × 110 cm., Year: 2023

At the base, carved by Mr. Petch, a Thai artist, there is an elephant's head presenting an iconic Lanna symbol of dignity and grace, honoring the majesty. In the upper part, above the elephant's mouth, the art piece presents an affirmation of a prompt promise. The upper part symbolizes the honor and glory in achieving a status through the head-hunting custom, using the feathers of the hornbill and the house's horn carved by Mr. Vikuolelie. This piece of work connects the two diverse nations, cultures, traditions, histories, and ways of life of Nagaland and Lanna, Thailand. As a result, the emergence of the artwork creates new friendships for a better new world.

The Hassadeeling bird (นกหัสดีลิงค์) is not merely a bird but a mythical hybrid creature in Lanna culture, bearing a body often described as a combination of an elephant and a bird. It symbolizes majesty and spiritual guardianship and is frequently seen in ceremonial contexts such as temple art or funerary traditions. The Hassadeeling is revered as a divine intermediary, connecting the earthly world to the spiritual realm, and its representation speaks to the harmony between humans, animals, and the sacred. The hornbill, deeply embedded in Naga tribal culture, holds sacred and symbolic value. Its feathers are used in ceremonial headgear, signifying honor, bravery, and tribal identity. It is often associated with warriors and elders; the hornbill represents strength, courage, and the intrinsic link between nature and culture. According to Lanna culture, the Hassadeeling serves as a means of transportation to the spiritual realm, accompanying the spirits of the deceased to the heavens. Similarly, in Nagaland, the hornbill's feathers, used in ceremonial attire, often represent the individual's ascension in status, honor, and communal memory—another form of transcendence.

Great Hornbills (Buceros bicornis) are found in northern Thailand and areas close to Nagaland, making it plausible that the hornbill, or similar avian species, inspired local myths in both regions. Both regions associate their respective birds with nobility, status, and cosmic significance. The Hassadeeling in Lanna culture is depicted in royal works of art, temples, and symbolic rituals, while hornbill feathers in Nagaland signify leadership and community honor. Birds, therefore, serve as universal cross-cultural symbols of transcendence and mysticism. This wood carving sculpture shows that the Lanna representative holds up Nagas's identity through this connection and relationship.

Wood sculpture in collaborations of Mr. Vikuolelie Keditsu (Nagaland) and Mr. Petch Wiriya (Thailand). Title: *"Hands of connection"*, size: 20 × 40 × 110 cm., year: 2023

"Hands of connection" could be extended to modern initiatives. Both Lanna and Nagaland face pressures of globalization but also exhibit strong efforts to preserve identity, crafts, and traditions in a modern global framework. Collaborative cultural initiatives, such as wood carving exchange programs, could serve as "hands," literally connecting historical regions that exhibit a deep relationship with nature and spirituality today. Lanna culture often connotes the intertwining of animist beliefs and Buddhism. Evidently, the Lanna rituals and reverence for sacred landmarks constitute a cultural framework that is mirrored in Nagaland, where traditional Naga spiritual practices align deeply with nature and community events. Critically, these partnerships shape community structures by providing chances for comparative study.

The metaphorical "hands" could also signify mutual recognition of spirit and respect between the two cultures. Traditional communal practices and festivals offer spaces for synergy. Lanna's Yi Peng Festival (the lantern festival) is celebrated with communal unity, much like Nagaland's Hornbill Festival, which showcases the vibrant traditions of various tribal groups. When taken as a whole, the festivals act as a social glue, bringing their diverse cultures together and providing possible glimpses into common ideals of celebration and harmony.

Wood sculpture in collaborations of Asst. Prof. Sutthisak Puthararak and Ms. Parinyapon Jaiodton. Title: "*Positive Energy*", size: 30 × 30 × 152 cm., year: 2023

Throughout history, Asian cultures have witnessed constant transfers, exchanges, and circulations. To represent the transmission of positive energy, this piece of art uses the symbol of hands touching with only the tips of their fingers. This symbol is comparable to the positive culture that maintains strong connections between India and Lanna. Artists from Thailand and India who have collaborated on projects and developed a friendship will continue to inspire the public with their artistic energy and beauty.

In Nagaland, one can find many festivals, songs, dances, and other performance arts. Furthermore, they are renowned for wearing vibrant, colorful traditional clothing. In addition, they have a long and illustrious legacy of wood carving. The Nagas are skilled wood carvers, clearly exhibiting the hidden symbolic significance of their carving culture. Nagas are talented craftspeople who imprint their ethnicity on most of their everyday items.

The celebration is a very complex ceremonial process since it involves the participation of the community from other villages as well as resources. The entire process starts with animistic rituals that only men may undertake. The men select the tree to be felled, carry the tree, and cut the wood. Simultaneously, the women cook for the men. To this day, no technology is utilized to carve these symbols and designs (Chishi & Lemtor, 2023). Large trees are needed to produce vertically positioned walls, which are only erected, allowing for the synchronized carving of the symbols. The distinct symbols and motifs that are carved on the wood and gifted to the Feast of Merit holders include *mithuns*, buffaloes, deer, flying foxes, and human skulls as well (because of the customary head-hunting practices). The existence of these expertly carved wooden skulls of different animals, birds, and people suggests that the owner of the home or a previous ancestor hunted various creatures. The Nagas generally have had wooden plates, for in the past, metals were scarce. Thus, they began eating on wooden plates rather than banana leaves. The idea of using wood-carved utensils has been practiced since early times and has been used until the present day; it enhances the flavor of the curry, which makes it taste better (Enok, 2023).

Families that have observed the feast of merit are clearly regarded with social significance in the villages since they inherit their homes from their forefathers, and the existence of such homes is a sign of a prosperous past. Even the numerous community members who gave their time and effort to build the houses are proud of what they did and enjoy the attention it garners in society. A fundamental change has occurred when headhunting practices have shifted to peaceful relationships, and the feast of merit has disappeared. Despite the decreasing number, the motifs, and symbols of wood carving are still regarded as sacred and are not used in commercial goods, even though animistic practices have declined and Christianity has become more prevalent in the villages, which has led to wood-carving as a dying craft (Metha et al., 2023).

According to a proverb, the wood-carving does not portray the life of the subject or the backstory. Conversely, the *Anghs* (chiefs) and their heritage drive the narratives. The stories about the triumphs or accomplishments of the *Anghs* become the inspiration behind every carving they create. Hence, wood carving serves only to document the rich customs and culture of the *Anghs* and to preserve their historical facts.

Despite the rapid growth of industrialization and technology, wood-carving enables people to continue learning about their own cultural and traditional beliefs as well as to keep track of the customs and culture of their ancestors (Tikerila & Razielung, 2023).

Wood log sculpture in collaboration of Mrs. Buakiang Kiwkaew and Mr. H. Akup Buchem. Title: "*The Strength and Prosperity of the Lanna Kingdom towards Nagaland*", size: 30 × 50 × 230 cm; year: 2024

The collaborative exchange event of a wood carving workshop held in Dimapur, Nagaland, in 2024, presented this log sculpture. The use of elephants as a motif in sculpture stems from their symbolic significance in the Lanna Kingdom, where elephants represent strength, power, stability, and good fortune, as well as being auspicious. Moreover, elephants play a significant part in Thailand's religious rituals. In Buddhist history, elephants are mentioned and are regarded as symbols of prosperity. According to a Buddhist legend, Queen Sirimahamaya, the mother of the Buddha, had a dream in which a white elephant presented a lotus flower to her before the birth of Prince Siddhartha. This narrative contributed to the elephant's status as a sacred animal.

The use of the "flower pot" or "Puranakhata" (ปูรณฆฏะ, flower pot of perfection) as an element in sculpture is based on the belief that it symbolizes a vessel of abundance, encompassing happiness, peace, tranquility, prosperity, and the flourishing of wisdom. The Puranakhata pot, often referred to as the "flower jar," is adorned with intricate and beautiful patterns and was traditionally used in religious rituals to honor Buddha, bringing prosperity, and abundance. In Lanna culture, this design is known as the "flower pot motif" or "flower pot pattern," and examples can be found on temple walls, such as in the vihara of Wat Pong Yang Kok and Wat Phra That Lampang Luang, Lampang province, in Northern Thailand. The pot, filled with water, represents prosperity, symbolizing the abundance of rice, fish, food, and wealth. The flowers, on the other hand, represent material wealth, including property and followers.

The carving integrates the imagery of elephants with the Puranakhata pot motif, with the intention of conveying strength, power, stability, good fortune, and auspiciousness, while also imparting prosperity. This includes happiness, peace, tranquility, and the flourishing of wisdom. The combination of these symbols, drawn from the Lanna Kingdom, is aimed at fostering harmony and permanence between the Lanna Kingdom in northern Thailand and the state of Nagaland, northeastern India, ensuring that both regions coexist in peace and lasting stability.

Wood log sculpture in collaboration with Mr. Vikuolelie Keditsu and Asst. Prof. Ronnakrit Buratemee. Title: "*The Sustainable Relationship Between Nature in Lanna and Nagaland*", size: 60 × 70 × 220 cm; year: 2024

The carving features of the culturally based animals are from both regions. The hornbill and bull's head symbolically represent Nagaland, India, and the elephant at the base symbolically represents Lanna, Thailand. These symbols signify the connection between nature, wildlife, and the abundant cultures living together in the same forest. The collaborative artists have decided to carve without turning the wood, choosing a huge teak stump that is the largest log available. Given the log's massive size and the difficulty of twisting or moving it, the carving technique conveys strength and stability. Through this wood carving workshop, the artists have presented their cultural symbols and beliefs, reflecting the harmonious relationship between nature and culture that continues to be preserved in the present day.

Lanna artists are known for their intricate designs that often depict scenes from Buddhist mythology and nature. From crafting temple structures like the iconic gables of Lok Moli temple to everyday utensils, their carvings embody harmony with the environment, which can be united with Naga artistic skills rooted in the traditions of their warrior-hunt culture and animistic spirituality. Lanna practices, influenced by Buddhist principles, often emphasize moderation and interdependence with the environment. On the other hand, the Nagas community traditionally emphasizes shifting agriculture, detailed forest management, and reverence for animals, which point to their harmonious cohabitation with nature. The concept of the cycle of life, involving water, animals, flora, fauna, and humans in wood carving highlights the continuity and sustainability of ecosystems. The form of a spiral or circular pattern can represent perpetuity; the essence of rich traditions can be unified to create art, not only reflecting shared heritage but also envisioning a sustainable future.

11.4 Conclusion

The COVID-19 pandemic has undeniably altered the normal life of artists, from the disruptions in work routines and material supply chains to the economic dimension. The eruption has reshaped global interactions, prompting re-evaluations of cultural expressions, cross-regional collaborations, and sustainable artistic practices. In the context of contemporary wood carving, a collaboration between Lanna (Northern Thailand) and Naga (Northeast India) artists provides a unique opportunity to unite regions with shared artistic traditions and philosophies under unprecedented global constraints. This academic exploration identifies the key elements of art that can guide such collaboration while emphasizing the transformative influence of the pandemic on artistic practice and its role in fostering resilience and sustainability.

The pandemic has severely impacted the woodcarver's everyday life. It has induced a shift in the target audience from seeking external sources to catering to local purchasers, as the globe faces a closure of international and tourist markets. In addition, people spent more time at home during the lockdown, which raised interest in home dĕcor and created a demand for carved ornaments and furniture. During the shutdown, artisans had more time to hone their skills and experiment with new designs or artistic expressions. This period saw the blending of traditional techniques with modern or minimalist designs, appealing to younger buyers. It appears that the lockdown's interruption of wood carving methods has highlighted the traditional arts' dependence on tourism and exports, as well as the lost skills brought on by fewer apprenticeship options. However, the pandemic has forced craftspeople and companies to adopt more sustainable business methods, such as digitizing marketing and sales, when physical interactions are limited. The use of digital tools, such as 3D modeling, online workshops, and virtual collaboration platforms, becomes central to crafting and sharing designs. These technologies allow artists to transcend geographic barriers, merging contemporary innovation with traditional craftsmanship. The lockdowns challenged the livelihoods and continuity of wood carving traditions in Northern Thailand, but they also created opportunities for innovation, adaptation, and new ways of reaching buyers. The crisis emphasized the importance of preserving and modernizing the traditional craft for its cultural survival in an ever-changing world. However, it has also prompted innovation, adaptation, and a renewed appreciation for local resources and community connections.

While diverse in cultural traditions, the collaboration must embrace unity by finding common ground through shared values such as respect for nature, spiritual symbolism, and a focus on community. The pandemic revealed the fragility of natural ecosystems, making sustainability a central element for contemporary collaboration. Using reclaimed or salvaged wood, the artists reimagine traditional carving practices while honoring both regions' cultural respect for forests and preservation of natural resources. The challenge extends beyond artists to engage local communities in both regions; the principles of collaboration and exchange can be achieved by fostering dialogues through art and wood carvings. The transformation is the act of carving into a universal language of resilience, a dialogue between nature and culture, and a testament to human adaptability in the face of adversity.

Both Lanna and Nagas have faced significant cultural disruptions during the pandemic, including economic challenges for artisans and limited avenues for creative output. The collaborations highlighted art as a form of resilience, allowing artists to express recovery, hope, and cultural continuity through their carvings. The COVID-19 pandemic has illuminated the ways in which art and culture can offer hope, healing, and a sense of connection in times of crisis. A collaboration between Lanna and Naga wood carvers in the contemporary context reflects this broader role of art in addressing global issues.

References

Bhirasri, S. (1961). *Thai wood carvings*. The Fine Arts Department.

Castro, L. (2020). *Woman between Two Kingdoms*. Cornell University Press.

Chanphosri, A. (2014). *Suññakala Art Catalogue*. Fine Art.

Chishi, A., & Lemtor, A. (2023). *A study on the challenges in woodcarving production and marketing in Dimapur, Nagaland St.* Joseph University. (Unpublished source).

Enok, K. K. (2023). *Diversity of tribal wood carving in Nagaland (with special reference to Mon District) (Unpublished source)*. Department of Management, St Joseph University Chumukidema.

Jacobs, J. (2012). *The Nagas Hill peoples of Northeast India*. Rivers Book.

Kunavichayanont, S. (2017). *Modern & contemporary art: The Western World & Thailand*. Faculty of Painting, Sculpture and Graphic Arts, Silpakorn University.

Leemuangpai, P. (2000). *Art appreciation*. Odeon Store.

Metha, Kevimenuo, & Asani. (2023). *Symbolism in woodcarving: A study under Phek district of Nagaland*. Department of Management, Nagaland University. (Unpublished source).

Punjabhan, N. (Ed.). (1994). *The charm of Lanna wood carving*. Rerngrom Publishing Co., Ltd.

Ratana, P., & Chaimuangchun, P. (2021). The role of creator and transmitter of Lanna intellectual cultural heritage by her highness Princess Dara Rasmi. *Journal of Thai studies, 17*(2), 51–72.
Tangnamo, S. (2004). *Research in visual culture*. Faculty of Fine Arts, Chiang Mai University.
Tayac, S. (2017). *Teaching's manual for the course 109100 man and art*. Faculty of Fine Arts, Chiang Mai University.
Teerapongvishnupon, P. (Ed.). (2006). *Silpa Ngodngam Songkramkradkreaw*. Jirasak Pattanapong's Fund.
Tikerila, & Razielung. (2023). *Wood carving and its trends in the Tuensang District of Nagaland*. Department of Management, Nagaland University. (Unpublished source).
Tourism Authority of Thailand. (Undated). *Ban Chang Nak*. https://www.tourismthailand.org/Attraction/ban-chang-nak

Chapter 12
Gods and Spirits in the Naga World of Woodcarvings

Iris Odyuo

Abstract This paper attempts to describe the involvement of gods and spirits in the creation of woodcarving arts among the Nagas. The Nagas holds great reverence for their ancestors and nature spirits, who embody the unwritten laws, traditions, and customs of their villages. Prayers were offered to these entities for good harvests, population growth, success in warfare, and protection against sickness, famine, floods, and earthquakes. The art of woodcarving, including the construction of the *morung* (men's house), log-drum, and village-gate, was accompanied at every stage by a series of taboos and ritual performances. Any deviation from the prescribed methods was believed to result in calamity or even death. Despite using simple tools such as chisels, adzes, axes, and machetes (*dao*), and facing the numerous difficulties and insecurities of life, the Nagas' artistic creativity thrived due to their strict adherence to rituals and taboos associated with these arts.

Keywords Naga woodcarvings · Men's house (*morung*) · Log-drum · Village-gate · Naga artists · Gods and spirits

12.1 Introduction

Wood carving is one of the most significant artistic practices among the Nagas. Important artworks, such as the main post in the men's house (*morung*), the log-drum, and the village-gate, are created with adherence to a series of taboos and ritual performances at every stage. Deviating from these prescribed forms is believed to result in calamity and death. For the Nagas, rituals are deeply integrated into all aspects of their lives. They hold the belief that everything possesses some degree of

I. Odyuo (✉)
Associate Professor, Department of History, Sao Chang College, Tuensang, Nagaland, India
e-mail: irisodyuo@gmail.com

Y. Santasombat (ed.), *Bio-Cultural Diversity, Ethnicity, and Local Knowledge*,
https://doi.org/10.1007/978-981-95-4242-0_12

supernatural and spiritual significance. According to Kanato Chophy (2020), there are three classes of spirits in the Sumi pantheon: *alhou, kungumi,* and *aghau.* However, to understand the functionality of magico-religious beliefs and practices, it is primarily the *aghau* that are actively involved in the daily affairs of humans. The sacred and the spiritual are seen as pervasive forces, influencing the physical, psychological, socio-economic, and spiritual realms—thereby affecting every aspect of life (Nienu, 2015:151). This paper explores how the gods and spirits are involved in the creation of wood carvings among the Nagas. V. Nienu elucidates:

> Rituals and ceremonies made the Naga supernatural world meaningful. The invisible became visible. What was mundane became sacred and what was ordinary became profound. Rituals and ceremonies provided a forum within which the Nagas developed a deeper understanding of their environment, both physically and cosmically, and helped them gain a greater knowledge of the role they played in the scheme of things. (Nienu, 2015:151)

Ceremonial wood carvings, prominently displayed in the men's house, as well as the log-drum, village-gate, and the residences of wealthy individuals, serve as vital expressions of the tribes' "value systems" (Lotha, 1998:2). These artistic representations convey important cultural messages within the community. The Nagas, despite facing numerous challenges and uncertainties in their daily lives, managed to cultivate a rich artistic tradition. They worked with basic tools such as chisels, adzes, axes, and *dao* to create intricate and meaningful artworks.

This creativity flourished in an environment that was often constrained by strict rituals and taboos surrounding various art forms. These rituals and taboos not only shaped the artistic practices of the Nagas but also underscored the significance of their cultural heritage. The careful adherence to these guidelines highlights a deep respect for tradition while allowing for a dynamic artistic expression.

12.2 Cultural Setting of Woodcarving

To appreciate the woodcarving of the Nagas, one has to understand the "socio-cultural context of their production" (Lotha, 1998:4). William H. Davenport writes, "most works of ethnic arts are statements of faith in and adherence to the beliefs and social rituals in which the highest traditional cultural values are expressed. It is these traditional values that give motivation and direction to the entire society. The art is a testament of these values and motivations" (Davenport, 1971:422–423).

In Naga society (Odyuo, 2013), an individual's social status is largely determined by their ability to host feasts. Although attending these feasts is optional, the pursuit of prestige compels every married man to compete for status through these merit-based celebrations. These ritual feasts not only require the increased production of goods like rice, grains, and livestock such as pigs, mithun, and chickens, but they also ensure that wealthy individuals share and redistribute their accumulated wealth among less fortunate villagers, thereby creating a balance within the community.

Headhunting was once a culturally institutionalized practice among all the Naga tribes, sanctioned by both social and religious beliefs. Individual bravery was highly admired, and each warrior sought to distinguish himself by taking as many heads as possible. Every Naga aspired to claim an enemy's head, as doing so not only brought him prestige and honor but also the right to wear the coveted warrior's paraphernalia. Additionally, by killing an enemy, a warrior believed he was contributing to the well-being and survival of his village (Odyuo, 2013).

The first step in establishing a new village (Odyuo, 2013) was to choose a central location for constructing the *morung*, or men's house. Naga villages are divided into khels, or wards, and each khel has its own *morung*. The *morung* serves as a vital social, political, and religious component of the village. The number and size of *morungs* in a village depend on its population and the vitality of the clan members who build them. The *morung* is the centre of social and cultural life for the Nagas and does not exist independently from the community. It functions as a guardhouse and armoury, a recreation club, and a centre for education, art, and discipline.[1]

The *morung* often served as the place for the log-drum[2] considered sacred, for the heads of enemies were kept or hung displaying the men's dominance in warfare. The Nagas believed that without the log-drum[3] there would be famine and an increase in the birth rate of unhealthy offspring. Panger Imchen (1993) states that each village, upon being founded, should consecrate the log-drum within 3 years of being established. The log-drum served various purposes, including conveying important messages such as announcements of victory, dangers of enemy advances, and notices of lunar or solar eclipses. It was also used to inform the community of the death of a warrior or feast giver and to announce festivals. Additionally, the log-drum played a crucial role in raising alarms about fires or the presence of tigers near the village (Odyuo, 2013).

Among the Tenyimia Naga group, the village-gate (*kharu*) is considered the most significant component of the village. It served (Nienu, 2015) not only for defence purposes but also holds spiritual significance, being regarded as a sacred embodiment of the supreme God. Vibha Joshi states that "In the past the *kharu* occupied a prime ritual position in the traditional set-up, serving as the point outside which the 'other

[1] It was here that debates and discussions concerning the interests of the clan and village, economic, and political rituals were held, and the art of warfare and skills such as woodcarving, basket making, songs, dances, and folklore of the tribe were communicated orally from generation to generation.

[2] The origins of the log-drum are somewhat unclear. According to the traditional oral history of the Khiamniungan Nagas, men used to bang bamboo sticks together to scare away the *pia goa*, an evil spirit believed to be about to consume the sun or the moon. If they did not create loud sounds, the evil spirit might devour the celestial bodies, leading to an increase in both accidental and natural deaths. Eventually, they discovered that a hollowed tree could produce a much richer sound, which is how the log-drum was first created (Lungo, personal communication, April 20, 2007).

[3] There are two varieties of log-drums found in Nagaland and among the Burma Nagas: open-ended cylinders and those that are closed at both ends. According to Oppitz (2008), drums with closed cylinders are primarily found on the western side of the northern tracts, among groups such as the Ao, the Phom, the Konyak, and the Wancho. These drums can also be found among the central groups of the northern Sangtam and the Chang, as well as among the Chagyik, Pounyu, and Macham in Burma. In contrast, open-ended drums are found from north to south among the Nocte, the Laju, the Wancho, the Rangpan, the Heimi, the Lainong, the Gongvan, the Yonkon, the Pounyu, the Khiamniungan, the Chang, the Yimchungrü, the southern Sangtam, the Tikhir, and the Makuri.

world' began. Traditionally any offerings suggested by the diviner to the spirits *(ter-huo-mia)* at the time of a personal illness would be left outside the *kharu*" (Joshi, 2008:61). The gate is seen as essential for the village's welfare and overall well-being. It acts as a vital spiritual connection to the village or khel community, functioning as the main religious centre and providing protection against physical threats and supernatural dangers. The gate is adorned with carvings (Lotha, 1998) of warriors, enemy heads, spears, *daos*, and stylized representations of hornbills and mithun heads. Additionally, the gate serves as a sanctuary for the village's prosperity stone, known as *Kiputsie* (Nienu, 2015). The history of the khel is also recorded on the village-gate.

Life-size wooden effigies dressed in warriors' custom were erected over the graves of successful warriors and feast givers by the Angami, Ao, and Konyak Nagas. The Nagas believed that the soul of the deceased person enters the carved figure into the earth, enriching it. Such effigies were also carved for victims of head-hunting. To summon back the soul of the victim "whose heads could not be recovered, in order that the soul may not become a wandering spirit between earth and heaven, a skull or hairs of the deceased person were placed on top of his effigy for a certain period of time so that the souls may pass into the wooden figure" (Lotha, 1998:14). W.G. Archer notes "the village subscribe rice-beer and two pigs, a member of the warrior's family cuts and carves a tree and after a general feast the village youth drag the effigy onto its site" (Jacobs, 1990:126–127).

12.3 The Ritual Tradition of Wood Carving

The selection of the tree for ceremonial woodcarving carries a significant sacredness, particularly in the processes involved in constructing, pulling, and installing the log-drum, *morung* post, and village-gate. For instance, among the Chang Nagas, a meeting is convened among village elders before heading into the jungle to select the tree for the *morung* post or log-drum. Four men, representing the four clans of the Chang Nagas—*Ongh, Lomou, Kangshou, and Honghang*—are chosen. These men must have living parents, be free from physical blemishes, and have never been struck by lightning or attacked by enemies, wild animals, or evil spirits. Once selected, the men proceed to the jungle to choose a suitable tree.

The chosen tree must be straight, tall, and of considerable girth. The best log-drums are made from closed-grained hardwood that grows long and straight, producing a rich sound when struck. Careful consideration is given to selecting the tree for the log-drum; trees that have been struck by lightning, have creepers growing around them, or whose roots are entangled with those of another tree are avoided. The Nagas believe that the environment surrounding a great tree should be clean, with surface roots that do not entangle with other trees, and without broken branches, bird nests, or termites present (Bümütyano, personal communication, April 6, 2006). According to the Konyak Nagas (Nienu, 2015), the log-drum can only be made from a special tree species known locally as *lanyangpu,* meaning "king of the forest trees." They believe that Tangma and Nyoya, the mythological King and Queen, reside in the forest. Thus, the log-drum is personified as embodying both male and female aspects and is considered to be deified. Once a tree is selected, the villagers return to obtain permission from the owner of the jungle where the tree is located. It is believed that unless the owner willingly agrees to donate the tree, the log drum will not produce a proper sound and may eventually rot.

12.4 Dreams and Omens

The Nagas place great importance on dreams and omens. According to Herschel B. Chipp, "Dreams are often considered more real than actuality, since the dream world is the abode of the ancestors and other spirits that control events in the real world. Since dreams and mystical experiences are the main avenues of contact with the spirits, it is natural that these should have been the immediate sources of much mythology and art" (Chipp, 1990:163–165). Among the Chang Nagas, the village shaman (*seabu*) ventures into the forest to appeal to the spirit of a selected tree. This spirit is invited to come to the village *morung*, reside in the log-drum, and become the village's Protector God. If the tree spirit is willing, it reveals its name and sex to the priest through a dream. (Impongso Chang, personal communication, May 5, 2009). In the Khiamniungan village of Lengnyu, both male (*Shilem*) and female (*Jonglio*) log-drums can be found in their *morungs.*

Omens are observed by closely watching the bird species that perch on the tree. If a scarlet minivet *(shupangshu)* is seen perched on the tree, it is considered a bad omen, as its red colour signifies blood. The Chang Nagas believe this could indicate that villagers may die at the hands of enemies or that there could be a fire outbreak in the village, prompting them to abandon the tree and select a different one. Conversely, the presence of drongos (*aukhumaishek*) on the tree is regarded as a good omen, since these birds are known for their bravery in fending off larger birds, such as eagles and hawks. Additionally, the appearance of a hummingbird (*angu*) is viewed positively, as it is swift, symbolizing that the *morung* youths would be able to run quickly, especially after confronting an enemy. However, if green leaves fall during the first symbolic strike to the tree, it is considered a bad omen, predicting famine or death in the village. If the weather turns windy on the day the tree is felled, it is believed that disputes may arise among the villagers (Mülong Chang, personal communication, May 4, 2009). Omens serve as warnings for all khel members, and the priest carefully observes any consequences at this stage. After interpreting the dreams, the tree is named according to its gender. The following day, an offering of ginger (*see*) and salt (*sachem)* is made to ward off evil spirits. The founder of the village or the owner of the jungle, who donated the tree, makes the first symbolic cut. Then, four clan members, wielding a *dao* together, deliver the first axe blow to the tree.

12.5 The Symbolism of Naga Wood Carving

For the Nagas, the *morung* serves as a testament to their culture and society, expressed through architectural elements. Ross Bowden, in discussing the architecture and sculpture of the Kwoma men's ceremonial house in Papua New Guinea, states, "Kwoma actually attributes to men's houses qualities which they admire in men, notably the attributes of homicidal aggressiveness and creativity in the contexts of yam cultivation and human reproduction. One consequence of this is that Kwoma collective representations of men are tacitly identified with ceremonial buildings. (Bowden, 1995: 68). Similarly, the Nagas attribute qualities they deeply admire in men, especially traits associated with aggressiveness in war, to the architecture and wood carvings of the *morung*. These qualities are particularly linked to men's roles as headhunters, providers, and maintainers of lineage. The trees used to construct the main post, side beams, and ridge pole of the *morung* needed to be strong, straight, and tall, without any blemishes; in the same way, the village men were expected to embody justice and honor. In the context of a society like the Nagas, where headhunting was a male-dominated activity, successful headhunters were held in high esteem.

The carvings of human figures and animals within the *morung* "communicate information about the society in which they are found by examining, in general terms, the relationship between these buildings and the men who construct and inhabit them" (Bowden, 1995: 77). The main post of the *morung* is known as the *kethung* among the Chang Nagas. The carvings on the *kethung* typically feature a pair of hornbills. On the second inner main post, called the *thaunyu*, there are carvings depicting a tiger or leopard approaching the head of a mithun or buffalo. Life-size human figures wielding spears, shields, and holding enemy heads adorn the side posts of the *morung*, referred to as *kamkhung*, representing the warriors of the village (Bümütyano, personal communication, April 6, 2006). These figures are meticulously carved from the actual wood, with their outlines initially sketched roughly using charcoal or chalk. Traditionally, the figures were coloured using natural pigments; for example, chicken and pig's blood provided red, soot created black, and chalk yielded white. The carvings of the human figures display a naturalistic style, showcasing prominent ears, broad noses, slits for mouths, and distinctive tattoo marks typical of warriors on their faces, chests, and arms. Headgear made from mithun horns adorns these human figures.[4]

[4]Woodcarvers from Yangpi, a village of the Chang Nagas, are regarded as the finest woodcarvers in the region. One common motif found in the Yangpi *morungs* is the depiction of two tigers facing each other on the outer main post, alongside carvings of pythons and hornbills on the inner posts. The horizontal crossbeams inside the front porch of the *Nokphu Chubang morung* are painted with red discs surrounded by thin black projections that resemble sun rays. These paintings symbolize the spiked brass armband known as *nam khengkap,* which is cherished as an heirloom and holds significant value as a symbol of bravery and wealth among the Naga people (Alan Chang, personal communication, May 5, 2006).

Bringing the enemy's head into the village was not only the best proof that an enemy had been killed, but also the head was believed to contain the soul, which ensured to increase in the crops, success in war and the well-being of the village. According to Nienu:

> When a head-taker returned with a head, regardless of the victim's gender or age or of the circumstances under which it was secured, all the men in the village, led by the old men and accompanied by other warriors, turned out to welcome the head-taker. The entire village population joined to celebrate the joyous occasion, for they believed that this head brought them prosperity, success and happiness. (Nienu, 2015:176)

Although decapitated, "the head was treated like a living person: it was spoken to, cajoled, and fed with rice beer, and food. The drink and food offerings were a magical rite and symbolized present victory and prayer for future success. Thus, the more heads that were acquired, the greater the power and virtue that accumulated, both individually and collectively as a community". (Nienu, 2015:172). The Naga's carvings of human heads on the *morung* posts, in front of wealthy individual houses, on the village gates and on the log-drum symbolize the strength, courage and victory of the male population in the village.

The mithun is the most "valued, prestigious and the biggest animal domesticated by the Nagas. Its confidence, grace, robust and meaty features, give a sense of well-being to the owner. It is also the animal of the *sikhingo* the spirit of human beings different from the soul" (Lotha, 1998:17). For the Nagas, the mithun was associated with all aspects of Naga's religious, social, political and economic life. Feasts of merit, negotiation of bride-price and celebration of marriages, and festivals, successful headhunting raids as well as negotiating political pacts, fixing of penalty and ramson for war captives. Nagas also believed that the mithun should be treated with consideration for it was similar to that of mankind. Only the best mithun skull with well-shaped evenly curved horns was displayed in front of wealthy individual houses and the morung.[5] The mithun provides the best meat; it is strong, and unpredictable and can even kill and maim humans. These characteristics of the mithun, prosperity, strength, and danger are the type of nature the Nagas ascribed to their gods.

The tiger holds significant meaning for the Nagas, symbolizing valor and strength. It is both feared and respected within their culture. According to Sema folklore, a mother had three children: a spirit, a tiger, and a man. When a tiger is killed, its head is brought to the village and displayed in the *morung* or hung near the village gate. The Angami Nagas observe a day of rest in mourning when a tiger or leopard is killed, as they consider it the death of an "elder brother" (Hutton, 1968:208). The Nagas hold a deep reverence for tigers, viewing the act of killing one as equivalent to taking the life of an enemy. All Nagas have a profound respect for tigers; they take oaths on the tiger's tooth and skull. Additionally, the

---

[5] It is not just in actual Mithun heads displayed on the front porch of wealthy individuals but warriors used the Mithun horns for drinking cups and to embellish their headgears and carrying baskets.

descendants of a person who has been killed by a tiger are prohibited from consuming meat from a tiger's kill (Hutton, 1968), as this would be akin to sharing a meal with a hereditary enemy. Killing a tiger grants the slayer the same status and decorations that are awarded to successful head takers. However, this act also necessitates a series of precautions and ceremonies to protect (Saul, 2007) the slayer from any harm that the tiger's spirit may inflict, either in this world or the next.

The hornbill (*Dichoceros bicornis*) is highly honored and respected by the Nagas, and it is regarded as the king of birds. "It is the largest bird in the country, of magnificent appearance, and makes a great impression soaring slowly overhead with very loudly whirring wings, audible at a distance and height at which the bird itself is barely visible"(Hutton, 2003:392). Its beak and feathers symbolize valor and wealth. Most Nagas believe that the hornbill represents the beauty and youth of the men in the village, leading to significant attention being given to its carvings. The hornbill's "body is represented by a rough rectangle, the posterior slightly narrower and the tail, which is an extension of the body itself, not attached to the beam or pillar, is simply a flat one painted with a black stripe in the middle obtained from soot and pig's blood or powdered charcoal" (Ao, 1968:63). Among the Konyak, the hornbill is connected (Mills, 1926) to the Angh as a decorative motif, and it is carved onto their coffins, stools, and large beds where they sometimes sleep and are laid to rest after death.

The python is described as being "carved in high relief" (Ao, 1968:66), with its entire body depicted in a zigzag pattern to create the impression of movement. The head of the python is "detached from the beam or pillar, and the two eyes are distinct. The mouth is represented either as a small slit or simply by a black line. The entire body is adorned with black spots that symbolize its scales" (Ao, 1968:66). The Ao Nagas attribute miraculous powers of attraction to the python, often comparing it to a wealthy individual. They suggest that "both can attract and draw to themselves anything they desire—one through its mysterious influence and the other through his wealth" (Mills, 2003:298). The Chang Nagas also view the python as a symbol of wealth. It is believed that killing a good python would lead to a bountiful harvest, while killing a bad one would bring misfortune. During a severe famine, a python may be killed to invoke its wealth. Additionally, the zigzag pattern found on the rich man's cloth, known as *Pi nei,* was inspired by the python. Great care is taken to weave this zigzag design uniformly into the *Pi nei* shawl.

The *morung* woodcarvings of the Konyak Nagas feature various representations, including concentric circles that symbolize the sun and the moon. According to Konyak beliefs (Jacobs, 1990), eclipses occur when an animal attempts to devour the sun or the moon. This event is depicted in carvings on the house posts of Konyak villages. The animal responsible for the eclipse is believed by some to be a frog, while others think it may be a tiger or a dog. Additionally, weapons such as spears, *daos*, rifles, and gongs are significant symbols of wealth and prestige. The *morung* of the Konyak Nagas also includes a considerable number of erotic themes, which reflect the power and virility of its members.

12.6 The Artist

In traditional Naga society, every man can carve simple wooden objects such as plates, bowls, mugs, pounding tables, and beds, as well as construct their own houses. However, some individuals possess greater carving skills and are therefore called upon for ceremonial wood carvings. These wood carving rituals are deeply rooted in spiritual beliefs and foster community solidarity. There is a profound understanding and spiritual connection between the artists, the wood, and the tools they use. The tools employed by the Nagas include the *dao*, axe, adze, and chisel. As noted by Ao "Considering the tools, the carvings are excellent" (1968:59). Anthony Forge writing about the Abelam artist of New Guinea states "The amount of time they spent helping with the actual painting and carving, as opposed to the many other activities necessary in the preparation of ceremony, is largely a matter of choice….the training and selection of artists are completely informal. A youth who shows aptitude will be encouraged and allowed to perform increasingly difficult task under supervision" (Forge, 2006:116). In a similar vein, within the Naga community, a young man aspiring to become an artist has the valuable opportunity to assist and learn from skilled and experienced artists. This mentorship allows him to observe their techniques, understand their creative processes, and gain insights into the art form. By collaborating with established artists, he can refine his own skills, explore various styles, and deeply understand the cultural significance of their art. And although no specialists are officially recognized and expert artists do not receive special status or rewards, individual artistic creativity is appreciated and acknowledged.

Large-scale wood carving projects, such as the village-gate and the log-drum, involve collaboration among community members. Archer, who observed the carving of a new village gate in Kohima in 1947, noted:

> First four and little later six men all began work on the plank together-while the others stood or sat around-there was no leader and it was as if everyone had the same idea of how the gate should be cut….In the first stage the rough forms were levelled down with the tool like an adze- the surface being hacked away in long parallel lines. It is this which gives the mithan horns one of their most beautiful effects…. Everyone was in complete command of his tools. Occasionally a line was marked in pencil to guide the cutting. In every sense a collective act. Altogether, 10–15 men at various times did the cutting—the basic pattern was cut in the fines—without a model, obviously relying on keen visual memory and a clear common conception of the fine carving. Only members of the Khel did the cutting; it was a Khel act. (Archer in Channa, 1992:98)

During the construction of the log-drum or the village-gate, all activities in the village are suspended. This allows the carvers to focus on their work. The wood carvers are prohibited from having any sexual contact with women and are required to build a temporary hut near the work site. They must prepare their own meals and adhere to a strict diet consisting of rice, chicken, and pork, as all other foods are taboo until the project is completed. Chipp writes, "Such an attitude towards the material and the technique of carving wood, even if the reasons were magical, produced a stringent discipline among the workers and elevated the technical process to a sanctified position" (Chipp, 1990:152). The artists once engaged in the process of making these items are believed to be in "a state of ritual purity…till the period of construction is on, the men working on the plank- in the rites of transition-do not visit their village….thus keeping the conceptual distinction between the settled community and the transitional period in which the carvers exist" (Channa, 1992: 98). In this context, "the work itself becomes the sole object of their vital energy" (Chipp, 1990:154).

Women were prohibited from approaching the site where the carvings were made or from touching the wood during its dragging and installation ceremony in the village. This restriction stemmed from the belief that a woman's presence would render the log-drum ineffective. The ceremonial wood carvings were symbolically linked to male prestige and the expression of male values. If women were allowed to witness the carvings or even touch them, it was thought to weaken men's ability to take the heads of enemies or defend the village.

Additionally, the Naga artists took great care to avoid any accidents while working, as such mishaps were interpreted as bad omens for the entire village. Chipp writes,

> the ancestors and the spirits that inhabit the other world are believed to control happenings to the physical one. Therefore, an accident meant that the workmen had incurred the wrath of a deity. They must then resort to magic rituals to attempt to regain the good will of the offended spirit. This ritual was closely bound up with the craft of woodcutting….the solution to difficulties precipitated by poor workmanship lay in the tapus of the craft itself; namely, that good workmanship precluded the possibility of the occurrence of bad omens. (Chipp, 1990:152)

In the village, it is considered taboo for the family of a deceased individual to participate in ceremonies or perform funeral rites following an accidental death. For instance, in Tuensang village during the log-drum dragging ceremony in 2005, an elder from the village of Bilashi Khel tragically lost his life when he was accidentally crushed by the log-drum. Although the log-drum had been carved and completed, it was left in the forest, and another log drum was made to replace it (Mülong Chang, personal communication, March 9, 2009).

> The selection of the tree for the log drum and its installation in the village inspired the Nagas to compose songs and dances celebrating the beauty and significance of the tree to their tribe. The Phom Naga serenade the selected tree by singing this song:O, ulong heyu jah pan yinghi long yuma?
> Nungmen huntang hungi hunglet phongjingpeth
> Nungbe chingu yipikah
> Nyu nyu oh, oh, meima yankei shekeja
> Chingsho longpa I nyuki lem nguke
> Longnyak shokphen nyuke
> Lomvai lomaong daihah phamching ongyanglei
> Jangmoi pangyak jontoh neihong la nglukah
> Nonipailang rar-u wannyang khumlu-a
>
> Tree from which amongst the jungle tree
> Selected you by knowing your name through the religious priest
> Tell goodbye to your neighbour

Depart from your family happily without feeling sad and come with us
Calling and welcoming you by village chiefs and elders
Longing to receive you by every village women
Let us now leave your beloved family
And go soon to the chief's house forever without hesitation
Every black and white joins in the long-live cane ropes and pull
Let us hold and pull lightly, such a long-life log-drum is.
(B.Tongshauh Phom, personal communication November 20, 2007)

The tree is personified as a young individual who is urged to leave the forest and his family in order to become the village's log drum. According to local beliefs, the tree selected for the log drum should not be forcibly taken from the forest.

While pulling the log-drum from the forest to the village, the Chang Naga cajoles the log-drum by singing:Ho hei, ho shei lengma hei hai
Ah saching pelong angkhi he, ho saching pelong angkhu hei,
Ho pushou puna, thu thük longei thu nyang nyang
hüh oh thu thuk longwi thu nyang nyang
Ah shangpong ali na hi ali nahe asemba, ho thüngto langto heima yo
Leng shang küling shang, punyüh küpongyüh
Aha ho, ho hou
Ah, kaku kayang mang suhta, ho kaku kayang mangsuhda
Ah melang kongto ansiji, oh melang khonto ansiji
Ah, pikhu pangkhu ankhu he
Ah, athok Laie ale hei, ho athok lai ei ale hei
Ah, yangpon onbu pu-nguh ha ho yangpon onbu pu ngyuh ha
A langnge pakbou angsiji, ho langnge pakbou angsiji.
Ah, thungpo lukbou angsiji, ho khunbo lukbou ansiji
Ah, hayang jangto ale, he ho haying jangto ale he
Ah, nyangbon ongbou punguh ha ho nyangbon onbou, pungah ha
Ah, hiao, hiao, hiao
Sangkhung lanei nokyen yanshoba sangkhung sang ho lei ho

Pull the log-drum with join hands
To hang the enemy's head on it, to hang the enemy's head on it.
Young trees surrounding the selected tree do not grumble
Among the trees, it has been selected for the ring-bell
Rest on the wooden sledge prepared for you
Log-drum do not resist and come
Ah! your tall strong healthiness is being wanted
You shall be adorned with precious ornaments
The heads of male and female enemy shall bejewelled you
Come now without obstruction
The jungle tree-like thunder lightning is being brought
The tree is brought where the hornbill perches
The tree is brought where the squirrel takes shelter
The log-drum is chosen from among the trees for the ring-bell
The log-drum is now approaching
The trees that are the home to the flying squirrel are brought
Now the log-drum is arriving in the village.
(N.C. Yimba, personal communication, May 3, 2009)

This song celebrates the beauty of the tree selected for the log-drum and encourages it not to resist but to lie down on the prepared wooden sledge and be taken to the village. The tree is transformed from a shelter for hornbill birds and flying squirrels into the chief of the village, where it will be adorned with ornamentation of enemy's heads, both male and female, around its neck.

Once the log-drum is installed in the *morung* it is smeared with the blood of the animals killed for the feast. For every new log-drum, an enemy's head was to be offered to seek the blessings and prosperity of the village. J.P. Mills writes:

> In the old days any stranger, whether friend or foe, who came to a village on the day when a new drum was dragged in was killed and his head placed on the drum, which was joyfully beaten to celebrate such a happy initiation. If as usually happened, the village did not have a piece of luck like this, the drum was fenced round as soon as it had been placed in position, and the fence could not be removed or the drum beaten till the young men been out and brought in a head. (Mills, 1926:80)

The arrival of the log-drum is welcomed into the village with prayer by the village priest to the new log-drum chanting:

Tsü phian khem o aya
Nyuk chethso sangcham
Nyuk chethso phachan
Pem lojao aota
Pem lopon aota
Pem losong aota
Pem lolaio aota
Nyu koleng aota
Nyu kotin aota

Warriors from another village
Their heads be brought under this log-drum
Chopped heads from many villages
Chopped hands from many villages
To wear a warrior dress
To keep the body strong
To keep the body bulkier
To keep the body longer
To decorate the village
To beautify the village
(Mulai Khiamniungan, personal communication, 3 June 2009)

When the tree is finally installed in the village *morung*, a prayer is offered to the spirit of the log-drum, asking for blessings of many enemies' heads, abundant harvests, livestock, healthy offspring, and overall prosperity for the village. From that point on, whenever an enemy's head is brought into the village, it is first taken to the *morung* and hung on the log-drum for the night. The log-drum is believed to be the dwelling place of the village's protective spirit. For the Chang Nagas, the high god *Shambüli Mühghaü* is thought to reside within the log-drum. As Ao states, "These log drums are the repository of the life or virtue of the clan or khel to which the morung belongs and are indeed treated almost as a village God" (Ao, 1968:66). The wood used for the log-drum is considered sacred by the Nagas, and they would never dream of carving stools, dishes, beds, or pounding tables from the remaining wood. Chipp notes that the artist:

> by his understanding of these precepts, the sculptor attuned himself to the flow of spiritual power to the tribe. Hence the commission of a technical error- a mistake by a priest in reciting a ritual or by a sculptor in observing traditional procedures of work- might interrupt or destroy the customary channels of communication with the spirit world and might be atoned for with the penalties of death…the incorporation of the carver into the discipline of the tribal work organisations bound him even more firmly to the community ideals of technical excellence. (Chipp, 1990:168)

No new carvings or paintings are done once the log drum is installed. To the Angami Nagas, the village gate, "Being an embodiment of the spiritual power, the gate is believed to be inhibited by spirits, and although touching it is not prohibited, any new carvings, paintings, or working on the carvings are not permitted on all days. An attempt to do so is to infuriate the khel and also invite the wrath of the Kharu spirits" (Channa, 1992:111). Among the Chang Nagas, old-log drums that are replaced are never destroyed or discarded. The rope used to drag the log-drum is never touched by a *dao*; instead, it is carefully preserved in the *morung*. Further smoothing, polishing, and repairs to the log-drum are only performed during the *Jeinyulum* festival in July (Bümütyano, personal communication, April 6, 2006). The Naga artists are aware that respecting taboos and omens leads to good workmanship.

12.7 Contemporary Naga Wood Carving

Among contemporary Naga artists, two individuals stand out for their innovative approaches to art: T. Veswuzo Phesao and Lepden Jamir. They excel not only in their use of new tools and techniques but also in their ability to integrate traditional belief systems and Christian faith into their creations.

T. Veswuzo Phesao is not just an artist; he embodies the rich cultural heritage and artistic expression of the Chakhesang Naga community. His work spans various forms of artistry, including folk singing, dancing, and sculpting, demonstrating his deep connection to the traditions and narratives of his people. In 2003, Phesao embarked on a significant project that showcased his commitment to preserving and promoting Naga culture—a traditional rich man's house known as *Kunemeche*. Located in United Village within the Chümoukedima district, this house is more than just a structure; it serves as a cultural symbol and a testament to the architectural traditions of the Chakhesang Naga. Phesao meticulously sourced traditional building materials and employed age-old techniques that reflect the skill and craftsmanship of Naga artisans. The intricate carvings on the front panels—featuring designs of warriors, stylized mithun heads, and depictions of women's breast—are crafted from large, single pieces of wood, showcasing not only his artistic talent but also the significance of these symbols in representing warrior culture and the fertility of the land.

Before commencing construction, Phesao honored local customs through the stone-pulling and installation ceremony, a vital ritual within Naga tradition. His commitment to these traditions was palpable as he hosted a pre-Christmas feast, illustrating his respect for both Indigenous practices and the Christian faith that many in the community embrace. Akshaya Tankha's (2020) observations highlight the seamless blend of these two worlds within Phesao's work. The integration of the stone-pulling ceremony—a ritual deeply embedded in traditional Naga religion—with the celebration of Christmas symbolizes a harmonious coexistence of Indigenous traditions and Christian beliefs in contemporary Nagaland. This coexistence indicates that for Phesao, Indigenous identity, and Christian faith are not mutually exclusive but are instead intertwined aspects of his present-day experience. He navigates this complex landscape by upholding traditions that define his culture while simultaneously embracing the transformative influences of Christianity, reflecting the dynamic nature of cultural identity. In this way, Phesao's work becomes a living testament to the resilience of Naga culture, showcasing how traditional practices can evolve and adapt within a modern context while remaining rooted in historical significance.

Lepden Jamir carved seven wooden pillars between 2000 and 2006 in Mopungchuket village, located in the Mokokchung district. These pillars, which include the Pillar of Time, the Eye, the Pillar of Love, the Pillar of Promises, Try, Peace on Earth, and the Pillar of Protector, were installed near the village lake known as *Sungkotenem*. The Pillar of Protector, standing sixteen feet tall, was inspired by the legend of how the lake spirit (*awatsüng tsüngrem*) saved the village from the invading Ahom army in 1832, appearing in the form of a giant snake. Lepden adorned this pillar with carvings of "humans and a snake representing the mythological account of how the lake was created" (Longkumer, 2018:19–20).

Lepden was aware of the stories surrounding the lake spirit, so he sought guidance and permission from the spirit before starting the carving process for the Pillar of Protector. According to Tankha (2020), Lepden faced challenges from the spirit while creating the pillar. At times, the spirit would hinder his assistants' work, and on other occasions, it would cast spells on those who did not behave respectfully in its presence. To avoid these obstacles and show his respect for the spirit before representing it, Lepden sought blessings from the male spirit. During the initial attempt to carve the image of the *awatsüng tsüngrem* into the wood, he encountered difficulties. As a result, he sought advice from a knowledgeable friend familiar with traditional practices. The friend explained that the problem was not caused by the spirit of the lake, but rather by the spirit of the wood itself. It was only after acknowledging the tree spirit as the rightful owner of the wood that he was able to continue his work without further issues.

In "Spirits of the Material World," Arkotong Longkumer argues that while some villagers have rejected their past deities due to their conversion to Christianity, the situation is more complex. Conversations with Lepden and other villagers reveal that the lake god is not merely a relic of the past; rather, he is actively invoked among the villagers, many of whom identify as Christian, including Lepden himself. Additionally, Lepden's discussion about creating the Pillar of Protector illustrates (Tankha, 2020) that the shared understanding of space and time among the indigenous woodcarvers is both pluralistic and layered.

12.8 Conclusion

The introduction of Christianity, colonization, and education led to the abandonment of many social and ceremonial practices associated with the traditional religion of the Nagas. It also significantly reduced their dependence on and loyalty to the *morung* system, as well as the clan and village structures. New sources of prestige, arising from education, professions, and livelihoods, have undermined the traditional prestige system that once supported ceremonial art traditions. Furthermore, with the increase in conversion to Christianity, certain traditional songs and dances that invoke the blessings of the village's benevolent spirits, performed by feast givers, are now considered un-Christian and are forbidden. Many Naga villages have destroyed their log-drums in the *morung*, and traditional ornaments, as well as game trophies and mithun heads stored in individual houses, have been burned or abandoned. Even today, whenever epidemics or unusual incidents occur in the village, people often blame the log-drums or the wood carvings found in the *morung* architecture. For example, in February 2009, influenced by the village church, the villagers of Bilashi Khel in Tuensang destroyed their log-drum and *morung,* believing they were responsible for the deaths of 14 elders in the village (Mülong, personal communication, February 14, 2009). Additionally, the Indo-Naga conflict in the 1950s led to the burning of villages, resulting in the loss of traditional wood carvings, including log-drums, village gates, and *morungs.*

However, some villages in the central and northern parts of Nagaland, such as the Ao, Chang, Sangtam, Yimkhiung, Khiamniungan, Makuri, Phom, Tangkhul, Konyak, Chakhesang, and Angami, continue to create new log-drums and village gates and observe ceremonies related to them. Omens and dreams are still taken seriously in these communities. For instance, Limalenden Longkumer (2007) discusses how, in 1981, the villagers of Aliba in the Mokokchung district discovered that the tree selected for their log-drum had five major branches and was nourished by five significant roots. This was viewed as a good omen, signifying the future of the five clans in the village. The Nagas perceive the log-drum as 'alive' and believe it has a mind of its own. In the 1940s, when the villagers of Longtang in Tuensang district attempted to drag a log-drum, it would not budge. In response, they conducted a divination by slicing a broomstick plant, which fell in the direction of the nearby Litem village. The log-drum only moved when people from Litem village joined the pulling effort (Impongso Chang, personal communication, February 24, 2010).

Naga artists who specialize in traditional wood carving possess a deep understanding of the intricate relationship between their craft and the spiritual realm. They recognize that each piece they carve is not merely an object of art, but a powerful expression of their cultural heritage and beliefs. Through meticulous attention to detail and a reverence for the materials used, these artists create works that are intended to bridge the gap between the physical world and spiritual consciousness, reflecting the rich traditions and stories of the Naga people.

Morung at Tuensang village. Photo Iris Odyuo 2018

Log-drum at Kuthur Village. Photo by Iris Odyuo 2021

Kiphoma Kharu at Kigwema village. Photo by Kezhalezo Yhoshii 2024

Wooden funerary effigy at Ketsapomi village. Photo by W.G. Archer 1947

Carvings of human heads at the village-gate. at Pfutsero Village. Photo by W. G. Archer 1948

Carving of a mithun head. Cheswezumi. village. Photo by W.G. Archer 1947

Carving of a tiger and hornbill at Yangpi village *morung*. Photo by Mono 2024

Carving of a python at Yemchung Haki at Tuensang village. Photo Iris Odyuo 2024

Carving of the log-drum. Photo Luthor Thang 2022–2023

Carving of a log-drum, Khiamniungan 2022–2023. Photo Luthor Thang

Phesao in front of his traditional house at United Village, Chümoukedima district. Photo Iris Odyuo 2024

Monolith commemorating the stone-pulling and installation ceremony and hosting a pre-Christmas feast for his village in both 1998 and 2021. Photo Iris Odyuo 2024

Pillar of protector at Mopungchuket village. Photo Lepden Jamir 2006

References

Ao, A. M. (1968). *The arts and crafts of Nagaland*. Naga Institute of Culture Government of Nagaland.

Bowden, R. (1995). Art, architecture, and collective representations in a new Guinea society. In J. Coote & A. Shelton (Eds.), *Anthropology art and aesthetics* (pp. 67–93). Clarendon Press.

Channa, S. M. (Ed.). (1992). *Nagaland, a contemporary ethnography*. Cosmo Publications.

Chipp, H. B. (1990). Formal and symbolic factors in the art styles of primitive cultures in Jopling. In F. Carol (Ed.), *Art and aesthetics in primitive societies* (pp. 147–170). E.P Dutton and Co.

Chophy, G. K. (2020). On worldview and witchcraft among the tribes of India. In M. Behera (Ed.), *Tribal studies in India*. Springer. Accessed 26 April 26, 2021, from. https://doi.org/10.1007/978-981-32-9026-6_14

Davenport, W. H. (1971). Sculptures of the eastern Solomons in Jopling. In F. Carol (Ed.), *Art and aesthetics in primitive societies* (pp. 383–423). E.P. Dutton and Co.

Forge, A. (2006). The Abelam artist. In H. Morphy & M. Perkins (Eds.), *The anthropology of art: A reader* (pp. 109–121). Blackwell Publishing.

Hutton, J. H. (1968). *The sema Nagas*. Oxford University Press.

Hutton, J. H. (2003). *The Angami Nagas*. Directorate of Art and Culture, Government of Nagaland.

Jacobs, J. (1990). *The Nagas: Hill peoples of north East India*. Thames and Hudson.

Joshi, V. (2008). Angami Village Gates kharu. In Richard & V. Joshi (Eds.), *Naga—A forgotten mountain region rediscovered* (pp. 61–63). Christoph Merian Verlag and Museum der Kulturen.

Longkumer, L. (2007, October 6). *Log-drum is a souvenir from the past, up for the records*. The Morung Express.

Longkumer, A. (2018). Spirits in a material world: Mediation and revitalization of woodcarvings in a Naga village. *Numen: International Review for the History of Religions, 65*(5–6), 467–498. Accessed October 16, 2022, from. https://doi.org/10.1163/15685276-12341509

Lotha, A. (1998). Understanding Naga woodcarving. (unpublished manuscript).

Mills, J. P. (1926). Certain aspects of Naga culture. *The Journal of the Royal Anthropological Institute of Great Britain and Ireland, 56*, 27–36.

Mills, J. P. (2003). *The Ao Nagas*. Directorate of Art and Culture, Government of Nagaland.

Nienu, V. (2015). *Naga cultural milieu: An adaptation to mountain ecosystem*. Dorylus Publishing Group (DPG). 2309 Noriega St, San Francisco, CA 94122–4239.

Odyuo, I. (2013). The various aspects of Naga art. *Journal of Humanities and Social Science, 9*(4), 13–22. http://www.iosrjournals.org/iosr-jhss/papers/Vol9-issue4/C0941322.pdf

Oppitz, M. (2008). The Log Drum. In Oppitz et al. (Eds.), *Naga identities: Changing local cultures in the Northeast of India* (pp. 169–197). Snoeck Publishers.

Panger, I. (1993). *Ancient Ao Naga religion and culture*. Har-Anand Publications.

Saul, J. (2007). *The Naga of Burma: Their festivals, customs and way of life*. Orchid Press.

Tankha, A. (2020). *An aesthetics of endurance: Art, visual culture and indigenous presence in Nagaland, India*. (Doctoral thesis, University of Toronto). Accessed March 23, 2024, from https://hdl.handle.net/1807/125976

Chapter 13
Understanding Naga Traditional Attires from a Semiotic Perspective

Temjensosang

Abstract The Naga traditional attires carry full of symbolic meanings this is to state that in traditional time ones' attires and ornaments was used to mark its identity. The entire concept of attires is centred on the principle of feast of merit and head hunting therefore the stratification of Naga society is based on these two. A dress for most human society is to cover its body, though it has its adornment, however for a Naga it is the social distinction of one clan from the other and most importantly attires signify the status and position of the wearer therefore attires are an identity marker. Each pattern and motifs inscribed in the shawls represent a semiotic meaning of the wearer. Tattoo forms an integral aspect of Naga culture and sets a benchmark in the social stratification. The use of hornbill feather signifies of being leader and symbolizes of high status. What is significant in this study is the use of objects such as animals, birds, human heads and many more all these carries sociological as well as semiotic meaning. Therefore, attires are not just for covering the body but reveal the wearer's identity in the context of naga culture.

Keywords Naga culture · Attires · Semiotic

13.1 Introduction

To a common person, attire is defined as a piece of clothing used for body ornamentation. As a result, any piece of clothing that is worn by members of a cultural society to adorn their bodies and whose beliefs and values are generally held by the group can be classified as attire. In contrast, semiotics is the science of signs. As a result, all signs and symbols employed by any cultural society have meanings that are mostly shared by its members to maintain the continuity of its interactions and acts. The Nagas primarily use animals such as tigers, elephants, lions, and mithuns as symbols. Moreover, the hornbills and cocks in the avian family, as well as

Temjensosang (✉)
Nagaland University, Zunheboto, India

Y. Santasombat (ed.), *Bio-Cultural Diversity, Ethnicity, and Local Knowledge*,
https://doi.org/10.1007/978-981-95-4242-0_13

planetary bodies like the sun, moon, and stars, are used. The human head is popularly used, as Nagas were once headhunters. Therefore, this essay will attempt in a modest way to comprehend Naga traditional clothing, particularly shawls and tattoos, and how their diverse patterns, motifs, and designs are widely used by society's members to convey meanings to both the signifier and the signified. The fact that the Naga clothing is more than just a piece of clothing worn for body ornamentation is what makes it worthy of a sociological and semiotic examination. The semiotic meaning is more important than the fabric or textile that the weaver used to create the final product or its quality.

The concept of semiotics has been used in a variety of contexts to comprehend different facets of culture. In fact, the Swiss linguist Ferdinand de Saussure's work was how semiotics was first used. He views semiotics as a closed system, a historical structural approach to signs. Meaning is derived from the sign, which is a function of the signifier and signified combined as a mental construct. G.H. Mead, a social psychologist, developed a popular symbolic framework. "How does a sign (a symbol) become significant?" he asks. According to Atkinson et al. (2009), they envisioned a small scenario in which individuals gestured and indicated objects that were then filled with meaning and action potential. This created a dialogue between the I, present action, the Me, reflection on action, and the significant and generalized other.

The most systematic approaches to semantics relevant to ethnographic work emerged in the 1960s, which were developed by anthropologists, influenced by biological taxonomies, cognitive psychology, and mathematics. This work shows how they organized key domains in culture, such as color, kinship, and ethnobotanical categories. Later work systematized native notions of mental illness, weddings, firewood, lesions, and law and mapped taxonomy, yielding Western, linear, hierarchical, exclusive, Aristotelian classification. One of the broadest definitions is that of Umberto Eco (1976), who states that "semiotics is concerned with everything that can be taken as a sign." It not only involves the study of signs in everyday speech but also of anything that stands for something else. In a semiotic sense, signs take the form of words, images, sounds, gestures, and objects (Daniel Chandler, 2017). Thus, Kathryn Sullivan Kruger (2000) argues for the link between the written texts and woven textiles. Encoded by pattern, symbol, and dye, textiles offer an important form of communication that has been ignored until now. It further asserts that before written texts could record and preserve the stories of a culture, cloth was one of the primary modes for transmitting social beliefs and messages. Anthropologists have long been intrigued by the various ways in which cloth embodies the unique ideas of a culture. They can trace the history of a culture through the record of its textiles, reading cloth like a written text. Indeed, cloth transmits information about the society that created it in a manner like a written language.

In all civilizations, dress is worn for warmth, protection, modesty, and to proclaim his place in society, including rank, occupation, age, sex, place of origin, marital status, and religion, claimed Ralp and Hoijer (2007). For instance, about 4000 years ago, in the warm Indus Valley, an assured little dancing girl from Harappa felt well-dressed, wearing only bracelets, knowingly massed, a necklace, and well-coiffed hair. The earliest forms of clothing seem to have been adornments such as body painting, ornaments, scarification (scarring), tolling masks, and often constricting neck and waistbands. According to Rampley (2005), many of these have deformed, reformed, or otherwise modified the body. Therefore, clothing is a significant factor in human development. Clothes are a way of showing social status, religious beliefs, and artistic aspirations. They are often easily identifiable and fiercely treasured as cultural heritage. As Galinsky and Adam said, clothes invade the body and brain, putting the wearer into a different psychological state. It influences not only how we think about ourselves but also how other people see us. Thus, clothing has become a symbol of an individual's identity. It represents and expresses the style of a person, and it is impossible to wear clothes without transmitting social signals.

Having stated the above, it must be mentioned that Nagas are a Mongoloid race who speak Tibeto-Burman languages and are indigenous inhabitants in many Indian states, including Assam, Arunachal Pradesh, Manipur, and Nagaland, and are spread across Myanmar (Burma). However, for this workshop's purpose, I shall concentrate on the Naga tribes, inhabitants of Nagaland state, India. Nagaland is the 16th state in the union of India. It is one among the seven sister states and one brother state in the northeastern region of India.[1] Consisting of 16 administrative districts, Nagaland is situated in the northeastern region of India, currently having a population of about two million with a literacy rate of 79.55%, which is above the national average of 74.04% (Census of India: 2011).[2] The state is inhabited by 17 major tribes and a host of sub-tribes.[3] People of Nagaland claim themselves to be a family of the same cultural community; hence, they share the same origins, belonging to the Mongoloid race. However, each of them is unique from the others due to certain innate features or cultural aspects. They all use the same name, Naga, even if these distinctions span language, attire, and customs. It is necessary to tell a legend that is well-known among the Naga tribes before continuing. The folklore claims that, in contrast to today, when they lack a written script, the Nagas formerly had their own script. Unfortunately, a dog ate the script that was written on a piece of animal skin. As a result, the Naga script was lost, and their history is not properly documented in writing. Despite not having written scripts, Naga cultures are rich in signs and symbols; therefore, it is believed that the lost scripts are manifested into signs and symbols that carry meanings and are shared among the members of the society. This indeed has attracted the attention of many scholars and researchers.

[1] Initially, northeast India was comprised of seven sister states, such as Assam, Nagaland, Arunachal Pradesh, Meghalaya, Mizoram, Manipur, and Tripura. Sikkim was officially integrated into the NEC in December 2002, making it the eighth state in the region.

[2] The 16 administrative districts in Nagaland are Kohima, Mokokchung, Tuensang, Mon, Wokha, Phek, Zunheboto, Dimapur, Peren, Kiphire, Longleng, Noklak, Tseminyu, Niuland, Chumoukidema, and Shamator.

[3] Seventeen major tribes in Nagaland include Angami, Ao, Chakisang, Dimasa, Kachari, Khiamniungan, Konyak, Kuki, Lotha, Phom, Pochury, Rengma, Sangtem, Sumi, Yimchunger, and Zeliang.

As stated above, the Nagas have no written historical record of their origin and the route of migration to their present inhabitation. Some writers believe that the Nagas have migrated from three directions: Northeast, Northwest, and Southeast Asia. However, it is a general belief that most of the Nagas immigrated from Southeast Asia through the corridor of the Indo-Myanmar border to the present-day Naga Hills. To add further, the Nagas have a rich and diverse cultural heritage. As stated by Adangla Changkija (2017, p. 5), the Nagas' self-sufficiency and sustainability were once established by their natural skills in basketry, pottery, cultivation, spinning and weaving, wood carving, dyeing colors, and metalwork. Naga folk songs are an integral part of their culture. They are, in fact, the most expressive form of communication and message delivery to an individual or group of people. Every song has a connection to a circumstance or event. The songs carry the poetic expressions addressing a person with reference to such symbols; for example, a brave warrior was referred to as a tiger or lion, while a vigilant person was referred to by the symbol of a cock. Therefore, it is crucial to understand the semiotic meaning in the context of Naga culture since it facilitates a more thorough examination of the social structure of Naga society.

G. S. Ghurye has advocated using an Indological perspective while trying to understand Indian society. The Indological perspective is a methodological approach involved in the close reading and analysis of ancient Indian scriptures like the Vedas, Upanishads, and the Puranas. It also included a sociological study of various Indian rituals, customs, and festivals. This is because without studying the ancient culture and texts, one may not understand the actual state of Indian society. He was critical of some aspects of Westernization in India that threatened Indian values and culture. Thus, in light of Ghurye's Indological approach, the best way to comprehend Naga society is to look at its traditional cultural elements, which include the feast of merit, headhunting, traditional Naga folk songs, the Naga worldview, attire and tattoos, governing institutions, family and kinship systems, and many other cultural elements that are essential to comprehending the true nature of Naga culture and society. This methodological approach, as I term it, *Nagalogy*,[4] can help us understand the basic meaning of Naga society that has not yet been addressed.

13.2 Rationale of the Study

The shawl, the wealthy man's home, and the village gate all bear cultural symbols and meanings that are deeply ingrained in the Nagas' way of life. These symbols and meanings take the shape of animals, planetary objects, and birds. In addition to motifs and crisscross patterns, shawls are made with an assortment of designs and symbols, including human heads, tigers, lions, and elephants, to mention a few.

[4] Nagalogy is a term used in this paper to explain the study of Naga society by carefully examining the cultural aspects so as to have a comprehensive understanding.

Besides these, tattooing is also an aspect of Naga culture, which is limited to head-hunters of the Konyak and Chang Naga tribes, whereas among the Ao Naga tribe, it was only women who got tattooed. Information regarding tattooing among other Naga tribes is either nonexistent or extremely limited. In fact, a woman would not be ready for marriage unless she got tattooed on her face and body. Studies pertaining to the semiotic analysis of cultural symbols into meanings that are valuable to society are particularly scarce. Among the Naga tribes, it might be said that a man who performed the feast of merit was customarily given the highest social honor. It takes a year to perform this kind of merit feast, and there are many taboos to follow. When all formalities have been performed and the final feast is successfully completed, it is socially acceptable to wear an ornamental shawl that is woven with multiple patterns and symbolism. A person not performing such a feat of merit was not qualified to use such an ornamental shawl. The importance of this study is highlighted by the fact that, in the modern world, semiotic meanings are becoming less significant, and we are failing to recognize their rich history and traditional art; as a result, there is a tendency to undermine traditional values.

13.3 Major Discussion

Every Naga tribe is distinguished from the others by its diverse traditional clothing. Additionally, the marks on the clothing help its clans and sub-clans remain identifiable. Notably, the clan system is the foundation of Naga society's social structure. The diverse ways of dressing distinguish the statuses of the elite and common people. The style and pattern of clothing, like shawls, can adequately convey one's identity without verbal communication. Therefore, a Chang Naga woman wearing the necklace below her navel is normally understood to have been from a rich family, while for those who wear it above the navel, it is an indication of being from a poor family. In the past, the ranks and statuses of individuals could be easily distinguished from the dresses they wore. A poor man's attire is simple; his shawls are without any designs, whereas a rich man's attire is complex, and his shawls are woven with different patterns and motifs with variegated objects and signs and colors. While dwelling on this discussion, a further relevant question arises as to who maintains the propriety of who should wear what. And what was the course of penalty in case of infringement? Since people lived in small village communities, everyone knew who was who; therefore, the question of moral infringement towards propriety in dressing was either very minimal or did not exist at all. Unlike modern society, parental role in educating their children was very minimal; however, the institution of *Tsüki*[5] and *Arju*[6] were strong enough that everyone learns the social etiquettes, moral responsibility, and norms and values of the society. This is to say that society itself was vigilant in moral policing. It is sufficient to mention that the structure of Naga society is based on a clan system; therefore, in the case of women, and especially with Ao women's loincloths (waist cloths), the daughters cannot share the same patterns and motifs with their mothers. This is true because Naga society is patrilineal and follows the principle of exogamy; thereby, children take the father's title. When talking about Naga clothing, it is important to note that the idea is based on two fundamental ideas: the warrior (headhunters) and the giver of a feast of merit. It was customary for Nagas to host feasts for the people in their society. The practice includes a sequence of sacrifices, rituals, and taboos. The community members pull a monolith stone to give the host social honor as the feast of merit ends. The Angami and Chakhesang Naga tribes, respectively, share a common practice of erecting monoliths. The Ao, Chang, and Konyak tribes, on the other hand, did not keep many monoliths. What made the feast of merit special was that a significant shawl with patterns and motifs was used to obtain social honor and status. The decorative shawl was reserved for the person who gave the feast of merit. The objects inscribed in the shawl, such as mithun, hornbill birds, and the planetary objects, are all marked as symbols of high status achieved in society through its riches. The most coveted shawls among the Ao tribe are *Tapensa-sü/Takar-sü.*[7] Fig. 13.1 depicts the image of this shawl.

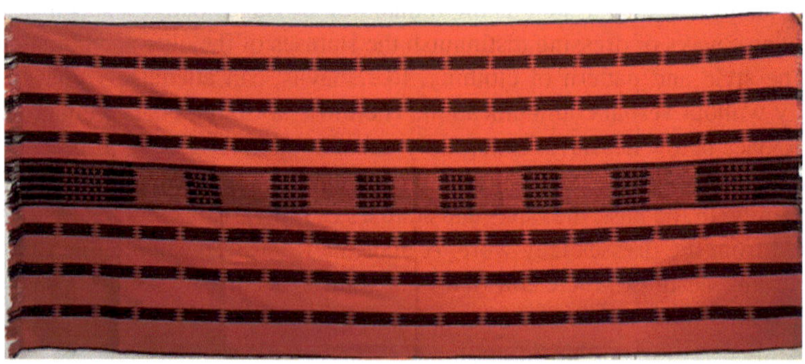

Fig. 13.1 Tapensa sü (The Ao Naga feast giver shawl). Photo by temjensosang

[5] Tsüki. It is a women's dormitory. Women of the same clan would sleep at night in the house of an elderly widow who belonged to the same clan as the girls. She acts like a guardian angel over the girls in life skills training, etiquette, and even in the selection of a life partner.

[6] Arju. It is an adult boy's dormitory; boys aged 13 years and above would sleep in the arju at night. It was used as an institution for learning life skills, etiquette, and others.

[7] Tapensa-sü/takar-sü. It is a decorated shawl awarded to a man for his performance at the feast of merit. Tapensa means complete in Ao Naga vernacular. This would mean the giver of the feast of merit has performed and completed all the series of rituals connected to this feast.

Naga shawls are also reserved for headhunters; these are earned through their bravery. A head was typically acquired as a prize during a village raid or when hunting wildlife. It was unthinkable to witness headhunting from a humanitarian standpoint. However, it must be noted that village raids were never conducted without any reason. It is sufficient to mention that Nagas adorned themselves with ornamental shawls, especially for feast givers and headhunters. Only their various shawl patterns need to be explained in this instance. For instance, the Ao *tsünkotep-sü*[8] is known as a warrior shawl. The shawl is depicted with the pictures of planetary objects like the sun, moon, and stars; the symbols of animals like lions, tigers, and birds like hornbills; and many more objects that carry significant meaning that are used normally as metaphors in the social usage of language. Figure 13.2 shows the image of this shawl.

Every Naga tribe has its spectacular and coveted shawls. For instance, *Shangbu Nei* is a warrior shawl worn by the Chang tribe; its members had taken their enemy's head as a trophy. It is also important to note that women weave these cloths. The women strictly adhered to the prohibition on eating or drinking in other people's homes until the weaving duty was finished. Figure 13.3 shows the image of this shawl. There is also a warrior shawl called *Moh Nei*, or cowrie shawl, which could be worn only by a man who has taken six or more heads. Figure 13.4 shows the image of this shawl. These demonstrate that Naga shawls are more than just items of clothing used to keep their bodies warm; they are a means of expressing one's social standing.

The use of hornbill feathers is another significant aspect of Naga culture. The hornbill feathers indicate a high status achieved through the feast of merit or being a headhunter. For example, an Angami warrior decorates his headdress with the hornbill feathers according to the number of heads taken. Konyak warriors put two feathers in their headdresses to show their warrior status. In contrast, the feast giver

Fig. 13.2 Tsüngkotep sü (The Ao Naga warrior shawl). Photo by temjensosang

[8]Tsünkotep-sü. It is a shawl awarded to a warrior for his bravery. The shawl is inscribed with objects like a tiger, lion, elephant, human head, hornbill birds, and cock, and planetary objects like the sun, moon, and stars with patterns and lines.

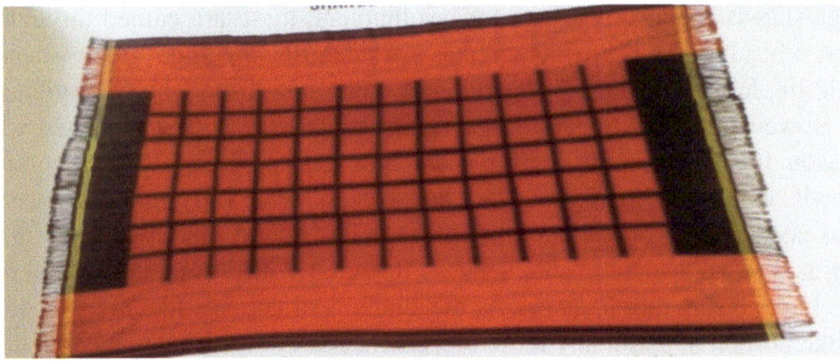

Fig. 13.3 Shangbu Nei (the Chang Naga feast giver shawl). Photo by temjensosang

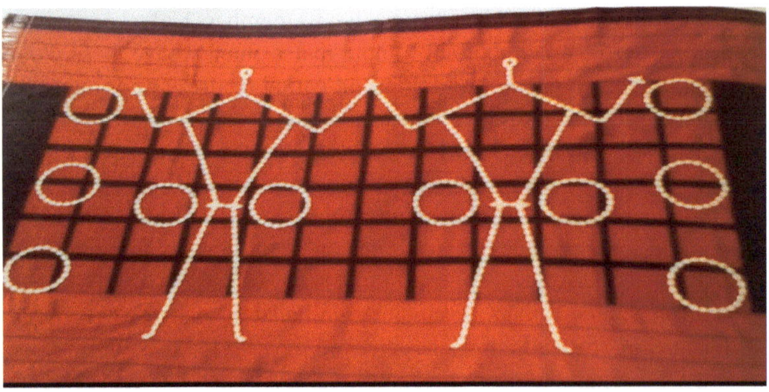

Fig. 13.4 Moh Nei (the Chang Naga warrior shawl). Photo by temjensosang

in Aos adds one feather each time the couple performs the feast of merit. The sociological significance of the hornbill bird in Naga culture is that it is regarded as the king or leader of all birds. As a result, the Nagas use this feather to symbolize a leader or king-like individual in their community who has achieved opulence or bravery. A relevant question is often raised as to who enjoys a higher status between the feast giver and the headtaker in society? It was always the headtaker who had the highest social standing. The rationale is that while the feast giver works for his personal wealth, the headtakers serve to safeguard the village and the community. Thus, this marks the social distinction between the elites of the community among the Naga tribes in general.

Besides shawls, tattoos occupy a significant space in Naga culture. The practice of tattooing and piercing the body is found in almost every subsection of Western popular culture and, arguably, throughout the world (Rachael et al., 2012). In today's mainstream culture, whether in the West or America, the concept of body art as a form of nonverbal communication has started to gain popularity. Celebrities, well-known figures in popular culture, and even regular people on the street are seen wearing tattoos these days as a statement of their cultural conformity. As a result, traditional Nagas wear tattoos on their faces and bodies as a nonverbal cultural representation of their status as a warrior, village chief, or even a symbol of their achievement as women. Among the Nagas, female tattoos typically have geometric designs that reflect different stages of life, whereas male tattoos are more figurative and associated with social standing. If a Konyak woman has a small tattoo behind her knee, it means she is married. This not only deters other men from wooing her but also proves that she is eligible to wait for her husband in the afterlife should she die prematurely. The tattoo serves as a sort of "divine passport" in this regard and is sometimes deemed significant enough to justify tattooing corpses.[9]

The girls, particularly those from the Ao Naga tribe, had tattoos on various body areas. The entire tattooing process for this tribe takes about 5 years. Since tattooing is prohibited on the body until puberty, the legs up to the calves were inked during the first year, in accordance with Ao Naga customs. In the second year, the chest, chin, and shoulder were tattooed. In the third year, the patterns of the calves were finished, and lastly, in the fifth year, the tattooing was completed with the wrists and stomach. It must be noted that there are certain innate differences in the pattern of tattoos between the *Chungli*[10] and *Mongsen*[11] speaking groups. However, there are four vertical lines on the chin, a chain of lozenges from the throat to the bottom of the breastbone, an inverted letter "V" shape on the front of the shoulders and stomach, lozenges and side squares on the wrists, lozenges on the lower part of the leg, and a sign of an arrow pattern on the knee (Mills, 1926, p. 30). This is why tattooing is based on the stages of a person's life. In the community, tattooing is a significant activity since the girls' body markings serve to both identify them as members of the society and set them apart from the "slaves." Individuals without tattoos are also seen to have less likelihood of finding compatible partners, while girls who want to remain single are perceived as a social embarrassment.

Sociologically speaking, those who tattooed their bodies were not the victims of fashion but active members of society who responded according to the social and cultural values. Tattoos and body piercings work alongside the brands and styles of clothes one chooses to wear to create a rhetorical statement about the kind of person one is. Therefore, if clothes constitute the "sentences" in the grammar of fashion, then it is reasonable to argue that body piercings and tattoos serve as the "punctuation" in those sentences. Thus, getting a tattoo can be viewed as encoding a cultural value. Figures 13.5 and 13.6 display the images of Ao Naga tattoos.

---

[9] http://web.prm.ox.ac.uk/bodyarts/index.php/permanent-body-arts/tattooing/169naga-tattooing. The Pitt River museum/Naga tattooing tools/University of Oxford, England.

[10] Chongli is a dialect spoken by the Ao Naga tribe. Hence this group is known as the Chongli-speaking group.

[11] Mongsen is a dialect spoken by the Ao Naga tribe. Hence this group is known as the Mongsen-speaking group.

Fig. 13.5 Chungli speaking group. Source: Adangla changkija (2017)

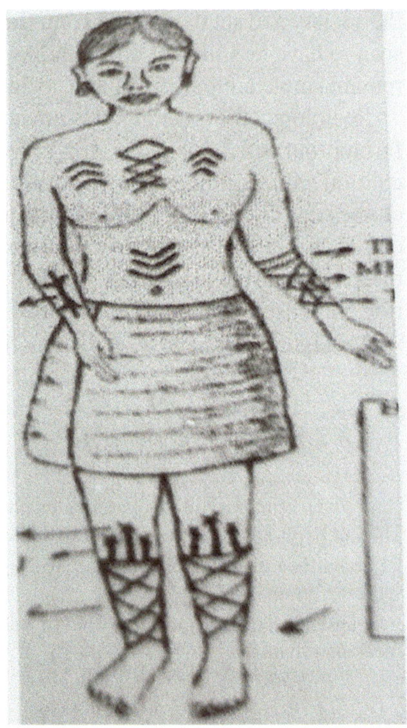

Fig. 13.6 Mongsen speaking group. Source: Adangla changkija (2017)

13.4 Semiotic Analysis

Semiotics assumes that visual images and their respective signs can be read like a text. In addition to having a language of their own, clothing and body decoration can be interpreted as a text and explanation of a person's status and personality. Cultural semiotics is a crucial component of visual rhetoric because it enables us to transform basic codes and signs into meaningful dialogue. Studying compositions, artwork, ads, movies, and a wide range of other produced materials is frequently a part of visual rhetoric research. The visual rhetoric of the objects around us can be used as a tool of evaluation, but the objects themselves and their presentation can also convey a message that is signaled and coded. Every individual has a unique personality and a set of beliefs and opinions. We utilize our bodies as a canvas to express these identities. According to Elizabeth Welter Wilson, the body is "dressed" in every society, and ornamentation and clothing serve aesthetic, symbolic, and communicative purposes everywhere. Dress is always unspeakably meaningful. The earliest forms of clothing seem to have been adornments such as body painting, ornaments, scarifications (scarring), tattooing, masks, and often constricting neck and waistbands. The body is modified, distorted, or reformed in many of these ways. Numerous factors can affect a person's rhetoric and look. There are multiple ways to use clothing to represent oneself, including jewelry, hairstyles, symbols, and many other rhetorical devices. In the Naga tribe, nonverbal communication through attire helps us comprehend societal structures and stratifications, while self-fashioning is the process of representing and exhibiting one's expression through nonverbal communication. The rhetorical expression exhibits its status and position one occupies in society.

Having said that, modernization- and globalization-induced changes are taking place rapidly in all aspects of life, including cultural values and ideas. As I continue this discussion, certain relevant concerns come up regarding whether the tribal clothing's symbolic significance evolves. Does the meaning remain the same across the different tribes among the Nagas? Does the meaning remain consistent across generations? These questions need further, deeper research for theorization. From an optimistic perspective, this discourse can imply that while the form may change in the present, the content would not. As a result, modernization and globalization are rapidly encroaching on all societies worldwide. However, in response to these changes in the context of Naga attire, especially men's shawls and women's waist cloths (sarongs), both young and old now prefer to wear modern fusion fashion (designs like modern Western wear created out of traditional designs and patterns) at any tribal event or, for that matter, at schools, colleges, and universities. The colors, patterns, motifs, designs, and symbols are all the same, so the meaning is unchanged. This may be understood as a positive impact of modernization and globalization. An additional argument in this context stems from the creation of the Hornbill Festival, a "festival of festivals" that was initiated by the Nagaland government. This has now grown to be a global occasion. Every Naga tribe celebrates a cultural holiday. All the tribal festivals come together during the Hornbill Festival. Besides establishing a forum for tribal contact with their rich heritage of culture, its introduction aimed to foster unity and a spirit of fraternity. As cultural troops take part and each tribe's morung presents their own ethnic culture—from food to clothing, dances, music, and much more—they both overtly and covertly convey the need to preserve and carry on the cultural meanings. Thus, our assertion that forms may change but meaning remain unchanged stands clear.

Without a doubt, it can be stated here that rural villages continue to use shawls in the old manner. This is borne by the fact that in any of the village council meetings or village annual meetings, members are expected to attend the meeting wearing their traditional shawl, if not the waistcoat for young boys. From an aerial perspective, the Naga tribes seem to be one and the same. Closer inspections and observations, however, show that sociocultural and governmental systems differ in certain fundamental ways. Therefore, even while symbolic markers and meanings may vary slightly among tribes at the micro level, the glory and social standing of headhunters and givers of feasts of merit remain the same at the macro level.

13.5 Encoding

Wearing a certain style of clothing can define a person as a member of a group as well as an individual. Therefore, the complete clothing code and the material items utilized by the Nagas are within their own cultural domain, which uses specific rhetorical cues to effectively transmit a particular idea of identity. Both the employment of encoded shawls and hornbill feathers provide examples. In the context of Naga culture, all of these are semiotically interpreted as expressing one's social status and portraying oneself as a leader.

13.6 Conclusion

One common definition of culture is described as the values, norms, institutions, and artifacts that are passed on from one generation to another by learning alone. For instance, no amount of separation and classification will diminish the rich and profound cultural milieus that surround "the American blue jean," the perfect pair of blue jeans, among Americans. In American society, these analogies are popular and deeply ingrained, even in people's minds. The analysis of American blue-jean culture provides a comparative example of Naga culture and traditional attire, demonstrating how the Naga people's cultural meanings and symbols have influenced every aspect of their existence. Therefore, the methodological approach of Nagalogy, which delves deeply into every facet of Naga life, can be used to facilitate any attempt at a deeper knowledge of Naga society and culture. The Nagas have indeed been supplanted by the cultural influences of the outside world. Famous anthropologist Prof. Furer Haimendorf, who has studied the Naga culture at different points in its history, from obscurity to popularity, claims that Christianity has been the main factor in the adulteration of Naga culture. Nonetheless, Naga civilization still has strong cultural ties; it requires the use of semiotics to explain its culture.

References

Atkinson, P., et al. (2009). *Handbook of ethnography* (pp. 149–150). Sage Publication.

Chandler, Daniel (2017) Semiotics for beginners. 3rd ed. Routledge, P: 8. Retrieved on June, 12, 2015, from www.wayanswardhani.lecture.ub.ac.id/files/2013/09/semiotics-theBasic.pdf

Changkija, A. (2017). Unpublished Ph.D Doctoral thesis. Nagaland University Hqrs Lumami.

Kruger, K. S. (2000). *Weaving the word: The metaphoric of weaving and female texture production*. Susquehanna University Press.

Mills, J. P. (1926). *The Ao Nagas*. Oxford University Press.

Rachael, A. C., Guitar, A. E., & Dillon, H. M. (2012). Ultimate answers to proximate questions: The evolutionary motivations behind tattoos and body piercings in popular culture. *Review of General Psychology, 16*(2), 134–143.

Ralp, L. B., & Hoijer, H. (2007). *An introduction to anthropology*. Surjeet Publication.

Rampley. (2005). *Exploring visual culture, definition, concepts, context*. Edinburg University Press.

MIX
Papier aus verantwortungsvollen Quellen
Paper from responsible sources
FSC® C105338

If you have any concerns about our products,
you can contact us on
ProductSafety@springernature.com

In case Publisher is established outside the EU,
the EU authorized representative is:
Springer Nature Customer Service Center GmbH
Europaplatz 3, 69115 Heidelberg, Germany

Printed by Libri Plureos GmbH
in Hamburg, Germany